Take This Stuff and Hack It!

About the Author

DAVE PROCHNOW is an award-winning professional writer, editor, and contributor to numerous technical publications and is the Contributing Editor for *Nuts and Volts*, and *SERVO Magazine*. He is the author of 27 nonfiction books, including the bestselling books *The Official Robosapien Hacker's Guide* and *PSP Hacks, Mods, and Expansions*, both published by McGraw-Hill. In 2001, Dave won the Maggie Award for writing the best how-to article in a consumer magazine. To learn more about Dave's books and other projects, visit his Web site: www.pco2go.com.

Take This Stuff and Hack It!

TRANSFORM EVERYDAY ELECTRONICS INTO MODERN TECHNO-WONDERS

Dave Prochnow

McGraw-Hill

NEW YORK | CHICAGO | SAN FRANCISCO | LISBON | LONDON | MADRID | MEXICO CITY
MILAN | NEW DELHI | SAN JUAN | SEOUL | SINGAPORE | SYDNEY | TORONTO

The McGraw·Hill Companies

Library of Congress Cataloging-in-Publication Data

Prochnow, Dave.
Take this stuff and hack it! : transform everyday electronics into modern techno-wonders / Dave Prochnow.
p. cm.
Includes index.
ISBN 0-07-147737-3 (alk. paper)
1. Household electronics. 2. Digital media. 3. Handicraft. I. Title.
TK9965.P78 2006
621.381—dc22 2006017237

1 2 3 4 5 6 7 8 9 0 DOC/DOC 0 1 2 1 0 9 8 7 6

ISBN-13: 978-0-07-147737-6
ISBN-10: 0-07-147737-3

The sponsoring editor for this book was Judy Bass and the production supervisor was Pamela A. Pelton. It was set in ITC Officina Sans by Cindy LaBreacht. The art director for the cover was Handel Low.

Printed and bound by RR Donnelley.

This book is printed on acid-free paper.

Contents

Disclaimer

Before you attempt to "dissect" *any* gizmo or gadget, please read, understand, and accept the following warnings, precautions, and disclaimers regarding the disassembly of electronics. Thank you.

PRECAUTIONS *Disassembling electronic devices will void your warranty.* There is no authorization for the disassembly or modification of any equipment. There could be a risk of electrical shock or fire by disassembling electronics equipment.

WARNINGS Monitors and LCD screens contain dangerous, high-voltage parts. Always remove the battery and disconnect any power cord(s) prior to disassembling *any* electronics equipment.

DISCLAIMERS The McGraw-Hill Companies and Dave Prochnow will neither assume nor be held liable for any damage caused to anyone or anything that is associated with the disassembly, modification, and hacking of any gadget, gizmo, or electronics equipment. The warranty for this equipment will be considered null and void if any associated warranty seal has been altered, defaced, or removed.

Dedication

On the morning of August 29, 2005, the lives of so many people changed forever. This book is dedicated to the plight of the residents of Southeastern Louisiana and the Mississippi Gulf Coast who suffered the most from the destructive force of Hurricane Katrina (see Fig. D-1).

Hopefully, this book will stimulate a dialog on how to empower people to survive in an austere environment with minimal or nonexistent governmental assistance by "hacking" their way to survival.

D-1 Young eyes behold the power of Mother Nature.

Foreword

BY HELEN GREINER

In 1977, my dad brought home a Radio Shack TRS-80. We called her "Tracy." It was supposed to be a family computer, but I wouldn't let anyone else near her. I was always hacking away at games—graphics characters make great space invaders—and helpful programs for things like sorting and lists. Then, I saw *Star Wars*—back when it was just plain *Star Wars*, before anyone called it "Episode IV." R2-D2 was more than a machine. He was a character with his own personality and agenda. I was enthralled, even after my brother told me that R2 was played by a man of short stature in a robot costume. And I have been hooked on robots ever since.

Tracy had all the components of a robot: motors (in the tape drive), sensors (keyboard), and a microprocessor. If I could just get to the Tosche Station to pick up some power converters, I knew that I could build an R2. Twelve years—and many trips to Radio Shack—later, after studying engineering at MIT, I co-founded iRobot Corporation with Colin Angle and Rodney Brooks: one step closer to that astro-mech droid.

Over the last 15 years, I've built many robots. Today, I find myself spending more time on the Dark Side concentrating on strategy, financing, management, and growth. But every now and then I am that 11-year-old girl, hacking away on a project, trying to bring us closer to R2 and machines with personality. That is why it has been one of my great joys to spearhead the iRobot Roomba Open Interface project, providing a platform for the hackers to make dreams into reality.

So what is the Roomba Open Interface (ROI)? Dave will give you the bits and bytes answer but in a nutshell, you can see the Roomba's sensor readings and control the motors over its onboard serial port. Although there have been Roomba hackers since its birth in the fall of 2002, it was not very accessible. You needed to open it up, figure out what the microprocessor was, solder cables onto the right pins, and then download entirely new code to it. This excluded all but an elite few. The ROI allows both hackers and developers to take control of the Roomba and make it into the robot that they always wanted.

Robot hacking is on the rise. It is at the same stage now as Tracy was in a family's wood-paneled, shag-carpeted den in the 70s. Robot clubs are springing up in major cities, with garage projects, classes, and contests in robotics all becoming more common. I hope that the ROI will give others the opportunity to hack robots, to learn by getting their hands dirty, and inspire creativity. Too much time is wasted in robotics by people duplicating work that has already been done rather than leaping to the new, creative stuff. Because of our investment and large-scale production, judged by price per robot capability, Roomba is off the charts. So use the effort we have put into the Roomba where you can...diverge where you need different capabilities. A few developers have already jumped in targeting the hobbyist and education community—from programmable single-board computers that control the ROI (such as Mind-Control™) to a blue tooth links–enabling control from a PC (such as RooTooth™). We encourage you to take advantage of this work as well.

You get the über-geek award if you: use the PSP to control your Roomba over the Internet, bring a beer to your buddy who is lounging on the couch, recognize her, and start podcasting her favorite show as the Robosapien stationed on top of the Roomba hands her the beer.

More seriously, if your Roomba project leads to the next great robot product success, inspires you to pursue an engineering career, or helps you get that cool new robot application up and running, we will consider it a success. Happy hacking!

March 16, 2006

Take This Stuff and Hack It!

INTRODUCTION

Designed to Be Hacked

We all have one somewhere in our house, office, or workshop—a "junk" box, drawer, or closet; a final resting spot for all of our outdated, obsolete, or broken technology. Actually, "junk" is too harsh of a descriptive noun for this repository. Maybe the residents of these techno-graveyards would be more aptly described as "disassociated inspirational components"; nah. Junk works just fine for me.

Others have coined catchy public relations–driven terms for describing our electronic refuse. The U.S. Environmental Protection Agency (EPA) has tried the hardest at putting a friendly face on our junk. Programs like **PLUG-IN TO ECYCLING** and **WASTEWISE** represent the government's attempt at stimulating interest in reducing, reusing, and recycling electronic debris.

Started in January 2003 as an element of the EPA's Resource Conservation Challenge (RCC) and Product Stewardship Program, Plug-In To eCycling represents the EPA's nationwide program for establishing, in the government's words, "flexible, yet more productive ways to conserve our national resources." More specifically, Plug-In To eCycling is attempting to migrate consumer electronics away from landfills and safely dispose of them in community-based recycling centers.

In order to reach this goal, Plug-In To eCycling is focusing on three primary areas of education and participation:

1. Educate the public about the virtue of safely recycling old electronics.
2. Develop a dialog between communities, electronics manufacturers, and retailers for sharing the responsibility for safely recycling old electronics.

3. Create pilot projects for testing innovative approaches to safely recycling old electronics.

Supplementing Plug-In To eCycling is another EPA program, WasteWise. Begun in 1994, WasteWise is a free, voluntary, program that enables participants to "design" their own waste reduction program. The ultimate goal of WasteWise is that organizations who participate in the program will reduce municipal solid wastes while, in the words of the EPA, "benefiting their own bottom line and the environment."

The solid wastes that are being targeted by WasteWise include: electronics, organic waste, paper, primary packaging and containers, and transportation packaging. According to the EPA, WasteWise partners who are interested in the disposal of electronics should consider these example activities:

- ♻ Donate reusable electronic equipment
- ♻ Buy recertified electronic equipment
- ♻ Lease electronics
- ♻ Recycle unusable electronic equipment

But many environmental activists suggest that this program fails to notice that the "Emperor has no clothes." Casting a sweeping proclamation toward "recycle unusable" electronic equipment fails to address issues like lead and mercury contamination.

In a *WIRED News* article ("EPA: Old Computers No Longer Junk" by Kendra Mayfield), critics of the EPA program were cited as stating that recycled electronics were "likely to be shipped to China, India, or Pakistan." Furthermore, this article stated that these overseas destinations dumped recycled electronics

Where Can I Dump Some Stuff?

You can discover your local used electronics recycling site from the Consumer Education Initiative (CEI) Web site which is sponsored by the Environmental Issues Council of the Electronic Industries Alliance (EIA) (www.eiae.org). These sites are a great place to start looking for your own junk to hack.

in waterways or burned them in rice fields rather than scavenging the parts for reuse in new electronics.

This dumping fear was exacerbated by the EPA's reclassifying of discarded mercury-containing electronics as universal waste (EPA530-F-05-010, July 2005). What sounds like an ominous impact on our environment is actually a modification of a previous EPA restriction. Specifically, the universal waste regulations modify the collection requirements for hazardous wastes found in batteries, pesticides, mercury-containing thermostats, and lamps (mercury-containing equipment is proposed).

As enacted in 2002, the universal waste regulations eased the regulatory burdens on businesses, promoted safe recycling, treatment, or disposal of these hazardous wastes and provided for sanctioned collection opportunities.

As defined by the EPA, federal universal wastes are (and remember, in order to be a universal waste, debris must be hazardous before it can be classified as a universal waste):

BATTERIES such as nickel-cadmium (Ni-Cd) and small sealed lead-acid batteries, which are found in many common items in the business and home setting, including electronic equipment, mobile telephones, portable computers, and emergency backup lighting.

Agricultural **PESTICIDES** that are recalled under certain conditions and unused pesticides that are collected and managed as part of a waste pesticide collection program. Pesticides may be unwanted due to being banned, obsolete, damaged, or no longer needed because of changes in crop behavior or other factors.

THERMOSTATS which can contain as much as 3 grams of liquid mercury and are located in commercial, industrial, agricultural, community, and household buildings.

LAMPS which are the bulb or tube portion of electric lighting devices that have a hazardous component. Examples of common universal waste electric lamps include fluorescent lights, high intensity discharge, neon, mercury vapor, high pressure sodium, and metal halide lamps.

MERCURY-CONTAINING EQUIPMENT is proposed as a new universal waste category. Mercury is used in several types of instruments that are common to electric utilities, municipalities, and households. Some of these devices include switches, barometers, meters, temperature gauges, pressure gauges, and sprinkler system contacts.

CATHODE RAY TUBES (CRTs), which are also proposed as a new universal waste, are the video display components of televisions and computer monitors. CRT glass typically contains enough lead to be classified as hazardous waste when it's being recycled. In the past, businesses and other organizations that recycled CRTs occasionally misunderstood the applicability of hazardous waste management requirements to computer or television monitors. The EPA proposed to revise the universal waste regulations to encourage more businesses to safely collect, reuse, and recycle CRTs.

Returning to the *WIRED News* article criticism of the various EPA electronics recycling programs, member organizations of the Computer TakeBack Campaign proffered a critique of the EPA Plug-In To eCycling program sponsored by the Basel Action Network (BAN). In this BAN document, the government recycling policy was chided as a "green-washing of a disastrous environmental policy and a band-aid over the growing global electronic waste cancer."

In an inventive spin on recycling, the Computer TakeBack Campaign, which is sponsored by the Silicon Valley Toxics Coalition, is attempting to make electronics manufacturers take a more proactive responsibility for their products—a cradle-to-grave responsibility. In other words, electronics manufacturers who give "birth" to a gadget should also offer to foot the bill for the device's "funeral" and take it back after it's outlived its usefulness.

Unfortunately, initiatives like the Computer TakeBack Campaign are quickly dwarfed by the sheer magnitude of tech rubbish that is cast out annually.

According to an article in *The Washington Post* ("'Digital Dumps' Heap Hazards at Foreign Sites" by Elizabeth Grossman), 20 to 50 million tons of electronics are pitched out every year. Of that amount of techno-junk, less than 10 percent gets recycled and a little over half of these recyclables are shipped overseas for disposal in Africa, China, and India.

Groups like BAN are quick to point out that this type of technological nose-thumbing toward developing nations doesn't bode well for bridging the digital divide, but rather it builds a digital dump.

Ironically, a group of former EPA administrators are taking the current EPA policies to task—specifically those policies related to global warming. Meeting

Exactly What Are The Universal Waste Regulations, Anyway?

According to the EPA, in exact government speak:

These regulations have streamlined hazardous waste management standards for the federal universal wastes (batteries, pesticides, thermostats, and lamps). The regulations govern the collection and management of these widely generated wastes. This facilitates the environmentally sound collection and increases the proper recycling or treatment of these universal wastes.

Additionally, these regulations ease the regulatory burden on retail stores and others that wish to collect or generate these wastes. In addition, they also facilitate programs developed to reduce the quantity of these wastes going to municipal solid waste landfills or combustors. It also assures that the wastes subject to this system will go to appropriate treatment or recycling facilities pursuant to the full hazardous waste regulatory controls.

States can modify the universal waste rule and add additional universal waste in individual state regulations.

These regulations are set forth in 40 CFR Part 273, which is an exciting publication for reading while curled up next to a smoldering pile of hazardous waste.

on January 18, 2006 at the Environmental Protection Agency 35th anniversary symposium, Carol Browner, Bill Reilly, Bill Ruckelshaus, Lee Thomas, Russell Train, and Christie Whitman jointly denounced the EPA's attempts at thwarting global warming.

It was Carol Browner, former head EPA administrator for President Clinton, who issued the most succinct call for action. She said, "If we wait for every single scientist who has a thought on the issue of climate change to agree, we will never do anything."

What's a Poor Technophile to Do?

So as a consumer of electronics, what can *you* do? Well, for one, stop throwing out your obsolete, outdated, and broken electronics in the garbage can each week. Rather, attempt to reuse, recycle, and repurpose your tech junk.

Consider this real-life problem: Following the release of Mac OS 10.3, "Panther," Umax Technologies, Inc., the makers of arguably one of the finest FireWire scanners ever manufactured, the PowerLook 1100, unilaterally elected not to support this new operating system with suitable drivers and software. Instead, owners of this scanner were curtly told to purchase third-party software. This blatant disregard for an installed user base outraged Mac owners. What should we do with these orphaned scanners?

Recycle This

OK, what's the real story behind that ringed, three-arrow recycling logo? Originally commissioned by Container Corporation of America (CCA), a graphics art student named Gary Anderson created the logo. Later a CCA employee, William Lloyd, revised the logo.Those funny numbers inside some recycling logos are a plastic resin identification coding scheme that was devised by the Society of the Plastics Industry. These numbers help in the post-consumer sorting of recyclable materials.

The Mac, Umax, and self-help online forums were filled with users who boasted about throwing away their beloved scanners and buying newer Mac OS X–friendly models. Jeez, that is exactly what *not* to do.

My solution was to dedicate an "outdated" Apple Computer G3 iBook notebook computer to being the scanner interface host. Formatted with an older version of Mac OS X, a version that Umax *did* support, and equipped with a Wi-Fi Airport card, this landfill refugee became my office's PowerLook 1100 scanner savior.

MY SCORECARD: 1 computer and 1 scanner saved from the junk yard.

Now consider a similar story about one of my neighbors, Mary. Mary claimed that she needed more computing power (whatever that means; all she does is surf the Web and read e-mail; go figure, eh?). So she buys a new computer. Now her old SCSI scanner won't work with this new PC because it doesn't have an SCSI port. So Mary also buys a new scanner. When I ask her what she did with her now obsolete hardware, Mary tells me that she threw them away. Oh brother, I would have gladly paid Mary for her old junk.

MARY'S SCORECARD: 1 computer and 1 scanner thrown away.

So the game is tied, now it's up to you. Make a big play and reuse, recycle, and repurpose *all* of your techno-junk.

I only wish that it was really that easy, but it's not. More and more manufacturers are actually making their products hacker-proof—so that you *can't* reuse, recycle, or repurpose them. From a Motorola Razr V3 to the Game Boy Advance, products are being engineered with an emphasis toward *not* being reengineered. In other words, tamper-proof fasteners, vendor-controlled digital diagnostics, limited lifespan, planned obsolescence, and restrictive warranties compel you to either pitch your old goods or seek professional repair services. Either way, you lose.

Now here's where it gets funny. The EPA WasteWise program actually encourages consumer electronic manufacturers to *consider* "redesigning an electronic product so that it can be more easily upgraded or remanufactured" and "establishing a take-back program for electronic products from customers, and remanufacture or upgrade these products for resale." Oh sure, some electronics manufacturers considered it, but most of them found it to be a financially poor idea for their bottom line.

Does that implied ambivalence make you mad? It should. You know what we ought to do? Take our stuff back and hack it!

And just like the arrows in the recycling logo, the solution to this growing techno-junk nightmare comes right back to you and dumps itself in your lap. If you want to save the environment, then it's all up to you. But there is some help for you. It's in this book.

In the following pages, you'll learn a varied assortment of techniques for breathing new life into your old electronics. For example, need some ideas for salvaging an old television set, then read Chapter 19. Are you looking for some ways to resuscitate a dated cell phone, then Chapter 23 will get your creative juices flowing. Maybe you're yearning for a new bicycle. Forget it, just repurpose your old bike frame. And Chapter 2 will show you how to make a "roadie" out of your old clod buster. Toys, TVs, computers, pagers, coffee makers, automobiles, furniture, and even clothing can *all* be reused, recycled, or repurposed by reading this book.

You'll learn how to treat your global neighbors with respect—by recycling your old electronics. But what about those neat gadgets sold in the stores today? What are you going to do with your latest iPod bauble next year?

Well, Part 2 provides four case studies featuring the latest and greatest gizmos that are quietly changing our daily grind into a digital lifestyle. Roomba, iPod, and PSP are today's bestselling electronic names that could become tomorrow's recycling challenges. Case Studies 1, 3, and 4, respectively, will show you how to make these cutting-edge products as useful tomorrow as they are highly sought after today. One other big-name, mouth-watering, pocketbook-draining product rounds out the case studies in this section.

"Great," you say. "All of this recycling effort is well and good, but I'm not a hacker. How could I possibly repair a broken iPod?" Actually, it's easier than you think. You just need the right tools and some helpful instruction. Luckily, this book gives you both.

Tools of This Trade

Oh sure, here it comes, the tool list that will cost a king's ransom to acquire. Rest assured (and stop rolling your eyes), this tool list is reasonably priced, easy to find, and worthwhile for tasks other than just hacking.

Your basic tool kit for aiding you in the recycling of your techno-junk includes:

- Soldering iron
- Screwdriver set
- Multimeter
- Solar-powered battery charger (optional)
- Tamperproof bits (optional)

SOLDERING IRON. This will be your preferred "weapon" of choice for resurrecting dead electronics. Accompanied by solder, desoldering braided copper solder wick, and a soldering sponge, your soldering iron will be your best friend. And like your friends in life, it pays to choose your iron very carefully.

I personally use an ISO-TIP® Quick Charge Soldering Iron Model 7700 by Wahl Clipper Corporation (see Fig. I-1). Why? I've got three great reasons for using the ISO-TIP: cordless, portable, and rechargeable. These are three marvelous attributes when you're working on one of your electronic projects out in the field, too.

Over the years my soldering iron has proven to be versatile and rugged. I've replaced the built-in lamp, installed a replacement rechargeable battery, and dropped it more times than I can count. Plus it fits in my back pocket so I can make a quick solder joint whenever and wherever I want.

But I didn't realize how valuable my ISO-TIP really was until I was stranded without power in the aftermath of Hurricane Katrina. You remember that storm, don't you? Well, my portable ISO-TIP soldering iron was able to kludge together several solar panels for supplying my family with some valuable electricity during the darkest moments following the storm. In fact, my soldering iron worked great for three days of heavy use. And it was like the loss of a friend when the battery finally failed. Luckily, when power was restored to our area, I was able to recharge my soldering iron and I'm still using it today.

I-1 Wahl Clipper portable soldering iron.

SCREWDRIVER SET. Basically, there are two types of screwdrivers that you *must* own: one regular, flat, slotted or straight tip and one star or Phillips head variety. With these two types of screwdrivers (in a couple of assorted sizes), you will be able to open virtually 90 percent of your electronics. For starters, you should own four slotted tip screwdrivers (sized 1/8, 1/4, 5/16, and 3/8) and three Phillips head screwdrivers (sized No. 0, P1, and P2). If you foresee a lot of exotic recycling work (repairing electric lawn equipment, for example), then you might opt for a couple of larger tip sizes *and* longer handles. Nothing beats a big, fat, long handle when trying to loosen a stubborn screw.

MULTIMETER. Unless you're some kind of digital Svengali, it's impossible to tell whether a voltage line is hot or not. Before you attempt to fix any broken electronics, you'd better know whether or not the juice is flowing. In fact, this little bit of information could be a life-saving amount of knowledge. Coming to your rescue is the digital multimeter (DMM). Also known as a multi-tester, multimeter, or a multifunction voltage meter, these instruments are generally capable of reading DC voltage, DC current, AC voltage, resistance, transistor test, diode test, and battery test. Best of all, a DMM will output these readings on a liquid crystal display (LCD) screen (see Fig. I-2).

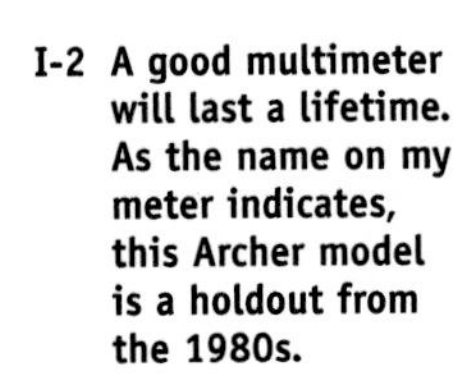

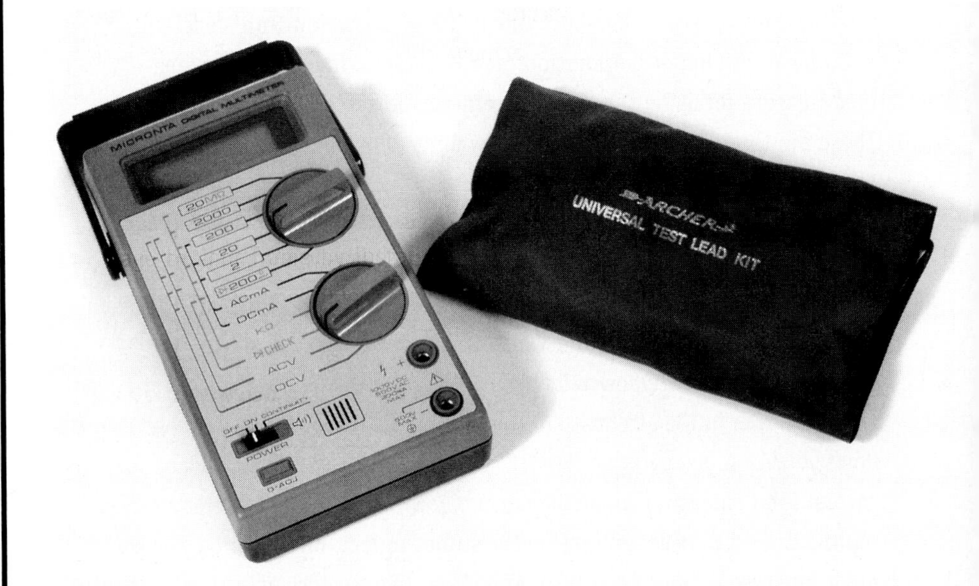

I-2 **A good multimeter will last a lifetime. As the name on my meter indicates, this Archer model is a holdout from the 1980s.**

One of the best deals in DMMs is available from Harbor Freight Tools. Item number 90899 is a seven-function DMM manufactured by Cen-Tech.™ Sold with a pair of 32-inch test leads, a low-battery indicator, and fuse and diode overload protection, this meter is able to test a variety of circuits with fairly accurate results. Best of all, however, is the price. Harbor Freight Tools sells this DMM for between $2.99 to $4.99 each! At that price is pays to buy a couple of them. In fact, I keep one in my car (along with a 35-mm nondigital, nonbattery camera) for testing surplus electronic equipment that I find at garage sales, auctions, and pawn shops.

SOLAR-POWERED BATTERY CHARGER. Dry cell batteries have become the main staple for powering our daily digital lifestyle. Unfortunately, they are also a primary contaminant of our environment. Enter the universal solar battery charger. Priced at under $15, the solar battery charger sold by C. Crane can accept two AA, C, or D size batteries at a time. The most common fault with these solar battery chargers is inadequate sunlight intensity. Luckily, a meter is built into this particular solar battery charger.

The intensity of sunlight can dramatically affect the battery charge time. For example, even with maximum sunlight, you will need about 9 to 18 hours for charging each pair of batteries. Having a meter will help you estimate your charging time. For example, a solar intensity of 120 mA will take approximately 12 hours worth of charging. In order to obtain the maximum charging capability, you can use the built-in prop stand on the bottom of this particular charger. This stand will help you achieve the best "angle of the dangle" for keeping your charger's best face toward the sun.

TAMPERPROOF BITS. Did you ever attempt to open a Game Boy Advance? If so, then you encountered a nifty little tamperproof screw. Known as a Triwing® screw, this fastener is meant to deter the owner from opening up their portable game console. Rookie hackers might think that they can simply wedge a slotted tip screwdriver into these screws and force them out. *Wrong-o, bucko.* Instead, purchase a set of tamperproof bits that feature a tip for handling Triwing screws. Sears sells the Craftsman 32-piece Security Screwdriver and Bit Set (Sears item #00947486000; Mfr. model #47486) for less than $25 which includes four Triwing bits, as well as a host of other tamperproof bits. Unless you regularly repair exotic digital electronics, you might want to consider a set of tamperproof bits to be an optional addition to your hacker's tool kit (see Fig. I-3).

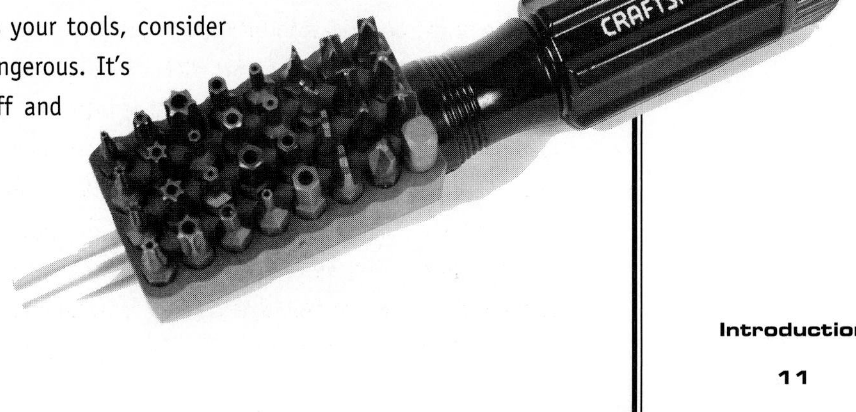

I-3 Tamperproof security fasteners are the bane of the hacker. This Sears tool will help you crack most security fasteners found on electronic gadgets.

Now that you have your tools, consider yourself armed and dangerous. It's time to take your stuff and hack it.

A SOLAR-POWERED BATTERY CHARGER GIVES FREE E-JUICE

1 A solar-powered battery charger is both fun and affordable to operate.

2 This charger is able to charge two batteries at a time.

3 The built-in meter helps you maintain the maximum sunlight strength for optimal charging.

4 Keep the charger's best face toward the sun—its PV array face.

1

2

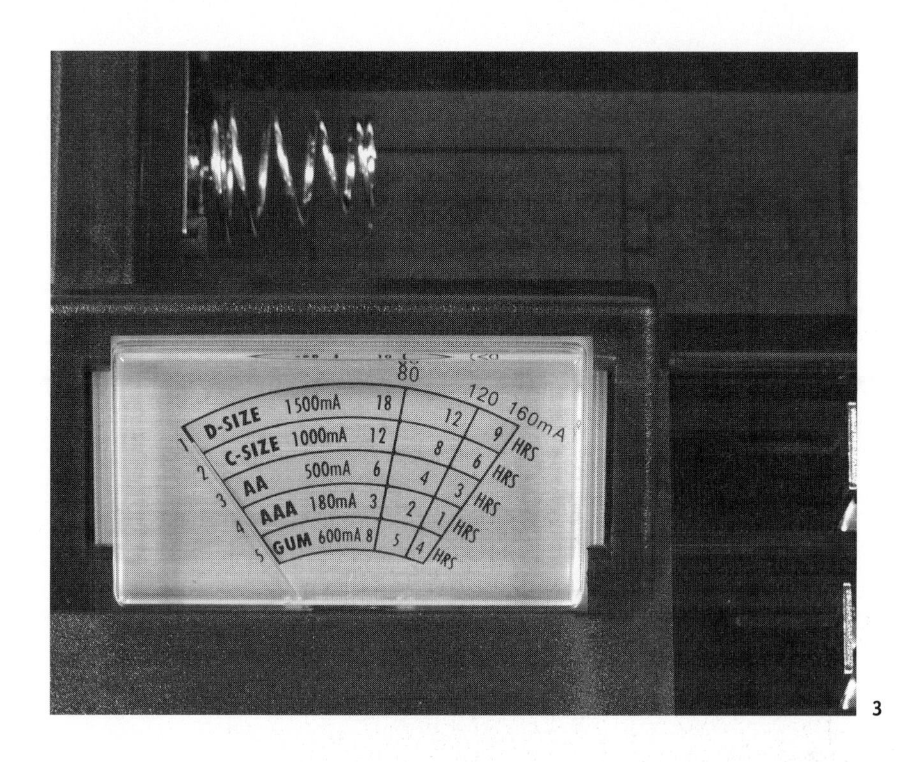
80
120 160mA
D-SIZE 1500mA 18 12 9 HRS
C-SIZE 1000mA 12 8 6 HRS
AA 500mA 6 4 3 HRS
AAA 180mA 3 2 1 HRS
GUM 600mA 8 5 4 HRS
1
2
3
4
5
3

4

PART 1

THE FINE ART OF HACKS, MODS, AND EXPANSIONS

CHAPTER 1

Baby You Can Mod My Rod

Can you name the first automobile that was hackable? OK, the Ford Motor Company Model T manufactured between 1908 and 1927 was a modder's dream rod. In fact, some models survived into the mid-1950s incarnated as surfer "hot rods" and "bucket Ts."

No, the car that I was thinking about was the 1972 Volkswagen Super Beetle. Now *that* was an auto that could enable any owner to "catch the hacking bug." Literally. Every owner of a "bug" quickly became a jack of all trades for repair, modification, and enhancement of their beloved VW—more out of necessity than artistic expression. Tuning the carburetor, replacing (or upgrading) the muffler, and installing a new cassette tape (or 8-track) deck were all "hacks" attempted by most bug owners.

The vehicle specifications were impressive for such a modest automobile:

ENGINE: Four-cylinder, 4-stroke O.H.V.-type rear-mounted engine with 96.66 cubic inches of displacement. The engine weighed 253 pounds, used 5.3 U.S. pints of oil, and culminated in a muffler with dual exhaust tips. There was a 10.6-U.S.-gallon fuel tank that was mounted inside the front luggage compartment. Fueling access was via an exterior latched door with a remote release located under the passenger-side dashboard.

TRANSMISSION: Four forward gear speeds with synchromesh and one reverse gear, all powering a rear axle drive shaft. A single-disc dry clutch was used for shifting.

PERFORMANCE: Fuel consumption mileage of approximately 26 miles per gallon with a "rated" maximum speed of 81 mph. This rating was beaten by me on one occasion in Kansas when I topped 90 mph.

DIMENSIONS: With a length of 158.6 inches, a width of 61 inches, a height of 59 inches, and a wheel base of 94.5 inches, a turning diameter of 34.2 feet was possible via a steering wheel that needed 2.6 turns to travel from wheel stop to wheel stop. The gross vehicular weight for this model was approximately 2,600 pounds with a "rated" maximum payload of 840 pounds. This rated payload was exceeded by me on several occasions at the University of Nebraska when I chauffeured seven fellow dorm rats between city campus and east campus.

Above all else, the Volkswagen Beetle was a reliable, dependable, and economical method of transportation. In 1972, the diminutive car caught the attention of the Smithsonian Institution when the 15-millionth Volkswagen Beetle, a Model 1131 Super Beetle, was donated to the National Museum of American History. And they continue to live on. I wish that I still had mine. Alas, in 1985, a microburst (i.e., a small intense downdraft of air) blew a garage down and it landed on top of my VW (see Fig. 1-1). Ironically, the superior engineering strength of the bug's egg-shaped design held the collapsed garage off of a neighboring Chevrolet Impala. Oh, the humility.

I iNstalled iPod® iNside my Car

Few of us drive 30+-year-old cars on a daily basis. So we must learn new methods for how to mod our modern cars. A classic demonstration of achieving this level of expertise can be obtained when attempting to install an iPod inside one of today's autos.

One of the most obvious product choices for integrating an iPod into a modern automobile is the Monster Factory-Linx™ iCruze® interface system. Although designed for supporting only Alpine aftermarket car stereos sporting the M-bus CD changer control, iCruze can be adapted to most other OEM stereo models with a supplemental interface module.

For example, in the case of trying to install iCruze inside a Dodge Caravan 2003 SXT, you would need the following iCruze products:

1

1-1 This is what it looks like when a garage lands on a 1972 VW Super Beetle—the horror.

- Monster Performance Car™ Factory-Linx™ iCruze™ OEM/CD Changer Interface Module, Model Name: MPC FX ICRUZ iPod Interface Module
- Monster iCruze™ Interface Module, Model Name: MPC FX IM-CHR1 Interface Module-Chrysler

As for the installation, beware of this note: The van's radio must be external CD changer–capable without factory navigation or an RBQ radio.

And, in case you're wondering whether this type of installation will void your van's warranty, here is the answer according to Monster: "No. It is against the law for your dealership to void your entire warranty because you installed an aftermarket component such as iCruze. Your dealership has the right to void a portion of the warranty that pertains only to the factory car stereo, but the remainder of your warranty remains intact."

Mod Rods with iPods

In 2006, the iPod was formerly accepted by the world's automotive industry as a product that they should support. Read this feigned support as: The automakers saw an opportunity to cash in by adding ten dollars worth of wire to a vehicle and charging a couple hundred dollars for the convenience of integrated iPod cachet.

During MacWorld Expo 2006 in San Francisco, the list of manufacturers supporting iPod included:

Acura
Audi
BMW; specifically models: Z4, X3, X5
Chrysler; specifically models: Pacifica, Sebring, Town & Country
Dodge; specifically models: Caravan, Grand Caravan, Neon, Ram, Stratus Sedan
Ferrari
Honda
Infiniti
Jeep; specifically models: Liberty, Wrangler
Mercedes-Benz; specifically models: C-Class, CLK, CLS, E-Class, SLK, M-Class, R-Class
Mini; specifically models: Cooper, Cooper S
Nissan
Scion; specifically models: xA, xB, tC
Suzuki; specifically models: Aerio SX, Grand Vitara
Volkswagen
Volvo; specifically models: S40, S60, S80, V50, V70, XC70, XC90

Others: in the Japanese market; Mazda, Daihatsu, smart, and Alfa Romeo

An easier option for installing an iPod inside your car is a car cassette adapter. This low-cost gizmo plugs into the line out port of an iPod and attaches to your car stereo via the cassette deck. In order to facilitate this attachment, the car cassette adapter is actually shaped like a cassette tape without the magnetic tape.

One of the best incarnations of this contraption that I've seen is the Sony CPA-9C Car Cassette Adapter. This adapter (which you can still purchase via the Apple Store® and amazon.com) was originally manufactured to integrate Sony Discman CD players with mobile cassette systems; like the one in your car. When coupled to an outdated iPod shuffle or similar MP3 player, the audio results are very good and the performance is acceptable. This is a great solution for recycling two obsolete products which most of us have lying around the house right now: a car cassette adapter and a castoff iPod shuffle.

Configuring this hack couldn't be simpler:

1. Turn on your car stereo cassette deck.
2. Configure your cassette deck by selecting "normal" or Type I with the Tape Selector button and turn OFF Dolby NR, Repeat, and Blank Skip switches.
3. Push the car cassette adapter into the car's cassette deck and press the Play button.
4. Plug the car cassette adapter's mini-jack into the line out port of the iPod shuffle or similar MP3 player (where the ear buds are connected).
5. Turn on the iPod shuffle.
6. Press the iPod shuffle's Play button.
7. Tuck the iPod shuffle away someplace safe and enjoy.

NOTES

- Use the car cassette deck's volume control for adjusting the sound output.
- If you don't hear any music in the car's speakers, press the Tape Change Direction button on the cassette deck.
- Disconnect the car cassette adapter by pressing the Eject button on the cassette deck.

A rap against this type of arrangement is that your salvaged iPod shuffle could lose its battery charge during lengthy trips. What you need is a cigarette lighter

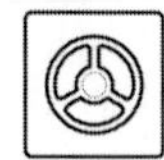

to USB port charger adapter. One example of this adapter is the Lenmar AI-DCU iPod Charger with USB Charge Port. Just plug the adapter into the car's cigarette lighter (must be 12 V) and plug your iPod shuffle into the adapter's USB port. Now your iPod shuffle will charge while it is playing music through the cassette adapter. Luckily there is a red charging LED built into the Lenmar adapter. This subtle warning light will help you remember to disconnect the iPod shuffle when you turn off the engine.

Who's Drivin' This Thing Anyway?

Can you imagine this scenario? You're driving along listening to your master-blaster iTunes™ playlist, talking on your cell phone, and checking the contents of your hard drive. What? Are you nuts? Talk about motor vehicle suicide. How could you check the contents of your hard drive while driving? Tee-hee, check this out.

First of all, yes, you are correct—you should *never* perform any ancillary tasks while driving. When you're behind the steering wheel, just drive. OK, enough warnings from your mother. Here's a super hack or mod for any car, truck, or SUV. Take a computer that you have lying around, load it up with some great MP3s, JPEGs, and MPEGs, attach an LCD screen, and connect a digitizing tablet. Assemble all of your goodies inside your vehicle and you have a terrific mobile entertainment platform.

Now throw in a potent Wi-Fi port, a massive storage system, and some competent support software and you'll be able to commute in computing bliss. Well, I wanted to test this hypothesis, so I repurposed a spare iBook G3 notebook computer and installed it inside my family van. I considered this test to be a "proof of concept."

Regardless of your preferred computer operating system, here was my first "hack" at a low-cost and painless method for integrating a computer into my automobile:

1. I first added a leftover Airport Wi-Fi interface card to my Apple Computer iBook. This notebook computer was formatted with Mac OS 10.3 "Panther." It also had iTunes, iPhoto, and iMovie installed on the 20-Gb hard drive. Any of the older versions of Mac OS X would work, as well. Just head over to Shreve Systems and purchase an operating system disc set for less than the cost of a current release DVD movie.

2. I set up a user account on this mobile system for accepting wireless uploads of music, pictures, and movies. This feature enabled me to sequester the main CPU and keyboard out of sight with only the LCD screen being visible.
3. I separated the LCD screen from the keyboard/hinge and installed both components inside my van.
4. A recycled digitizing tablet was connected to one of the iBook's twin USB ports. Power for this tablet was drawn from the USB port. I used a Wacom Graphire3 Blue 4" x 5" Pen Tablet.
5. The van's power supply was bridged from the cigarette lighter to the iBook with a Kensington 70-watt DC Power Adapter for Apple® Notebook (Model: 33194).
6. I fired it up and tested it. And it worked, kinda.

LESSONS LEARNED FROM THIS TEST: Well, as you might have expected everything didn't go exactly as I had planned. First, when the iBook was placed into "sleep" mode, the digitizing tablet was unable to "wake" the computer back up. Only a key press on the keyboard was able to reactivate this mobile system. Rats. That was a major disappointment.

Next, the screen was way too big for easy installation inside the van. Likewise, the main CPU and keyboard were still tethered to the LCD screen making the installation even more awkward. There had to be a better way.

ON TO PLAN B

And Plan B was spelled Bluetooth. First I had to make a trip to PowerMax, the Apple Store, and my techno-junk box. Additionally, there were a couple of more parameters that had to be accommodated by this new "Plan B" mobile entertainment system.

First of all, the final installation had to be neat and tidy. No cables or bulky displays would be tolerated. Next, I was expressly forbidden from messing with the vehicle's wiring, radio, and speaker system. The thinking there was, "If it ain't broke, leave it alone because we might still need it later after this harebrained idea flops." Finally, this revised system must be easily operated by *any* occupant *anywhere* inside the van. Luckily, that's where Bluetooth came to the rescue.

So, here's the laundry list needed for making Plan B work:

Apple Computer Mac mini

- 1.25-GHz PowerPC G4
- 512-MB DDR333 SDRAM – 1 DIMM
- 40-GB Ultra ATA drive
- Combo Drive (not really needed)
- Internal Bluetooth + AirPort Extreme Card
- 56-K v.92 Modem (not really needed)
- Mac OS X 10.3.9 "Panther"
- ATI Radeon 9200 with 32-MB DDR video memory

Apple DVI to Video Adapter

- connecting the DVI port on Mac mini to A-V composite input of display

Logitech Cordless Desktop LX 501 keyboard, mouse

Wacom 6 x 8 Graphire Bluetooth (if you don't want to use the wireless mouse)

SFBags Waterfield Designs Mac mini Sleevecase

- a durable ballistic nylon exterior protects against scuffs and scrapes, and a soft lined 6-millimeter neoprene interior cushions the Mac mini from bumps

JBL Encounter™ 2.1 speakers

- a compact high-performance sound system that connects to the audio out port with separate subwoofer
- 1/8-inch (3.5 millimeter) mini stereo jack
- Satellite: 13 watts per channel
- Subwoofer: 34 watts

Audiovox PVR1000

- 32-Mb on-board memory; SD/MMC memory; 1-Gb maximum
- USB interface
- 2.5-inch color TFT LCD
- Speaker
- Microphone: Built-in condenser
- Quality settings: Voice recording, audio/video recording: low, high
- TV system: NTSC/PAL
- Clock, Calendar, and Games (tic-tac-toe and puzzle)
- A/V input: 1 minijack 1/8-inch composite–video + audio
- Headphone jack: Yes (.125-inch)
- Hold switch: Yes

More Rod Mods

Here is a list of other projects that could be used for tricking out your ride:

Where R U? Install a Garmin GPSMap 60-Cx inside your car.

Mail Me. Schedule your onboard wireless computer to log onto a local Wi-Fi network and grab your e-mail—great for reading in those lengthy traffic jams.

Back OFF! Equip your wheels with a motion detector coupled to a digital camera. Then you can wirelessly e-mail a photo of your car thief to yourself. SMILE!

Remote Lock Down. Refer to your car's wiring diagram and connect a USB data acquisition interface relay controller such as one from Ontrak Control Systems between your electric door locks and your onboard wireless computer system's USB port. Just send a Wi-Fi (or Bluetooth) signal to your computer for relay to the door locks and "Open sesame."

- Power source: Battery or AC/DC power adapter (supplied)
- Batteries: Rechargeable lithium-ion x1 (internal/supplied)
- Battery life: Information not available
- Dimensions: 3 × 3.7 × .78 inches
- Weight: .25 pounds (without battery)

NOTE: I discovered the PVR1000 while I was writing my bestselling book, *PSP Hacks, Mods, and Expansions* (McGraw-Hill, 2006). If you can find one cheap (e.g., under $80), grab it. The parts and features are great for repurposing with other hacks and projects. Oh, and if you ever see the message "PVR ERROR" appear on your LCD screen, this product is totally dead. A doorstop. I had one PVR1000 display this informative screen right out of the box. Lovely. Get this: Those kind folks at Audiovox actually replaced the dead PVR1000 at no charge and returned it to me in less than one week. Now that's customer service. I wonder if they smile a lot.

The moment of truth was hooking everything together and then flicking on the switch. Hey, it worked. How, you ask? Well, in a nutshell, the Mac mini and speakers went in the back, tucked inside one of the storage areas that populate vans like ticks on a dog, while the PVR1000, keyboard, and Wacom tablet were housed up front on the dashboard, underneath the passenger seat, and in the lower front storage bin, respectively. Ideally, a center armrest storage console similar to the one found in a Dodge Ram 1500 would be the perfect housing for most of these components. In this case, the adage "your mileage may vary" is apropos.

Now I didn't have to carry my iPod with me in the car. I could just play my music through the iTunes interface that had been preloaded on the Mac mini. Likewise, I could add new songs/albums via the Airport card and a Wi-Fi connection.

One exciting fringe benefit that I didn't even realize I had was that I had inadvertently transformed my family van into a powerful, mobile, wardriving machine. Wardriving? Wardriving is a Wi-Fi "sniffing" or search activity where you locate and identify wireless networks without the knowledge of the network's administrator. The Airport utility built into the Mac OS X makes this capability a snap to use. Just drive around and watch the utility display all of the wireless networks that it can read. Believe me, you'll be amazed at what you'll discover.

HOW TO ADD A WI-FI NETWORK TO YOUR CAR

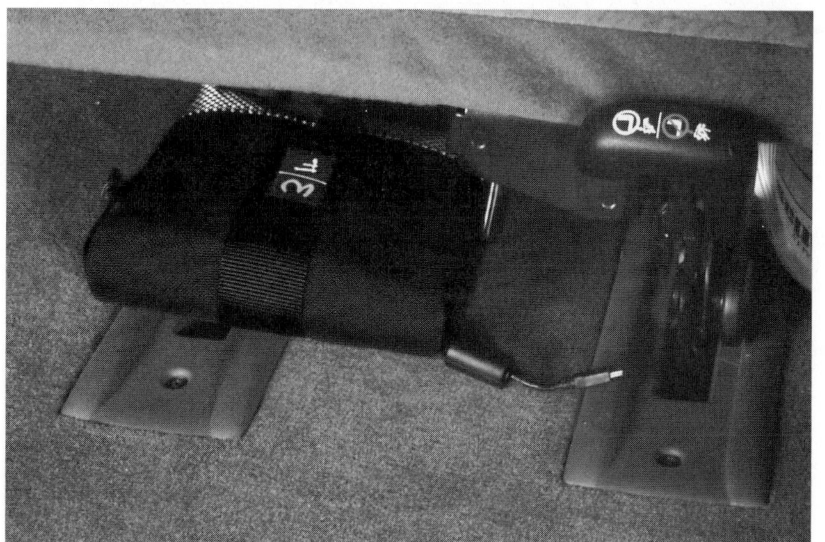

1

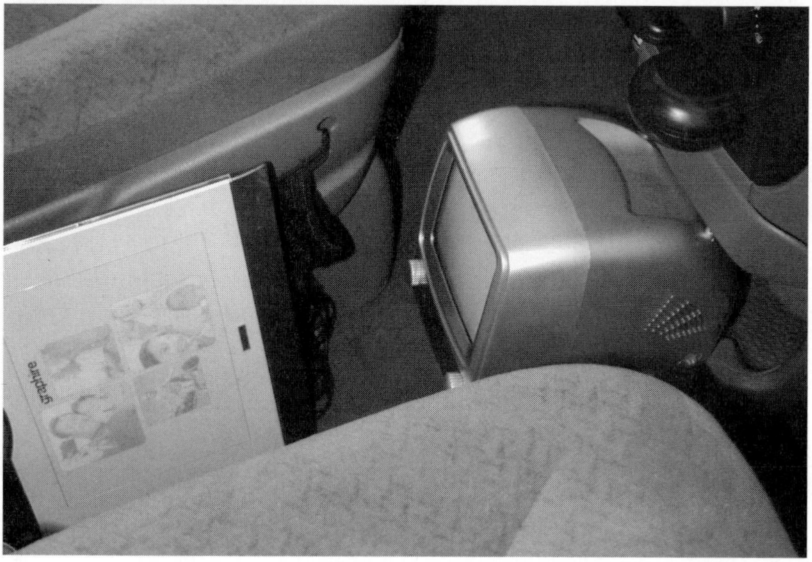

2

1 An Apple Computer Mac mini fits great under the back seat safely ensconced in a Waterfield Designs case.

2 Add a Wacom Bluetooth tablet and 12-V portable TV.

3 Wacom Graphire, 6 x 8 Bluetooth tablet.

4 Before using your Wacom tablet, you must insert the battery.

5 Likewise, the battery must be charged outside of your vehicle.

6 The Mac mini must recognize the Wacom tablet. Press the Connect button to begin this handshake process.

3

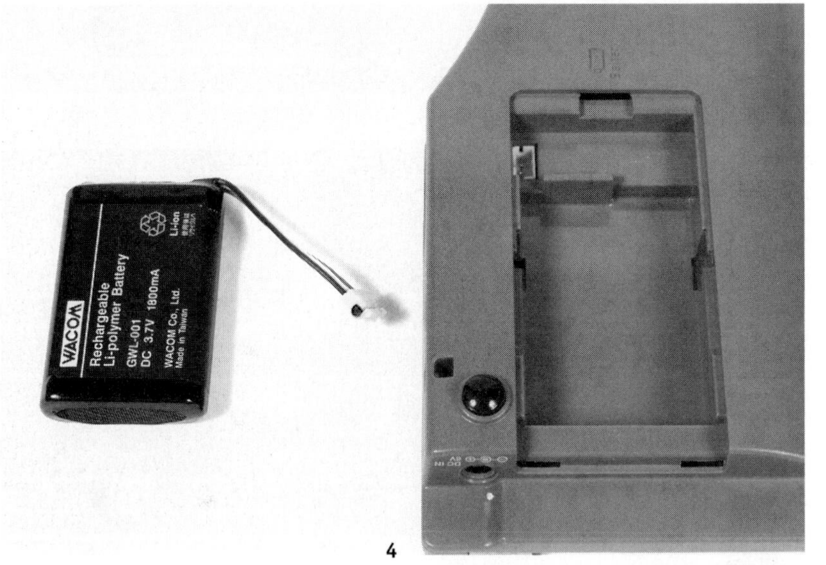

4

WACOM
Rechargeable
Li-polymer Battery
GWL-001
DC 3.7V 1800mA
WACOM Co., Ltd.
Made in Taiwan
Li-ion

5

6

7 This Waterfield Designs case is made for holding one Mac mini in dirty, dusty spots like a trunk or under a car seat.

8 A pull tab on the bottom of the case helps with removing the Mac mini.

9 Store your Wacom Bluetooth mouse and pen in the glove compartment.

10 The Bluetooth tablet is ideal for navigating around your OS desktop and it will wake up a sleeping Mac mini. You will not need a separate keyboard for waking this computer.

11 A wireless keyboard can also be used for navigating a PC desktop.

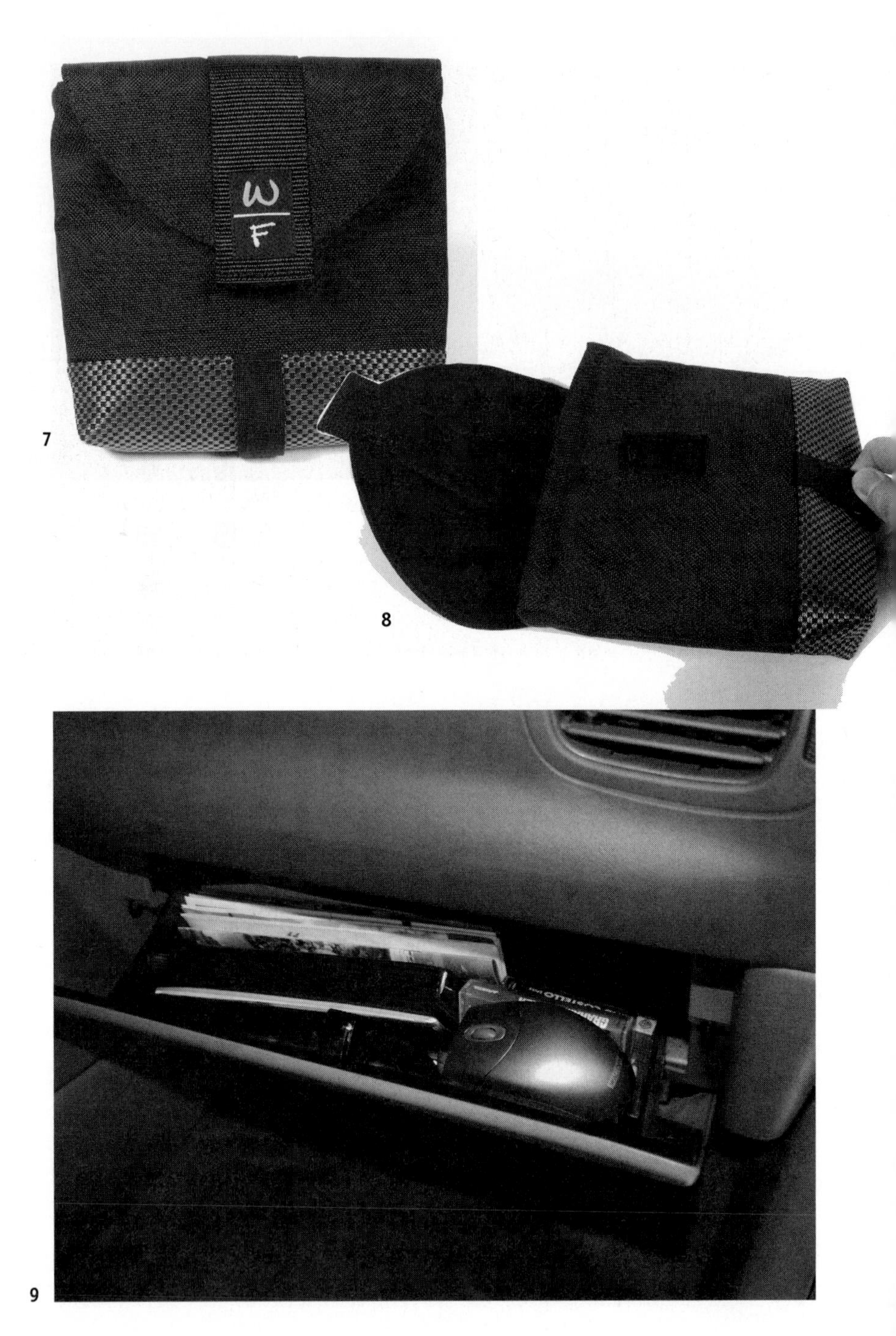

7

8

9

10

11

12 If your PC doesn't have Bluetooth connectivity, Logitech has a series of keyboards for you—LX Series Cordless Desktops.

13 LX Series Cordless Desktop for PC.

14 You can use your PC serial connection with the LX Series.

15 Or, you can use a USB port.

16 A special receiver is attached to your PC for talking with the Cordless Desktop.

17 The antiquated cigarette lighter port in your car is a great source for tapping power.

18 Boxwave makes a cigarette lighter adapter, VersaCharger, that can convert DC to AC voltages.

19 A retractable conventional AC plug is available with VersaCharger.

20 Boxwave also makes a Firewire + USB charger.

17

18

19

20

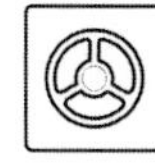

HOW TO BRING YOUR PDA ALONG IN YOUR CAR

1 Boxwave makes a Mobile Mounting Kit for holding a PDA inside your car.

2 A flexible shaft is attached to the Mobile Mounting Kit's base.

3 The base holds your PDA.

4 Move your PDA to a convenient position.

5 One final nifty Boxwave product is an all-in-one touch stylus, pen, laser pointer, and blue flashlight called OmniPen Pro. It beats looking for that PDA stylus while dodging traffic.

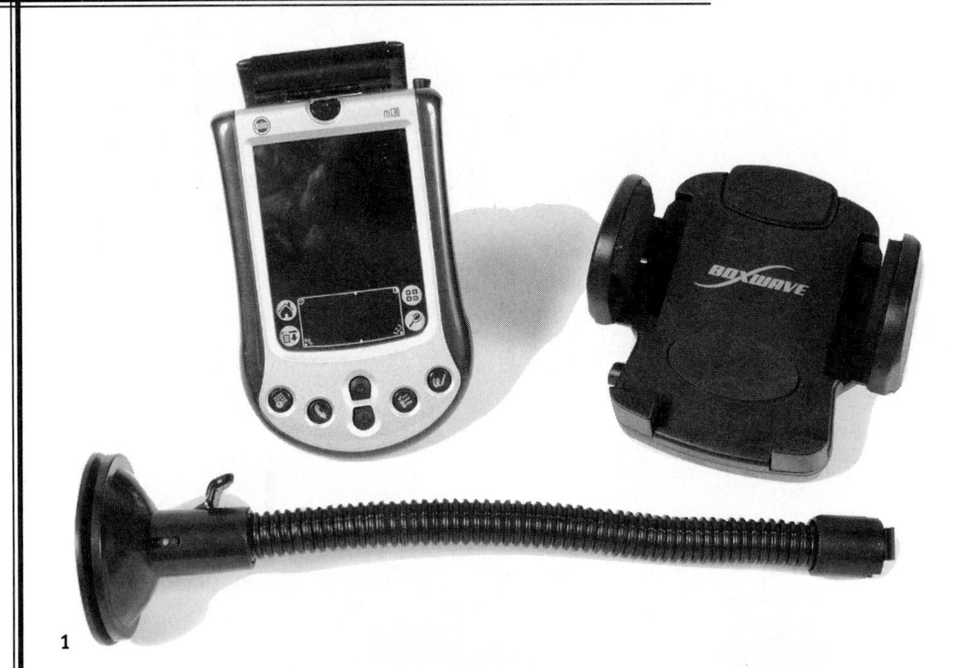

1

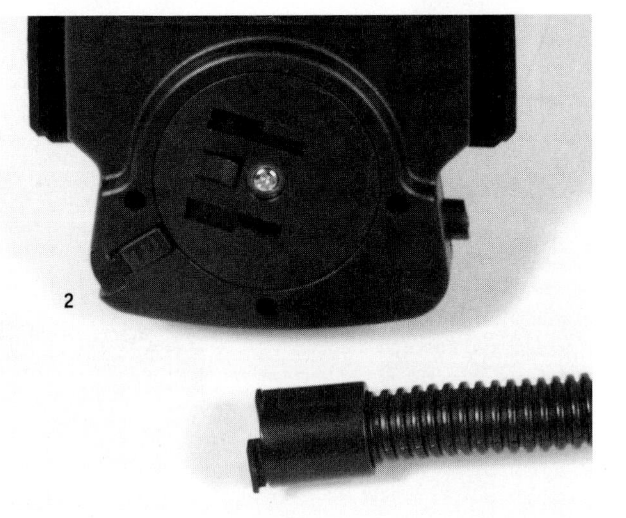
2

3

4

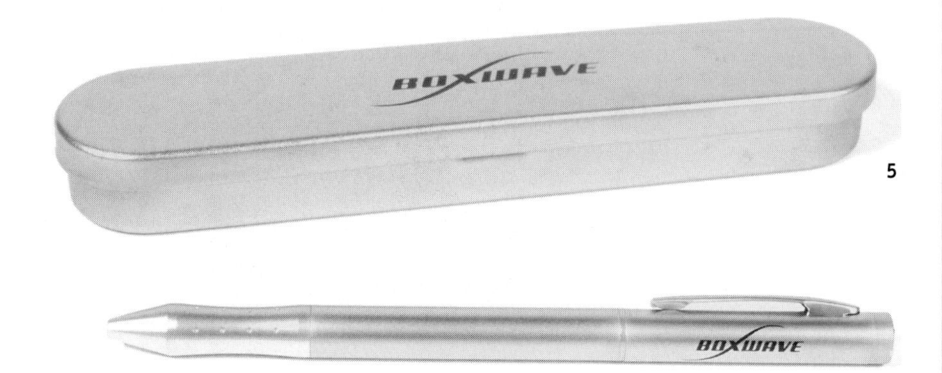

5

CHAPTER 2

Twist Dem Nuts

In February 2005, California Assembly Member Betty Karnette, D-Long Beach, introduced a bill (AB 1103) that would add a $7 disposal fee surcharge to the price of every new bicycle sold in California. Later the bike's owner would receive a $3 rebate when turning the bike into a state-certified recycler. According to Karnette the impetus behind her introduction of AB 1103 was threefold: reduce landfill waste, reuse and recycle bicycles, and reduce pollution.

Reaction to the introduction of AB 1103 has been varied. For example, not surprisingly, the Environmental Caucus of the California Democratic Party recommended support for this legislation. At the other end of the spectrum, on April 1, 2005, the California Manufacturers and Technology Association formerly announced that they would "either independently or through coalitions, actively oppose this bill."

An editorial opinion in *Silicon Valley/San Jose Business Journal* (April 22, 2005) labeled the surcharge a "tax" and suggested that the bill might be a cure without a problem. Claiming, "AB 1103 is not backed by any new studies on recycling patterns. It would punish people" who purchase bicycles.

Oddly enough, an article in *Bicycle Retailer and Industry News* (February 25, 2005) indicated that a professional lobbyist with Capitol Advocates, Ravi Mehta, had contacted an unspecified number of bicycle manufacturers trying to form a coalition to torpedo the bill but reported that "no companies have signed on."

Oh, in case you're wondering, sales figures published in *Bicycle Retailer and Industry News* indicate that between 3 and 4 million bicycles are sold each year in California. You do the math.

AB 1103 is not the only way to form a bike recycling program, however. The Arcata, California Bike Library loans bicycles for a six-month period at a cost of a $20 refundable deposit. According to the program's mission statement, this volunteer-based organization hopes to "inspire people to bicycle more often."

In Bethlehem, Pennsylvania, Coalition for Appropriate Transportation regularly accepts donations of used bikes at the Bethlehem Bicycle Cooperative. These recycled bikes become mainstays in shipments for the "Bikes for Africa" campaign with the International Transportation Development Program.

Meanwhile over in Boston, Massachusetts, Bikes Not Bombs gathers up about 3,000 used bikes each year. From this inventory, about 2,400 are distributed annually in various African nations through an economic development program.

In fact, there are lots of bike recycling programs out there. The International Bicycle Fund sponsors a list of organizations that recycle bicycles. While most of these organizations are based in the United States, they regularly ship large numbers of recycled bicycles to disadvantaged nations around the world.

It Is About the Bike

One of the best places to start a bike recycling program, however, is in your own garage. Just because you're switching bike sizes (e.g., your six-year-old needs a bigger bike) or changing your riding habits (e.g., cycling on dirt trails instead of driving along paved bike paths), you *don't* have to throw away an old bike and buy a new one.

First of all, let's look at some numbers. During 2004, through the month of August, according to the U.S. Department of Commerce, 661,031 bicycles were imported into the United States for sale. This number translated into $322.19 million in sales which, in turn, drove a market of $1 billion in components, apparel, parts, and accessories.

That's a lot of new bike stuff. Trying to nail down similar numbers and statistics for bikes that are dumped into landfills is very difficult. Returning to California, Assembly Member Karnette yielded some plausible figures. According to Karnette, California landfills accept approximately 250,000 used bikes per year. That's a lot of old bike stuff, too.

Other than dumping them in landfills, what can we do with these used bikes? Well, one man's junk is another man's million-dollar art business.

Founded in 1991 on a flat bicycle tire by Graham Bergh (he thought that the limp inner tube was an interesting product, took it home, and fashioned it into a stereo speaker hanger), Resource Revival has become the largest manufacturer of recycled gifts and home furnishings in the United States. We're not talking about some shoddy welded bike frames that have been thrown together in the name of art, either.

Joining forces with artists Andy Seubert and Kif Scheuer, Bergh and company design bottle openers, candle holders, desk accessories, and even furniture fashioned from recycled bike parts. The museum stores at the Whitney Museum of American Art and the Guggenheim Museum began selling their products in 1996. Even the *New York Times* is impressed with this recycled art form and exposes the talents of Resource Revival to the world. Resource Revival had become the darling of the art world.

OK, what about that bike in your garage—is it trash or treasure? Probably a little of both.

Other than dumping your bike, there are two great options for getting some new life out of those old wheels. You can either upgrade it or you can modify it.

A Betta Beta

The bike world today is buzzing with a new type of bike: the Women's Specific Design or WSD bike. Sounds great, doesn't it. A bike made specifically for a certain type of rider. If you're a woman, however, and you're trying to locate WSD products, forget it.

Which is the best bicycle part for my bike? Where can I find the cheapest cycling apparel? What is the best gear system for riding in city traffic? How do you calibrate a cyclocomputer? In today's WSD cycling world, all of these types of questions are answered by trial and error; or, word of mouth. There is no other real source for this type of information.

In contrast, the market for WSD bikes, components, apparel, and accessories is growing at an unprecedented rate. Growth rates of 38 to 44 percent have been reported by the Outdoor Industry Association's 2004 Outdoor Recreation Participation Study for all facets of cycling which feature women's involvement—road, off-road, and cyclo-cross. Gone are the days of step-through frames, tassels, and low-grade components; today's WSD bike is a high-tech, performance model that rivals the male models.

Just what makes a WSD bike different from another bike? Well, in the world

Complete Amended Text for AB 1103

BILL NUMBER: AB 1103

AMENDED BILL TEXT

AMENDED IN ASSEMBLY APRIL 12, 2005

INTRODUCED BY Assembly Member Karnette

FEBRUARY 22, 2005

An act to add Section 39006.5 to the Vehicle Code, relating to bicycles.

LEGISLATIVE COUNSEL'S DIGEST

AB 1103, as amended, Karnette. Bicycle recycling sales: notice. (1) Existing law requires each bicycle retailer and bicycle dealer to supply to each purchaser a preregistration form provided by a county or city that has adopted a bicycle licensing ordinance or resolution and to include, on the sales check or receipt given to the purchaser, a record of certain information. A violation of this requirement is an infraction pursuant to other provisions of law. Existing law, the Personal Income Tax Law, provides for the application of specified provisions of the Internal Revenue Code, relating to itemized tax deductions. This bill would require a bicycle retailer and bicycle dealer who sells, furnishes, or gives a bicycle to any person to provide a disclosure to the purchaser or recipient that states that a bicycle may be recycled and that state law permits a tax deduction for qualified charitable contributions of bicycles, as specified.

Because a violation of this requirement would be an infraction, this bill would impose a state-mandated local program by creating a new crime. (2) The California Constitution requires the state to reimburse local agencies and school districts for certain costs mandated by the state. Statutory provisions establish procedures for making that reimbursement. This bill would provide that no reimbursement is required by this act for a specified reason.

THE PEOPLE OF THE STATE OF CALIFORNIA DO ENACT AS FOLLOWS:

SECTION 1. It is the intent of the Legislature to support the recycling and reuse of bicycles sold, possessed, used, or disposed of within the state. The Legislature finds and declares all of the following: (a) The state supports qualified charitable organizations that refurbish bicycles for youth programs or alternative transportation programs by allowing tax deductions for donating bicycles to qualified charitable organizations. (b) Bicycle recycling or bicycle donations are the preferred methods to dispose of bicycles. (c) It is in the best interest of the state to support bicycle recycling and reuse programs.

SEC. 2. Section 39006.5 is added to the Vehicle Code, to read: 39006.5. (a) A bicycle retailer or a bicycle dealer who sells, furnishes, or gives a bicycle to any person shall provide a disclosure to the purchaser or recipient that states that a bicycle may be recycled and that state law permits a tax deduction for a qualified charitable contribution of a bicycle. (b) The disclosure required under subdivision (a) shall meet both of the following requirements:

(1) Be printed in not less than 14-point boldface type on a single sheet of paper that contains only the disclosure.

(2) Include the following statement: "THE STATE OF CALIFORNIA OFFICIALLY SUPPORTS RECYCLING OR CHARITABLE CONTRIBUTIONS AS THE PREFERRED METHOD FOR THE DISPOSING OF BICYCLES. AT THE TIME OF DISPOSAL, BICYCLE OWNERS ARE ENCOURAGED TO DONATE THEIR BICYCLES TO A QUALIFIED CHARITABLE ORGANIZATION TO INCREASE THE AVAILABILITY OF AFFORDABLE ALTERNATIVE TRANSPORTATION IN CALIFORNIA COMMUNITIES."

SEC. 3. No reimbursement is required by this act pursuant to Section 6 of Article XIII B of the California Constitution because the only costs that may be incurred by a local agency or school district will be incurred because this act creates a new crime or infraction, eliminates a crime or infraction, or changes the penalty for a crime or infraction, within the meaning of Section 17556 of the Government Code, or changes the definition of a crime within the meaning of Section 6 of Article XIII B of the California Constitution.

All matter omitted in this version of the bill appears in the bill as introduced in Assembly, February 22, 2005 (JR11).

of the WSD road bike, the basic differences can be distilled down to the following short list of tangible traits.

> **GET A GRIP.** Change the bike's geometry for centering the rider's weight over the bike which reduces your body weight from resting on your hands. Also, shorter reach brake levers are built into the handlebar.
>
> **THE WEIGHT OF THE WORLD ON YOUR SHOULDERS.** At least the weight of your body is more evenly distributed via a shorter top tube and narrow handlebar.
>
> **KISS MY SEAT GOODBYE.** There is nothing worse than riding a bike that gives you saddle sores. A different bike geometry and different saddle shape minimizes pressure on your pubic bone making your seat better to sit on.
>
> **YOU'RE A PAIN IN THE (LOWER) BACK.** A shorter top tube results in a less aggressive riding posture and a more comfortable ride.

Whether your ride is a WSD or another type of bike, you can add some extra mileage to your frame by installing a simple cycling computer. Typically, I would never recommend, let alone review, a computer product with the word "beta" in its name; but the Echowell Beta 1 Bike Computer is a rarity in the current crop of cycling computer products. It's inexpensive and it works—right out of the box. In fact, the Beta 1 is terrific for an "under $10" bike computer. Brought to you by the same people who make "Electronic Dart Games," the Echowell Beta 1 remarkably shares the same electronics as found in some higher priced "brand name" bicycle computers.

More commonly known as cyclocomputers, these devices are multifunction speedometer, odometer, chronometer gauges with digital displays and battery-powered operation. Consisting of three basic components: sensor (attached to your front fork), magnet (fixed to a front wheel spoke), and main digital unit (mounted on your handlebar), the original cyclocomputers used a cable for connecting the sensor to the main digital unit. Second-generation cyclocomputers were able to eliminate the need for a connection cable and relied on a wireless communication system between the sensor and the main digital unit. The most recent, or third-generation, cyclocomputer employs geosynchronous GPS systems for precisely determining the speed and distance traveled for your bike.

According to Echowell, the first-generation Beta 1 cyclocomputer provides five functions: current speed (in either kilometers per hour or miles per hour), trip distance (in either kilometers or miles), bicycle cumulative odometer (permanently stored until battery reset in either kilometers or miles), current time on a 12-hour clock (devoid of AM or PM), and a dubiously claimed "function" that scans through trip distance, bicycle odometer, and current time while continuously displaying current speed. Furthermore, an automatic start/stop operation activates the Beta 1 as soon as the magnet rotates past the sensor. Likewise, the Beta 1 will power down after approximately 10 minutes of wheel inactivity; although the current-time 12-hour clock is continuously displayed even when the power is off. All of these features are moot, however, if both the installation and calibration procedures are not followed religiously.

And the Beta 1 installation is as easy as 1-2-3. However, there are two points regarding this installation procedure that bear elaboration. First of all, you must be absolutely certain that the distance between the spoke magnet and the fork sensor is not more than 6 millimeters. Yes, even though the

A WSD Bike Sampler

If you're in the market for a new bike, two of the leading bicycle manufacturers offer several WSD models from which to choose:

Fuji

- WOMEN'S
- Finest RC
- Finest 1.0
- Finest 2.0

Trek

- Madone SL 5.9 WSD
- Madone 5.2 WSD
- Pilot 5.2 WSD
- Pilot 5.2 s.p.a. WSD
- Pilot 5.0 WSD
- Pilot 2.1 WSD
- Pilot 1.2 WSD
- 5000 WSD
- 2200 WSD
- 1500 WSD
- 1000 WSD

Echowell documentation recommends a gap distance less than 4 millimeters, I found in my ten different installation test cases that a distance less than 6 millimeters will yield reliable results. Please let me caution you, though, that the smaller the magnet/sensor gap, the better. In most cases, however, a gap distance of even 6 millimeters can be impractical, for example, with suspension forks. In this type of difficult installation, strive for a minimal gap and worry about calibrating your Beta 1 later.

My other installation procedure that needs to be highlighted is the use of clear packing tape for attaching the Beta 1 cable to your bike's brake and

Building a Roadie

Here is a laundry list for the components that would be used for building your own economical road bike:

- Mavic CXP22 Black/Shimano 105 9-Speed Black Wheelset
- Shimano 105 STI Levers
- Nashbar Road Frame
- Kestrel EMS Pro Carbon Road Fork
- Nashbar ISIS Road Crankset 53/39
- Nashbar Racing Saddle
- Hutchinson Fusion Long-Distance Road Tire
- Shimano 105 Rear Derailleur
- Shimano 105 HG70 9-Speed Cassette
- Nashbar Jail Brake Road Calipers
- Shimano 105 Front Derailleur
- Cane Creek C-1 Threadless Headset
- Nashbar ISIS Bottom Bracket
- Nashbar Road Seatpost
- Nashbar Ergo Road Handlebar
- Nashbar Threadless Road Stem
- Shimano HG53 9-Speed Chain
- Nashbar Deluxe Handlebar Tape
- Nashbar Small Presta Road Tube

shifter cables. Basically, short strips of clear packing tape are used for binding the Beta 1 cable to your bike's front cabling. This is a departure from the Echowell instructions. In those instructions, Echowell uses two supplied cable zip-ties for securing the Beta 1 cable slack to the front fork. I felt that the packing tape makes for a cleaner installation, but use whichever method suits your bike and your taste. Just remember to leave adequate slack in the Beta 1 cable for allowing steering movement and, in the case of suspension forks, shock absorption compression.

After this painless installation, it is time to calibrate the Beta 1. Now before you groan with an anticipated measurement of your front wheel's circumference and its subsequent input into the Beta 1, I have streamlined the Echowell instructions for ensuring that your calibration is both simple and painless. In this three-step calibration addendum to the supplied instructions, you can be guaranteed that your Beta 1 will accurately display speeds and distances. Here's how:

1. Use the supplied Beta 1 wheel size table and enter your stated calibration Setting Value into your Beta 1. For example, if you installed your sensor on a 26- × 1.5-inch wheel, then enter 2026 into the Beta 1. Don't worry, this setting will probably be wrong.

2. Now find a track that has well-defined and accurate distances calculated for it. A college or university outdoor track is ideal for this step. [Note: Most colleges dislike bicycles hogging their track. Make sure that you obtain the track manager's permission before you begin your calibration tests.] Now begin your first test drive. I found that using distances greater than 2-mile increments are ideal for a Beta 1 calibration. Just remember to forget about your speed and concentrate instead on your trip distance.

3. After cycling your prescribed track distance, check the trip distance reading on your Beta 1. If they match (or, are close enough for your tolerance), then your Beta 1 is correctly calibrated and ready for use. Depending on whether your stated trip distance is above or below your track distance, you will need to adjust your Beta 1 with a new Setting Value. As a starting point for your recalibrations, measure the sensor's height above the ground in inches, multiply this value by 25.4, then multiply that resulting value by 3.14. Enter the final product as your

new "beginning" Setting Value. Then redo Step 2. Repeat these steps until you get a track distance–trip distance match.

During my calibration tests of the Beta 1, all ten of these test cases required numerous tests/retests and calibrations/recalibrations. Here is a test/calibration sampling:

Starting Setting Value: 2026
First adjustment: 1555
Second adjustment: 1722
Third adjustment: 1802
Fourth adjustment: 1882
Fifth and final adjustment: 1962

One thing that these test cases showed me was that a Setting Value of 80 is roughly equal to one-tenth of a mile. You can use this equivalence as you fine tune your Beta 1 calibration.

While reading this calibration process might seem daunting, once you've labored through this relatively simple series of tests and adjustments, you will have a fully functional speed and distance gauge for your bike that works. The Beta 1 looks great, works great, and, best of all, it's priced great.

Just Let It Ride

More times than not, you will just want to modify your current bike frame for better performance or to make it more comfortable. These are easy mods that can save an otherwise good bike from the landfill.

Modifying your current bike is usually a two-step process. First, you have to identify the components that are needed for giving your old wheels a new lease on life. Secondly, once you have the components in hand, you have to install them on the frame. Both of these tasks could be incredibly daunting and challenge even the most knowledgeable bike repair shop if it wasn't for the leading online bike component shop in the United States—Bike Nashbar.

Bike Nashbar is a former mail-order business that has struck it rich in digital gold with the opening of its Web-based e-commerce site. Their inventory is enormous, ordering is secure and painless, and, best of all, their prices are unbelievably low. So low that you might buy all of your components, accessories, and cycling clothing from them rather than your local bike shop.

If It Doesn't Fit, Then You Must Assemble It

Have you ever been frustrated with finding the "perfect" bike for you? Or, have you had to settle for a stock manufactured bike that a salesperson assured would be "perfect for you"? Then after a couple of days worth of riding your neck ached, your back ached, and your hands ached; not to mention that your butt was a total disaster.

There is one common alternative to this painful dilemma, however. The custom-built bike has long been sought after as the Grail for cycling purists. Commanding prices that start at $1,500, for just the frame, only the most wealthy hobbyist or the professional competitive cyclist can entertain the thought of pursuing this bike option. That doesn't leave many options for the rest of us. Or, does it?

There is one final option for finding the perfect bike—assemble it yourself. Steeped in the same tradition as the homebuilder's do-it-yourself credo, an assemble-it-yourself bicycle begins with a pre-built frame. From this pur-

Building an MTB

If you're more interested in an off-road experience, then use the following list for building your own economical mountain bike:

- Answer Manitou Axel Elite '05 ATB Fork
- Sun Rynolite/Shimano XT Disc Silver Wheelset
- Nashbar Aluminum ATB Frame
- Ritchey V-Pro ISIS ATB Crankset 44/32/22
- Shimano PD-M737 SPD Pedals
- San Marco Azoto MTGEL Saddle
- Shimano 9-Speed Rapidfire Set
- Shimano 8-Speed Cassette
- Nashbar Carbon Riser Bar
- Shimano Deore LX Rear Derailleur
- SRAM 9.0 Linear Brakes
- Shimano Deore LX Front Derailleur

chased "foundation," select components are added to the frame which contribute to your specific body measurements and riding requirements. In short, with a matched set of bicycle components and a handful of common tools, the assemble-it-yourself bike quickly and easily becomes the perfect bike for you.

Using only simple tools, you can learn how to locate and then add the components to your custom bicycle. And best of all, when you're finished you will have saved hundreds, if not thousands, of dollars over the cost of the same bike from a name-brand manufacturer. The final result is the perfect bike for your ride.

Just ZIP It

In 2005, the hottest thing on two wheels was the Razor™ Pocket Rocket™ electric motor scooter. Able to cruise along at about 10 mph for 30 minutes worth of riding, these "toys" also earned the ire of numerous safety councils. Considered too small to be seen by gas-powered motorists, Pocket Rockets were frowned upon for use by unsupervised children. Under the proper guidance and in controlled driving environments, the Pocket Rocket is a gas to ride.

HOW TO TRICK OUT YOUR BIKE

1

2

1 Any bike can be improved with just a little ingenuity.

2 You can add a low-cost speed-ometer, odometer, chronometer to any bike. These devices can cost as low as $10.

3 A transmitter is mounted on your bike's front fork.

4 Attach your wiring with clear tape to your brake and shifter cabling.

5 Fix a small magnet to one of your front wheel's spokes. This magnet must pass in front of the transmitter. Now calibrate your cyclocomputer as described in the text.

6 Electroluminescent (EL) wiring can really enhance your night riding.

3

4

POWERED

5

SEE REVERSE SIDE FOR DETAILS
TECHNO FLASH™
EL LIGHT WIRE
10 AVAILABLE COLORS!
Aqua
White
Blue
Lime Green
Pink
Purple
Orange

6

7 Mount the battery pack under the seat.

8 Extend the EL wires down and around your seat post.

CHAPTER 3

Power to the People

Who can forget those events that started on Monday August 29, 2005? It was on that morning that Hurricane Katrina left its permanent mark on the Gulf Coast of Mississippi and the surrounding parishes of New Orleans, Louisiana.

Living in Hattiesburg, Mississippi, I had always thought that I would be isolated from the forces of the hurricanes that regularly churn up the Caribbean every summer. Not so with Hurricane Katrina. Granted, storm surge isn't a problem in Hattiesburg, but the hurricane-force winds and driving rain left my yard in complete shambles. But that was only the beginning.

Getting a cold shoulder from the insurance company didn't help either. I was left totally out in the cold and I had to fund all of my recovery out of my own pocket. And that hurt, but it didn't hurt nearly as much as it would have if I hadn't been prepared for a disaster.

How did I prepare for a disaster that no one forecasted accurately? Well, luckily I have a great supply of solar panels at my disposal. Furthermore, my ISO-TIP portable soldering iron was capable of soldering all of the circuit joints that I needed for fashioning several solar panel power generators.

One of the uglier aspects of the post-disaster recovery period was the enormous dependency that people had on gasoline. Long gas lines that stretched for miles and lasted for hours were a common sight as far north as Canton, Mississippi. Most of these people were actually trying to purchase gasoline for fueling portable gasoline power generators rather than for transportation. Even worse than this gasoline dependency

was the noise and air pollution that was produced by a neighborhood full of power generators running all day and most of the night. Curiously, the appliances being supported by these gas-guzzling generators were TVs, fans, and lights. And these are the exact types of appliances that are ideal candidates for a small, simple, solar power generator like the ones that I built.

Give Me the Power

The notion of generating power for electrical appliances through photovoltaic solar panels is not a new or revolutionary idea. What is interesting is a new initiative that brings solar power generators to about 1 million California rooftops. As reported by United Press International on PhysOrg.com, this will be a 10-year-long program (i.e., 2007 through 2017) that will consist of homeowner rebates to those families who install and use photovoltaic or solar panels.

By using just a handful of parts, you can easily grab enough free energy from the sun to power a TV, fan, and light all day with another 4-hour time allowance when used for charging a deep-cycle lead-acid battery. Follow these simple instructions and you'll be able to save your gasoline for a well-deserved trip instead of running a generator.

PARTS LIST

- 12-V solar panel
- 12-V deep-cycle lead-acid battery
- DC voltage controller
- 12-V DC meter
- DC-AC inverter
- Connection wire

SOURCES

You'll probably have a couple of these items lying around your house. In case you don't already own a DC panel meter, for example, here are some places that can supply you with everything you'll need to build a solar power generator.

- All Electronics Corp.—www.allelectronics.com
- The Electronic Goldmine—www.goldmine-elec-products.com
- Jameco Electronics—www.jameco.com

Many of the parts used in the construction of a solar power generator use different names. Following is a list of some of the more common names for acquiring the needed parts.

- 12-V 50-ma solar panel
- 12-V solar panel/charger
- Solar cell or alternative power battery charger regulator
- DC voltage controller
- 12-V 12-AH rechargeable battery
- 12-V 17-AH sealed lead-acid battery
- 25-Amp DC current meter
- Solar meter edgewise voltmeter
- 15-V DC panel meter
- 75-W DC-AC inverter
- 100-W inverter
- 150-W DC-AC inverter
- Cigarette lighter "Y"
- 4-foot coiled lighter extension cord

Another Story

The tale of my family's survival during and after Hurricane Katrina was documented by MAKE ("Eleven Days Gone"; Web Extra September 12, 2005). One noteworthy, yet invisible, point regarding this story is the date of its publication. While some people tend to remain inactive during a disaster, I juggled solar panels, pedal-powered generators, and a limited rechargeable battery supply around so that I could energize a limited number of electrical and electronic resources. The result was the publication of an eyewitness article five days after the disaster, while still living in it.

Go Pedal Some Power

If you're feeling more energetic, you can always make your own electricity. While not as efficient or powerful as a solar power generator, a pedal power generator can easily light some lights, operate a TV, or recharge a battery for short periods of time.

The basic principle for operation of a pedal power generator is very similar to the alternator found in most automobiles. An alternator consists of a rotating magnetic field coil, or rotor, which generates magnetic lines of force. These lines of force are intercepted by the outside concentric stationary windings, or stator coils, which are fixed in the alternator's frame. Now the rotor turns and the magnetic poles of N and S poles flip positions.

As current flows through the field coil (this current is typically supplied by a battery) an electromagnetic field is produced. The waves of electromagnetic force traveling over the stator coil make the electricity which serves to recharge the battery. Since the poles of the rotor's field coil flip their polarity, an alternating current (AC) is produced.

This AC is rectified through a diode bridge which results in direct current (DC). The DC is then used for recharging the battery. Since the current flowing through the field coil from the battery is DC, the alternator can supply its own field coil current *after* it has recharged the battery.

A Big Bag of Hot Air

According to the Tropical Cyclone Report published by the National Hurricane Center (December 20, 2005), Hurricane Katrina had weakened to a Category 3 storm before making landfall. This hurricane weakened further to a tropical storm about six hours later northwest of Meridian, Mississippi. Furthermore, this report claimed that the strongest sustained wind gust from an official reporting station was 99 kt, recorded at Grand Isle, LA.

Now if you are really adventuresome and a gifted cyclist, you can easily convert a used bike into a pretty good electricity generator.

PARTS LIST

- Bike
- Bicycle indoor trainer
- Alternator
- Switch
- Battery
- Three 22-ohm 10-W resistors
- Pulley and drive belt
- Electric clock

Be advised that your alternator might have some different lettering on its terminals. While the alternator case is always GND and should be connected directly to the battery's negative (–) terminal, the other connections might be called B, IG, S, and L or battery positive terminal (+), field, regulator, and lamp, respectively.

Most of these items are self-explanatory, but the bicycle indoor trainer might need some additional explanation. A good place to look for a trainer is Bike Nashbar. Three of the better models for building a pedal power generator are:

- Minoura Hypermag 1200 Trainer with Remote—The folding steel frame is durable, rigid, and fitted with nonmarring rubber feet to dampen noise and vibration. The hypermag resistance unit features a heat-dissipating top cover, a 1,200-gram flywheel and large-diameter roller for the smoothest action. Choose from seven different resistance levels with the handlebar-mounted controller.
- Nashbar 2006 Mag Trainer—A powerful, quiet, and smooth 850 series magnetic-resistance unit with a durable, folding steel frame, five resistance levels, 850-gram flywheel for exceptionally smooth action, and crank-style axle clamp mechanism.
- Ascent Magnetic 3 Level Trainer—Sure-footed wide, steel base with vibration-absorbing feet, quick-release skewer for secure bike mounting, and three levels of magnetic resistance.

Complete Text for R.04-03-017

The California Solar Initiative
R.04-03-017
Created by the California Public Utilities Commission on January 12, 2006

WHAT IT DOES

The California Solar Initiative (CSI) provides $2.9 billion in incentives between 2007 and 2017, divided as follows:

1. The California Public Utilities Commission (PUC) will oversee a $2.5 billion program for commercial and existing residential customers, funded through revenues and collected from gas and electric utility distribution rates.

2. The California Energy Commission (CEC) will manage $350 million targeted for new residential building construction, utilizing funds already allocated to the CEC to foster renewable projects between 2007 and 2011.

SPECIFICALLY, THE CSI WILL:

- Provide incentives to customer-side photovoltaics (PV) and solar thermal electric projects under 1 MW capacity.
- Authorize a pilot solar water heater (SWH) incentive program for customers of San Diego Gas and Electric Company. If successful, the PUC could offer SWH incentives statewide.
- Set initial PV incentive levels at $2.80 per watt effective January 1, 2006, to be reduced by an average of approximately 10 percent annually. Incentive levels for solar thermal electric projects and solar heating and cooling will be determined in 2006.
- Allocate 10 percent of program funds for low-income and affordable housing.
- Develop a pay-for-performance incentive structure to reward high-performing solar projects.
- The CSI will be coordinated with energy efficiency, advanced metering, demand response, and building standards programs at the energy agencies.

COST AND RATE IMPACT OF THE CSI

The estimated average cost to a residential electric customer will be approximately $12 a year; the average residential natural gas cost will be $1.40 per year. However, the total impact on a residential customer's monthly bill is expected to be minimal in most cases, because the cost of this program will be largely offset by the expiration, at the end of 2007, of a surcharge on utility bills to repay rate reduction bonds authorized in 1996 for electric sector restructuring.

ADDITIONAL INFORMATION

The agencies' existing solar programs will be consolidated into the CSI by January 1, 2007.

The PUC Self Generation Incentive Program funding and technology categories for 2006 will be modified as follows:

- Level 1—Solar technologies: $340 million
- Level 2—Other renewable fuel projects: $42.5 million
- Level 3—Nonrenewable projects: $42.5 million

The CEC Emerging Renewables Program will provide incentives for residential and small commercial solar projects through 2006. After 2006, the CEC will focus on residential new construction as part of the CSI.

And if you'd rather buy a kit for building a pedal power generator, Convergence Technologies sells plans and a kit for The Pedal-a-Watt Stationary Bike Power Generator.

If you're like me, buying an alternator for this type of generator is not very practical. A better idea would be to take the same principle and apply it to a more common induction motor—like the motor from an old furnace blower system or washing machine.

In order for this project to work, we have to master two issues: You have to supply a DC pulse to get the motor to make electricity and you must rotate the motor faster than its stated RPM.

While the first issue isn't too tough to figure out (alternators use this same principle to charge an automobile's battery), why the devil do we have to spin the motor *faster* than its rated speed?

Basically, induction motors have no physical connection between the stator and the rotor. The electricity flowing in the rotor is created by transformer action because the magnetic field in the stator is revolving at 1,800 rpm while the rotor is revolving at 1,725 rpm. The 75 rpm difference (4 to 5 percent) causes a current to be induced into the rotor and the motor behaves like a motor.

When this type of motor is used as a power generator, the motor must be driven 4 to 5 percent faster than the 1,800-rpm speed. This comes to about 1,880 rpm. When the rpms are timed correctly, the generator will be churning out 60-cycle power.

In order to "kick-start" your generator, you have to pulse the coils with DC to get it started. A switch connected to a 12-V battery will supply this pulse.

Some motors will need an 8- to 100-mF capacitor. You will have to experiment to determine the exact size. The capacitor must be rated for 250 to 300 V AC.

You can determine the frequency of your pedal power by connecting an electric clock to your generator. Calibrate your generated power by comparing the electric clock's sweep second hand with a known battery-powered time unit, like a wristwatch. When the two clocks synchronize, you will make close to 60-cycle AC.

This generator can easily drive a small 120-V 60-cycle appliance for as long as you are able to pedal the bicycle. Better yet, make the kids pedal for powering the TV. You'll have a fit family and probably not a lot of wasted hours viewing TV.

I Want a Hand Job

If riding a bike for making electricity sounds way too taxing, then how about cranking a generator with your hand? Yes, it's possible to generate electricity with a hand crank, but you probably won't be able to run 120-V 60-cycle appliances with it (see Fig. 3-1).

3-1 Use your bike to give you power—electrical power.

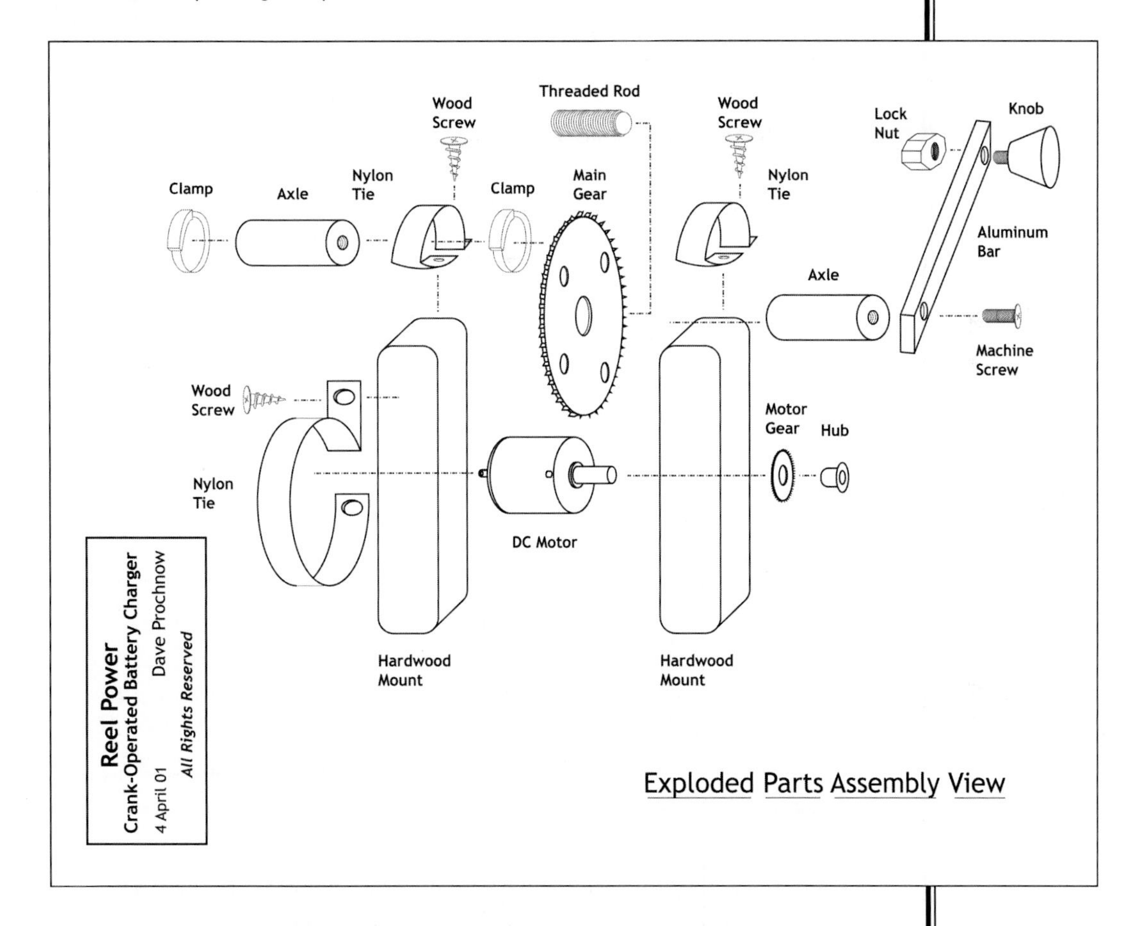

One nice feature of building a hand-crank power generator is that you can find almost all of the parts lying around your house. Even better, if you have an Erector® set, you can build an extremely efficient and professional-looking model.

One of the best sources for finding simple parts that can be quickly fabricated into a wide variety of projects and designs is the Erector set (the exclusive trademark of Meccano® S.N. of France). You remember Erector sets, don't you? What you might not know about today's Erector set (unless you inhabit the local toy store, like I do) is that these terrific metal building sets now come equipped with small electric motors and great plans for building up to 50 different models. These Erector Motion Systems,™ as well as the more traditional Erector sets, feature easily assembled pieces that can be readily adapted to any hacking requirement.

If you'd rather just start cranking out amps without all of the muss and fuss of building your own generator, then you might want to think about buying the Hand Crank Generator (Product No.: EL-GENHAND) from Home Training Tools, Ltd.

For less than the cost of a new DVD, you can quickly and easily be cranking your way to power generator happiness. Claimed to produce 6 V and a maximum output of 0.2 amps, the Hand Crank Generator comes with two small light bulbs for instant self-gratification.

We're In for Some Stormy Weather

Black & Decker has an all-in-one solution for your short-term power outages. The Storm Station (Model No. SS925; available at Lowe's, Target, and amazon.com) is a $100 answer to the question: "What happened to the lights? Complete with V recharging port, rechargeable flashlight, and power outage indicator lamp, the Storm Station is kept continually plugged into your household electrical system until it is needed. Then with a mad dash through the house, you will have all of your Storm Station goodies at your disposal. And if you have to let everyone know that you're safe, you can use the recharging port for powering your cell phone. That is, provided you have the right connector. Oops.

HOW TO BUILD YOUR OWN SOLAR-POWERED GENERATOR

1 Less than $150 will build your own 12-V solar-powered generator.

2 A 12-V solar panel.

3 Connect the solar panel to a DC voltage controller.

4 Simple 1/4-inch plugs make this DC voltage controller a snap to set up.

5 An inexpensive DC panel meter enables you to monitor your power generator.

6 A 150-W inverter will take your battery's DC power and convert it into AC power.

3

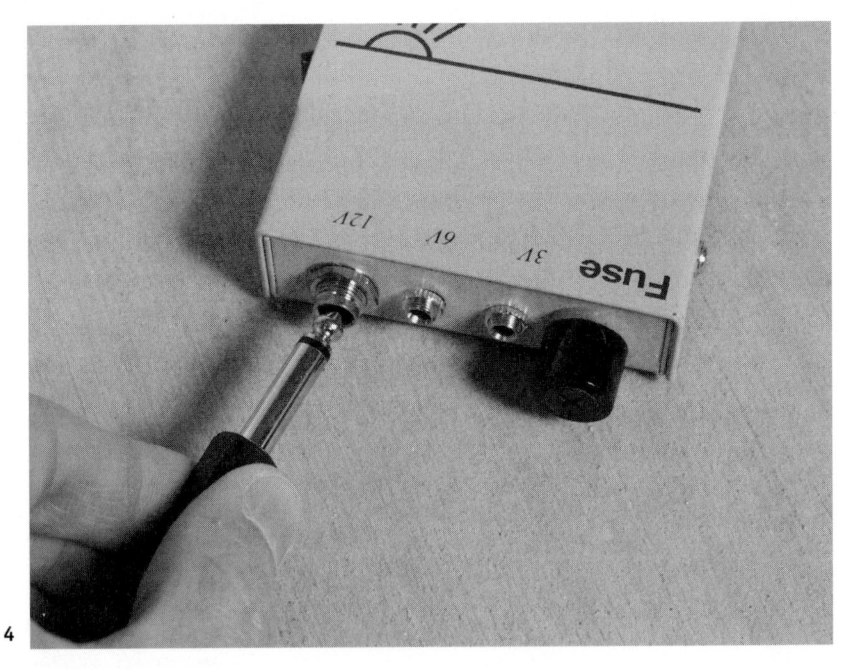

4

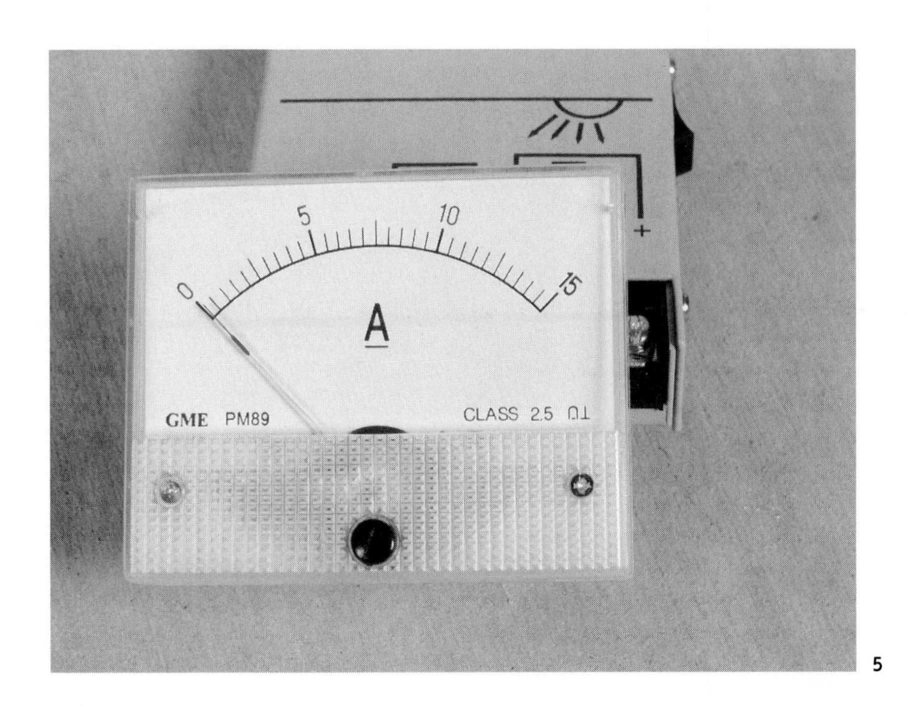

5

6

7 Most inverters feature a cigarette lighter plug.

8 Remove the cigarette lighter plug and replace it with a 1/4-inch plug.

7

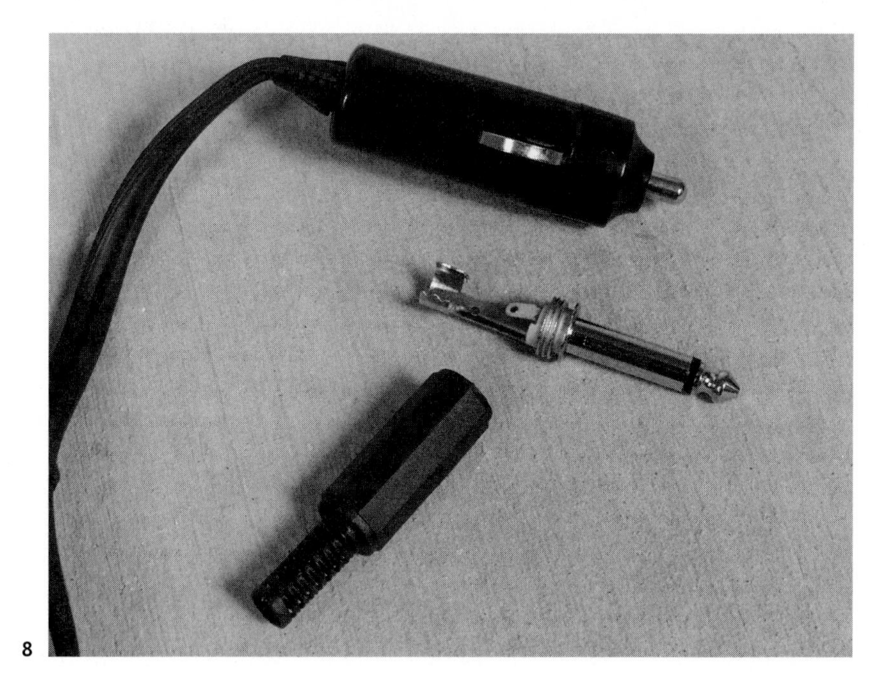

8

CHAPTER 4

Shuttered But Not Shackled

It was a dark and stormy night. Suddenly, a dog barks, a wailing car siren goes off, and you're awakened from a sound sleep. It's not your dog; it's not your car, so roll over and get back to that dream. Hey, thanks neighbor.

Sound familiar? If it does, then you should move. Seriously, alarms, dogs, and security systems are only as good as the people who respond to them. Although most perpetrators (or, perps) have a visceral flight response when confronted with a barking dog or a screeching alarm, if you want to stop a crime spree, then you've got to catch a thief.

Nothing says "gotcha" to the local police better than a nice clear photograph of a suspected criminal. And what better way to take this photograph than with an autonomous robotic watch dog? Behold, I give you *Beware of Bot* or *BoB*.

STOP, in the Name of BoB

In order to be a viable crime deterrent, BoB must be able to go anywhere, as well as "know" how to respond to a variety of different situations. For example, you wouldn't want your crime-fighting robot to try to arrest your lawnmower. Likewise, BoB won't be very effective in thwarting crime if it's stuck against a tree or spinning around in circles. What you need is a bot with brains.

In the bot brains department, there are two flavors: simple and advanced programmed "thinkers." I prefer the simple brains variety. If you need greater precision or sophistication in your robotic sentry, then you should opt for some higher brain activ-

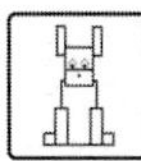

ity. If you own an iRobot® Roomba® or a WowWee Robotics Robosapien, then you already have access to a very low-cost and extremely powerful bot brain. Refer to Case Study 1 and Case Study 2, respectively, later in this book for detailed information on extracting these powerful brains.

On Brains of Bicore

At the opposite end of the robot brain spectrum lies the simple bug-like nervous system known as a bicore. What the heck is a bicore? The *bicore* is a common robot circuit that was originally designed by the same fellow who birthed Robosapien, Mark W. Tilden—Mr. BEAM (Biology, Electronics, Aesthetics, and Mechanics) himself.

While you can buy a ready-to-go bicore printed circuit board from Solarbotics, it is so simple to build one that you can easily wire the whole "brain" together in the same amount of real estate occupied by a postage stamp. Furthermore, only a couple of components are needed for building a bicore.

What sounds like a bunch of science fiction mumbo-jumbo is actually the basic working description of a bicore. The bicore is a relatively simple electronic circuit that is used to simulate a dual nervous (Nv) neuron network.

Delving into the history of simulating animal neural networks, you'll notice that Tilden originally coined the term *very slow propagation artificial neural system* (VSPANS) for networks of Nv neurons. Rather than VSPANS, many robotic physicists have adopted the term "nervous networks."

Regardless of the term, inside these networks a single Nv neuron is described as the "core" with a prefix defining the number of neurons inside the network. For example, two Nv neural cores are called a bicore. Just add more cores and you can design tricores, quadcores, hexcores, etc. So the bottom line is that a bicore consists of two Nv neurons or cores linked to each other. The product of this linkage is an oscillator that can drive lots of circuits from LED flashers to motors.

Understanding the theory of bicore operation is the hard part; building one is the easy part. Just slap a couple of garden variety capacitors on a 74AC240 Octal Buffer Inverter IC with a resistor and you have a perfect bicore. Specifically, a suspended bicore.

A suspended bicore is a form of bicore circuit that uses a resistor that is "suspended" between the Nv neurons or cores. This suspended resistor pro-

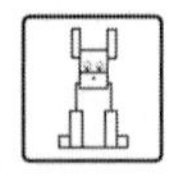

vides a "virtual ground" which swaps circuit action back and forth between the two Nv nodes or cores.

In the case of BoB, you could design a special suspended bicore made from a set of two photodiodes connected together in series (i.e., connected cathode to cathode). These dual photodiodes would then act as a replacement for the suspended resistor. One terrific attribute of using these twin photodiodes is that they will alternatively turn on and off each of the bot's motors while trying to locate the strongest signal of IR light.

By alternating the power (e.g., +5 V to 0 V to +5 V to 0 V, etc.) to each of BoB's motors, one bicore 74AC240 controls the movement direction of your robotic sentry. Another bicore could be installed inside the robot for pulsing (e.g., +5 V to 0 V, etc.) LEDs or firing a motion detector which, in turn, would activate a camera. Unlike the motor control bicore, these other bicore functions are managed with an actual resistor for making it into a suspended bicore.

As a recap, BoB consists of one suspended bicore that drives the motors using a pair of photodiodes connected in series, while another suspended bicore inside flashes an LED and a third bicore controls a motion detector. Beware crooks, you're about to be caught on robot candid camera.

With the brains taken care of, it's time to design and build our robotic guard. A BoB proof of concept "wish list" might consist of the following attributes (assembled in no particular order).

- Four-wheel drive
- Two sets of solar panels
- Phototrophic
- Touch sensor teverse
- Motion detector
- Digital camera with flash
- GPS
- Perimeter avoidance
- Data logger
- Water-resistant
- Dust-proof
- Shock-proof
- Rugged, durable
- Autonomous

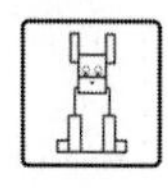

Picture This Hack

What started as an innovative disposable digital camera quickly exploded into one of the most widely publicized hacks of all time. John Maushammer took a CVS Pharmacy PV2 LCD disposable digital camera and ably rendered it into a reusable digital camera (i.e., you didn't need to return it to the pharmacy for processing). While this hack is exciting to re-create, it can be daunting to a beginner. You have to be part "solder head," "bit buster," and "net hound" to get all of the pieces and parts to work correctly in this hack. But Maushammer has made it possible for everyone to duplicate his feat.

Here is a very general outline of Maushammer's detailed instructions:

1. Display the special diagnostic page by pressing the shutter and display buttons simultaneously.
2. Open up the camera.
3. Build a USB interface.
4. Modify the camera's firmware.
5. Install software drivers for extracting image data from the camera.

If this sounds like something you can do, then go to Maushammer's Web site (www.maushammer.com) for all of the gory details.

Best of all, BoB can be built almost exclusively from toys and electronics that are languishing in your family e-junk drawer, box, or closet. For example, do you have a disposable digital camera (e.g., Walgreens' single-use camera)? Then you have virtually everything you'll need for building the camera portion of BoB. Luckily, other cameras also can be adapted to this design.

Seemingly, the four-wheel drive, shock-proof, dust-proof, water-resistant, rugged, and durable chassis sounds like an impossible problem to surmount with a discarded piece of techno-junk. But look no farther than your kids' old RC trucks.

I have had tremendous success with a New Bright 1:6 Scale Tricked Hummer (Item #6654), Traxxas E-Maxx, and Tamiya 1:12 Scale M1025 Hummer—all of which were salvaged from the back of my daughter's closet. Just discard the body, remove the radio control electronics, and install the new robot components. BoB is now ready to defend your home.

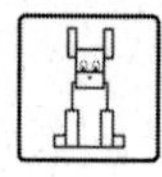

HOW TO CONVERT A TOY INTO A SECURITY ROBOT

1 Toys like the *Sewer Spewer* can be innocuous robotic security systems.

2 Radio-controlled toys can be converted into robots with simple bicore brains.

3 Squirt your intruders with some green "Sewer Spew."

4 Separate the toy's body from its chassis.

3

4

5 Disconnect the radio control circuitry and replace it with your bicore brain.

6 Add a sensor (e.g., photodiode) to your robot toy for triggering its security system.

5

6

CHAPTER 5

Smile, Say 43-68-65-65-73-65 (That's Digital "Cheese")

If you follow the world of hackers very much, then you have probably heard all the buzz about the "disposable" video camera manufactured by Pure Digital Technologies and distributed by CVS Pharmacy that had been hacked into being a "reusable" video camera. Once you study the quality of the video from this hacked camera, this $30 camera might not be worth the effort that was expended to make it into a reusable video camera. Its specs are appalling and its performance disappointing. A much more acceptable solution to low-cost video is the Aiptek IS-DV.

This palm-sized camera packs an impressive punch for under $160. Rather than using video tape for recording its video, the IS-DV uses SD/MMC media cards. Therefore, you can shoot until you fill a card, swap it out, and keep right on filming. Try that with the CVS video camera.

Then there's the camera hack heard 'round the world—a disposable digital still camera also sold by CVS Pharmacy that was hacked into being a reusable camera (see Chapter 4 for more information about this hack and its hacker). Once again, the quality of images from these reusable digital still cameras was not worth the cost of driving to your local drug store. Plus you had to fork over $10 to $20 for attempting to hack each camera. All it would take is one little soldering mistake or a firmware flashing screw-up and you were left with a truly disposable camera—throw it in the junk drawer.

What if I told you that you probably have a high-resolution digital camera hiding in your house right now? But it's not really a camera, you have to make it into a camera. Say what? Would you claim that I'm full of bologna? Even better, anyone can quickly and easily perform the steps necessary to hack this object into a mega-pixel photo-making monster. This camera is actually a flatbed scanner.

Don't give up on this hack just yet. By converting your scanner into a digital camera, you will have salvaged one of the ugliest pieces of digital detritus from littering our landfills. Furthermore, the quality of artwork produced by a scanner camera has been elevated to an art form.

Michael Golembewski has created an exciting portfolio of beautiful black-and-white imagery that explores an invisible tension between motion and time. This site, known as the The Scanner Photography Project, is definitely worth seeing.

I started experimenting with using flatbed scanners as digital cameras during the research and development of a book I wrote about Erasable, Programmable Read-Only Memory (EPROMs) (*Experiments with EPROMs*; McGraw-Hill, 1988). At first I toyed with using the clear window in an EPROM as a recording film plane in a conventional camera. By today's standards, the results were awful, but for that day the image was thoroughly enlightening and the concept, to me, was eye opening (see Fig. 5-1).

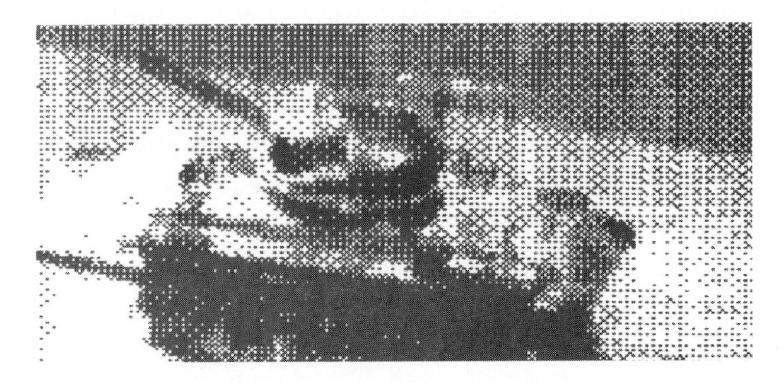

5-1 An early scanner-derived photograph, circa 1980s.

On a lark, I adapted a broken scanner to that same camera. Whoa, it was like a light bulb exploding in my head. There on my computer screen was an extremely clear, high-resolution photograph. My first work of digital art. Over the next couple of years I became a photocopier/scanner junkie. I was able to do things with these copiers and scanning digital cameras that digital photography experts can only dream about doing today. However, there were some distinct problems with scanner photography in those days.

First, scanners were heavy, bulky, and tethered to both a computer and a power supply. Another fault was that most early scanners were very sensitive creatures that didn't tolerate standing on edge very well. Finally, my initial scanning digital cameras were incredibly s-l-o-w machines. Exposure lasted several minutes and light would leak into the film/scanner holder causing streaks and reflections which marred the final scanned image.

Blow Your Friends' Minds with These Scanners

Before you begin this project, remember that you will need access to both a power supply and a computer for capturing your scanner digital photographs.

First of all let's get some common ground on scanner terminology. These definitions from Hewlett-Packard are applicable to our experiments in scanner photography:

- **CHARGE-COUPLED DEVICE (CCD)**—solid-state analog electronic sensor commonly used in scanners
- **CONTACT IMAGE SENSOR (CIS)**—sensor used in smaller, low-cost scanners; has limitations on resolution
- **COMPLEMENTARY METAL OXIDE SEMICONDUCTOR SENSOR (CMOS)**—sensor used in scanners and digital cameras based on a semiconductor process designed for digital electronics
- **OPTICAL RESOLUTION**—resolution of a scanner calculated by dividing the width of the scanned area by the number of pixels in the sensor

This is Only a Production Test

If you would like to run a series of diagnostic tests on the Mattel Juice Box, hold down the top button (rewind) and the bottom button (play) while you turn the power switch on. You will be greeted with "Production Test Ver 2.6." From this test menu, you can perform an LCD Test, Mem Test, Key Test, Audio Test, Battery Test, and Sleep Test. You can conduct a test by pressing the bottom button (play). When you're finished testing, just turn Juice Box off.

Your PSP Brag Book

Want to take your photos with you on a Sony PlayStation® Portable or PSP™? Well, first of all, your selected imagery must be up to snuff with the PSP's JPEG rendering engine. And the best tool for all digital image editing is the venerable Adobe Photoshop.

Here's a simple PSP-specific procedure for getting a photograph into perfect proportions for PSP viewing:

1. Select a suitable digital photograph.
2. Open the image in Photoshop.
3. Select the Crop tool and crop the photograph's dimensions to 480 x 272 pixels. These dimensions represent the size of the PSP screen.
4. Complete the cropping action by double-clicking inside the cropped selection of the photograph.
5. Save a copy of this image as a JPEG. Select the maximum image quality option and make it a standard JPEG.

Once you have a properly dimensioned JPEG image, you are ready to load it on a Memory Stick PRO Duo media card. In order for this image to be displayed by your PSP, all photographs must be saved in the proper location. This location involves a funky file and folder hierarchy structure that you can either "roll" yourself or let your PSP create (i.e., with the "Memory Stick Format" setting) the correct hierarchy. In either case, your photography must be copied into this folder:

Memory Stick PRO Duo _ PSP _ PHOTO _ YourFolderName _ Your Photos

Your photographs can now be viewed, zoomed, rotated, and panned while viewing them on the PSP.

There's more to the PSP than just this simple photo-viewing stuff. If you want to learn about Power User Mega-Hacks, as well as more great PSP info, then you should read my book, PSP HACKS, MODS, AND EXPANSIONS (McGraw-Hill, 2006).

- **PIXELS PER INCH (PPI)**—number of pixels captured per inch by a scanner
- **TWAIN**—software driver interface between a scanner and computer support applications

If your old scanner was made within the last 5 to 10 years, then most of these annoying difficulties will not prevent you from repurposing the retired scanner into being an incredible digital camera. Here's how:

1. Open your discarded scanner, remove the glass, and disconnect the lamp. I've had success with old Epson, HP, and UMAX scanners.
2. Locate the CCD, CIS, or CMOS sensor. The sensor may be hidden underneath a lens array. If so, this array can be removed and discarded. Your camera lens will focus the image on the sensor.
3. Modify the sensor carriage for maximum exposure of the sensor throughout the entire length of its travel. This is not referring to the exposure for the image. Rather it is an attempt to ensure that a clear line of sight exists between the sensor and the projected image from the camera's lens.
4. Put the scanner back together.
5. Fix the scanner to a "film holder." This film holder should be removable from the camera so that you can focus your lens prior to making an exposure.
6. Load your appropriate scanning software and TWAIN drivers.
7. Run a test of the scanner *before* you attach it to the camera.
8. Hook it all up to your camera and start making some art.

If you have trouble making your scanning software and TWAIN drivers work with your hacked scanner, refer to the SANE Web site. Scanner Access Now Easy, or SANE, is an open-source software repository and an application programming interface (API) for building a universal scanner interface. Most of this work has already been done for the major computer operating systems (e.g., Linux, MacOS, and Windows).

So get out there and start lugging around your new scanner digital camera and learn what it must've been like snapping glass plate photographs back in the late 1800s. Can you say: heavy, arduous...yet rewarding?

Show Me What Ya Got

Once you have some great scanner digital photographs, you'll want to show them to your friends (see Fig. 5-2). Well, after hauling your computer around with your scanner digital camera, you'll probably opt for a lighter, more portable method for bragging about your artwork. Enter the digital Brag Book.

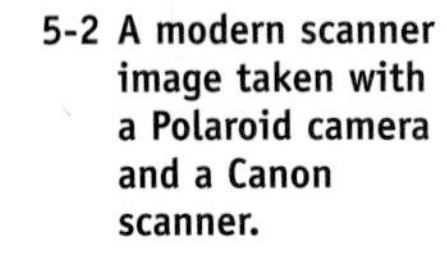

5-2 A modern scanner image taken with a Polaroid camera and a Canon scanner.

First of all you'll have to raid your kid's toy box. You'll be looking for the Mattel Juice Box.™ Based on an embedded Linux design made by Emsoft, the Juice Box is a 2.75-inch color LCD music, photo, and video player. All is not bliss in the Juice Box world, however. A special Juice Box MMC/SD media card reader program and USB card reader are needed for getting MP3 audio and still images onto the Juice Box. And even with this special software/hardware option (marketed as the Juice Box MP3 Starter Kit), you still can't load your own video onto the Juice Box.

Alternatively, you can purchase Mattel videos. Known as Juiceware™ these videos are supplied on a removable flash media card. By using the optional Juice Box MP3 Starter Kit you can rip CDs (using CoffeeCup® software) and load digital pix (converted into Juice Box Picture format or .jbp) into your own personal MMC/SD "juice ware," but you can't encode and transfer videos to the Juice Box.

If you'd like to try your hand at hacking the Juice Box, make sure to begin your research at elinux.org. Tim Riker has established a fledgling project for exploring, programming, and hacking commercial-embedded Linux toys like the Juice Box.

By doing just a little soldering, you can easily adapt your kid's Juice Box into an MMC/SD reader/player. Just what can you put on these removable memory cards? Well, music and images. Therefore, this hack makes a discarded Juice Box into a great Brag Book.

Unfortunately, Mattel tried to make the Juice Box as difficult to use as possible. The images on a Juice Box must be written in a weird image format called JBP. By using this format, you can't just take your scanner digital photographs and drop them on an MMC/SD card. You must first convert your JPEG digital photographs into JBP format files. Mattel does include a converter program inside the Juice Box MP3 Starter Kit, but this path can be a bit pricey. A better solution for converting your JPEGs into JBPs is with a little Java application called jpbConverter.jar.

While Juice Box might not be the perfect mobile multimedia player for your kids (hey, they did throw it into their toy box, didn't they), Mattel has ensured that at least you won't go (digitally) thirsty. The slurping noise on starting up the Juice Box is priceless.

HOW TO TAKE PHOTOGRAPHS WITH A SCANNER

1 A Polaroid 800 Instant Land camera.

2 This Polaroid camera is a vintage 1957–1961 camera. The original list price for this camera was around $120. I purchased two used ones for under ten bucks, total—with cases and accessories.

2

3 No matter what brand of camera you use, make sure that you can hold the shutter open during exposure. This setting is called B, or Bulb. Another setting that is equally useful is T, or Time. The difference between these two settings is that you must hold the shutter manually open with B, whereas the camera holds the shutter open with T.

4 This Polaroid model used instant roll film.

5 You must either remove the film back or mount the film plane directly onto the scanner.

6 Remove the scanner's lid or cover.

3

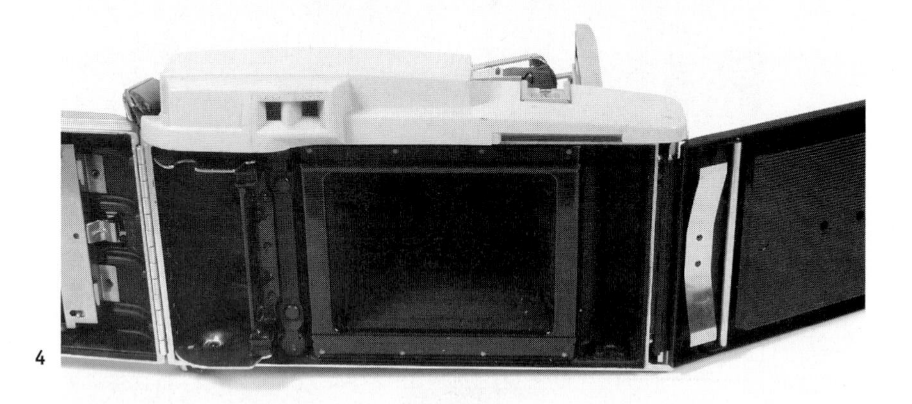

4

5

6

7 Remove the scanner's glass focus plane.

8 Disconnect the scanner's lamp.

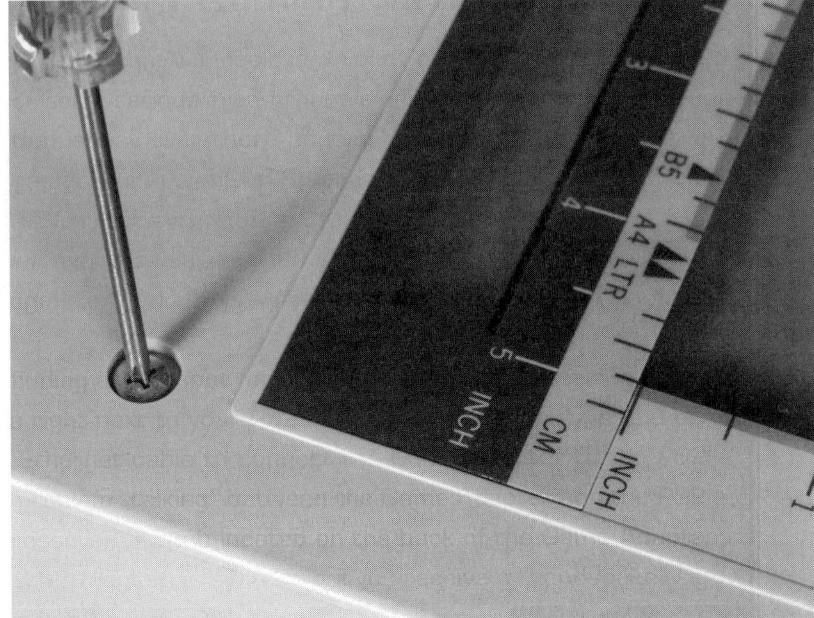

7

8

CHAPTER 6

Eat to the Beet

It started in 1997, at the Conference of Parties III (COP3) held in Kyoto, Japan, the Kyoto Conference on Climate Change. Following endless debate, the world's developed nations established a framework of specific targets for cutting the emission of greenhouse gases. This framework became known as the Kyoto Protocol.

In this framework the United States proposed to stabilize greenhouse gas emissions and not cut them. Conversely, the European Union proposed a 15 percent cut. Rancor aside, a compromise was the real result of the Kyoto Protocol and the majority of the planet's industrialized countries were committed to an overall reduction of greenhouse gas emissions to 5.2 percent below 1990 levels for the period 2008 through 2012.

If you personally want to get on board with the complete Kyoto Protocol, then just look to recycling your old refrigerator. Back with our friends from the EPA, an initiative for disposing of old "white goods" has been implemented. This initiative, for example, helps recycle approximately 95 percent of the parts inside an old refrigerator.

How does a dead frig relate to the Kyoto Protocol? That answer lies in ozone depletion. According to the EPA, some of the stuff found in older refrigerators contains ozone-depleting substances (ODSs). In other words, if these ODSs leak into the atmosphere they can contribute to the destruction of the earth's ozone layer.

Ozone depletion occurs when chlorofluorocarbons (CFCs) and hydrochlorofluorocarbons (HCFCs) are released into the atmosphere. Most of the refrigerators manufactured since 1996 contain an ozone-friendly replacement compound derived from hydrofluorocarbons (HFCs). By 2003 all new refrigerators contained no ODSs.

Another nasty aspect about refrigerators is that they are one of the biggest electricity hogs in your house. Even a 10-year-old frig can burn up to $400 per year in electricity costs. Conversely, a modern refrigerator that bears an ENERGY STAR label can use between 20 to 25 percent less energy than one of these older models. Initiated in 1992 by our good EPA buddies, ENERGY STAR has become a high-visibility symbol of energy conservation. In fact, the EPA claims that a $10 billion energy cost savings was delivered to businesses and consumers in 2004.

Now it's time to pay back on that savings—recycle an old frig into something more useful than pollution—something like a smoker cooker.

When Smoke Gets in Your Meat

Please don't make the same mistake I made when I converted an old Norge refrigerator into a smoker cooker. I forgot to remove all of the plastic parts from inside the frig (i.e., one of the plastic posts that held a wire tray) and when I heated up the inside (with a 20-pound turkey) a horrible black smoke filled the smoker (truly a "smoker") and ruined my Thanksgiving dinner plans.

There are two general types of meat-smoking techniques: cold smoking and hot smoking. The cold smoking process cooks cured meats at temperatures of around 90° F. Hot smoking, on the other hand, is the more generally accepted process for cooking meats. Capable of generating 250° F temps, a hot smoking cooker is the ideal choice for an older, recycled frig.

Before you attempt to build a refrigerator smoker cooker, make sure it is *metal throughout the inside*. Absolutely *no plastic* should be inside the cooker. Also, remove all circuitry, tubing, and the compressor assembly. Ideally, a two-compartment frig with a lower freezer or vegetable compartment makes the best smoker.

In this case, all of the smoke-generating equipment is housed in the lower compartment, while the meat is hung in the top compartment. You will need a conduit between the two compartments (generally 8 inches in diameter) with a baffle for channeling the smoke up and into the meat compartment.

In order to ensure the proper flow of the smoke and heat, cut a 4-inch-diameter hole in the top of the frig leading into the meat compartment. Attach a short 1-foot section of stovepipe with a damper control to this hole.

The smoke is best generated by an electric hotplate holding a pan filled with flavored wood chips. In my cooking tests, I found that adding a bed of

Eye of Newt, Sprig of Moss

Most of your traditional cooking recipes can be easily adapted to solar cooking. Just to make sure that you savor the flavor, here are some solar cooking tips:

Complete meals—When cooking with several containers (yes, it's possible), first add the food that takes the longest time to cook. Then when that container is boiling, add another cooking container. An average solar cooker like the one illustrated in this chapter can easily accommodate three cooking containers.

Noodles and rice—My kids love noodles and rice. Presoak the noodles or rice in water prior to cooking. Use a separate container for this presoaking and not your cooking container. I use two cups of water to one cup of noodles or rice. Add the water to the cooking container without the noodles or rice and place the container inside the cooker. Once the water is boiling (about one hour), remove the cooking container, top off the water level, and add the presoaked noodles or rice. Cook until ready to serve.

Vegetables—Most kids hate vegetables, but not when they've been cooked in a solar cooker. Why? Because it's more fun. Just leave corn in the husk, zucchini whole, and potatoes in their skins. Your meals will never be the same again.

Bread—This is my favorite solar-cooked food. Not because the bread tastes that much better, rather it's the enjoyment of taking a hot loaf of bread out of a silly foil box that I have sitting in my driveway and showing it to the neighbors. The best way to reliably bake bread is to employ a used metal coffee can (with a replacement metal lid; not the plastic disc lid), grease the inside of the can, mix your dough, drop the dough in the can, and add the lid. The dough/yeast must rise prior to baking. Once the dough has risen, set the coffee can inside the cooker. You can use the old toothpick test (insert the toothpick into the loaf, if it comes out with some bread dough attached to it, then the bread isn't done yet) for testing your bread's baking time. When the bread is done, just turn the can upside down, and give it a tap on the bottom. Out drops the loaf and let the good times roll.

lava rock underneath the wood chips helps to produce a more even heating of the wood. You also should include a pan for holding water inside the smoke-generating compartment.

If you would like a more detailed plan for building your own smoker cooker from a discarded frig, SmokeHouse Plans sells a complete set of diagrams and instructions. In no time, you will be the toast of your neighborhood.

Soup's Up... In the Sky

What's for dinner? An easy enough question, right? Well, what if you don't have electricity? Now what? Yes, the backyard cooker might be a viable option, but most of these grills require a fuel source. Build your own solar cooker and you will never have to scratch your head to answer any of these questions ever again.

You have two general choices when building a solar cooker: the passive oven-style and the blowtorch version. Whereas the passive oven model will cost you next to nothing to build (high-end options like a glass cover could make your costs skyrocket to near $5 for parts), the blowtorch model will require the purchase of a Fresnel lens. You can find this magnifying lens at Edmund Scientific for less than $5.

You can find plans for many different types of solar cookers at The Solar Cooking Archive. Sponsored by Solar Cookers International, this Web site will show you how to build almost every conceivable type of cooker or dryer that uses the power of the sun to cook a meal.

Before you begin to build your cooker, however, please don't confuse a solar cooker with the photovoltaic cells used in solar power panels. Yes, they both share the similar solar noun, but they are as different as, well, night and day. While photovoltaic cells are sophisticated manmade bits of magical silicon, a solar cooker is much more basic—relying on a couple of sheets of reflective foil that are held in place around your food.

Yup, you guessed it, that fancy-sounding reflective foil is nothing more than garden-variety aluminum foil—easily scored from any grocery store.

In building the Fresnel lens "blowtorch" cooker, all you need is a simple box (determine the dimensions of your box based on the focal length of the Fresnel lens—hold the lens in the sun and measure the distance needed to form a nice, tight, bright pinpoint of light), the lens, and some foil. Just mount the lens over one open end of the box, line the box with foil, and cut a square opening (about 3 inches square) in the end of the box that is opposite the Fresnel lens.

Now hang a small hot dog, smoked sausage link, or marshmallow on the outside of the box over the 3-inch-square opening. You can use toothpicks for skewering your meal. Then orient the cooker with the lens opening facing the sun and come and get it. This cooker will have a small hot dog ready for feeding your troops in under a couple of minutes (depending on the intensity of the sun and the focal length of the box).

As an enhancement to this cooker, you might want to rotate your food as it cooks. For example, a marshmallow will burn very quickly in intense sunlight. You won't need to worry about rotating your food with the passive solar oven.

At the other end of the solar cooker spectrum, this passive oven model is great for cooking larger meals, boiling water, and baking bread (see Fig. 6-1). Once again, your parts list is very modest— two boxes that can nest inside each other with a 1-inch air gap around them, two sheets of cardboard, foil, a cooking container, and a plastic bag or pane of glass.

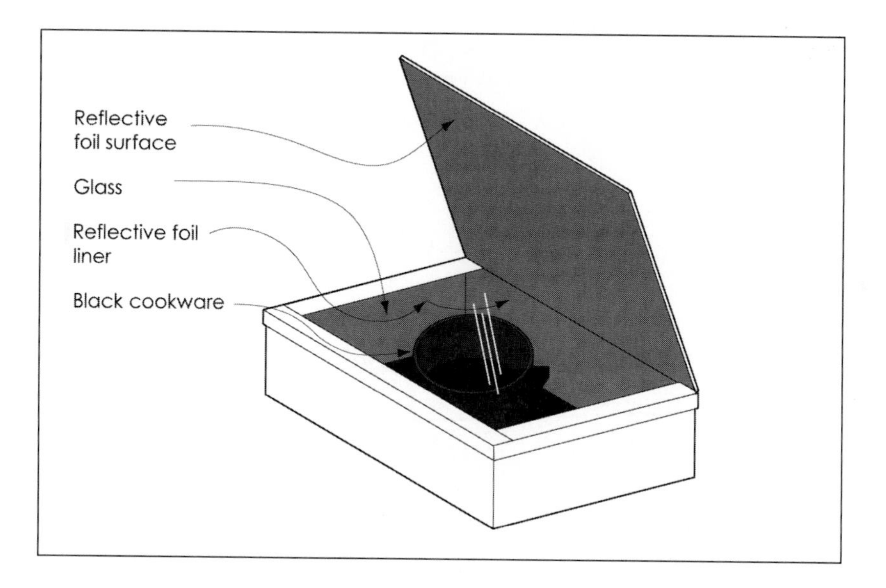

6-1 You can easily make a solar cooker from a couple of boxes, foil, and a sheet of glass.

Begin by trimming the boxes so that they nest together with a 1-inch air gap around all four sides. Next line the inner box with foil, glue two 1-inch spacers to the inside bottom of the outer box (these spacers can be either wood or metal washers and serve to support the cooking container when it is placed inside the inner box), and nest the two boxes together.

One of the cardboard sheets will be used as the inside cooking surface. Cut this sheet to the same dimensions as the bottom of the inner box. Paint this cardboard sheet flat black. Use a heat-resistant spray paint that has been approved for use on outdoor cookers. Drop the cooking surface cardboard sheet into place on the bottom of the inner box.

Your cooking container must be painted black. If you already have a black cooking pot, that also will work. Your food is added to this container. Be sure to include water, seasoning, etc.; just like your standard household cooking techniques. Now carefully wrap the cooking container inside a high-temperature plastic bag (e.g., oven bag) and lower the cooking container into the inner foil-lined box. If you are using the optional glass pane, just place the cooking container inside the inner box and lay the glass on top of the boxes.

Use the sheet of cardboard that has foil glued to its face to add a reflector to the cooker. Prop the nested boxes up toward the sun and adjust the angle of the reflector for maximum saturation of sunlight around the cooking container.

Check your meal, periodically. When it's been properly cooked, remove the cooking container, and enjoy.

HOW TO INCREASE VITAMIN C INSIDE YOUR REFRIGERATOR

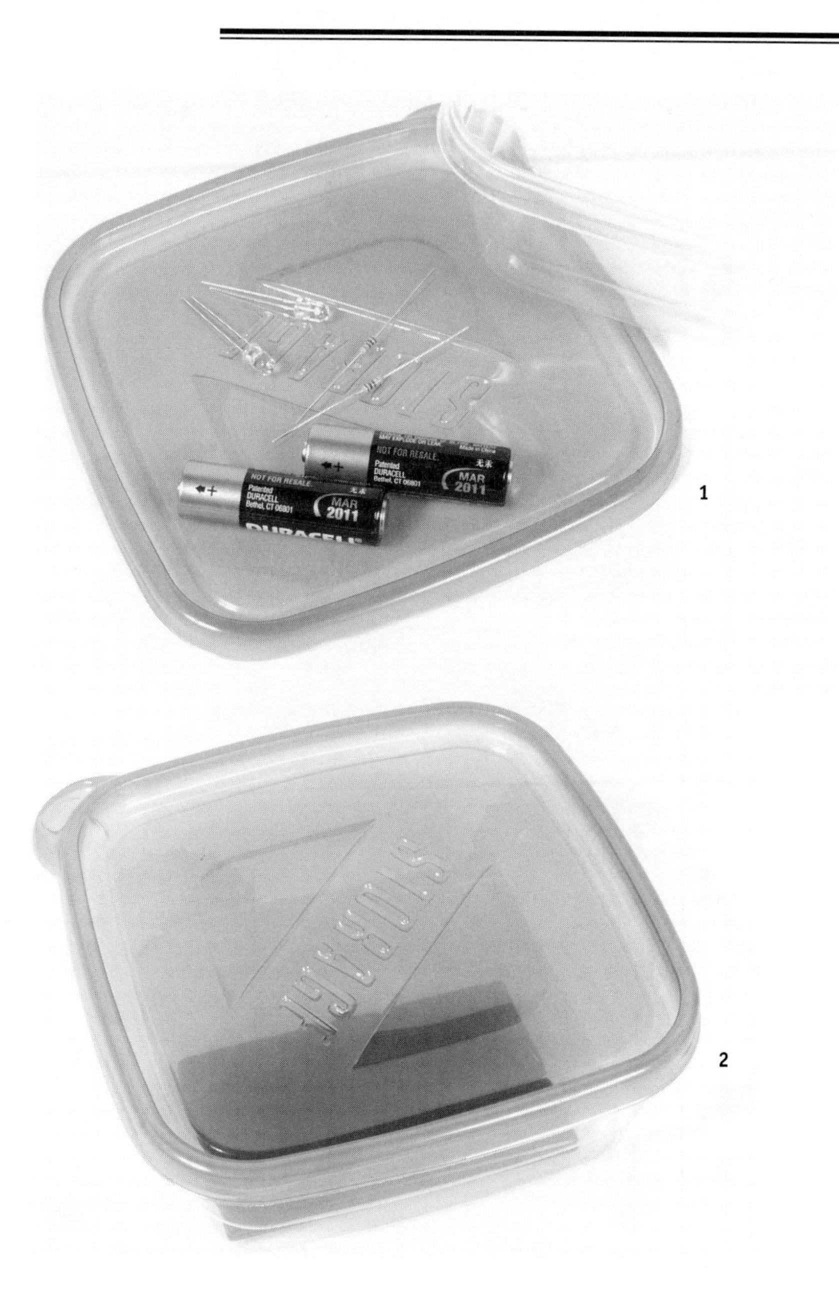

1 The Japanese manufacturer Mitsubishi claims that you can increase the vitamin C content of your vegetables by shinning orange light on them inside a refrigerator. You can add this same capability to any refrigerator with a sandwich container, some batteries, and a couple of orange LEDs.

2 Assemble an LED circuit and install it in a plastic sandwich container.

3 Test your circuit.

4 Place your LED-equipped container inside your refrigerator's vegetable drawer. Yum, tasty. Oh, the Mitsubishi frig with this feature costs around $2,600. Look at all the money you just saved.

3

4

CHAPTER 7

Put Your Finger on This Pulse

It was 50 years ago today, that the president taught us all to play. Sorry, Sgt. Pepper's fans. The President's Council on Physical Fitness and Sports (PCPFS) celebrated its 50-year anniversary in 2006 (see Fig. 7-1). Started in 1956 by fitness guru and president, Dwight D. Eisenhower, as the President's Council on Youth Fitness, the PCPFS brings more visibility to the importance of physical activity, fitness, and sports for improving and maintaining health. This high-profile program is a nationwide attempt at countering the staggering rates of obesity that continue to plague this country.

7-1 PCPFS Chairman, Lynn Swann (left), PCPFS Executive Director, Melissa Johnson (middle), and Secretary of Health and Human Services, Mike Leavitt (right) on March 1, 2005. (*Photograph courtesy of the President's Council on Physical Fitness and Sports.*)

OK, so drop that burger fatso and listen up. Stop living a sedentary life. It's time to get fit. Derived from the recommendations found inside the 1999 *Surgeon General's Report on Physical Activity and Health*, the PCPFS takes aim on three alarming statistics:

1. More than 60 percent of American adults are not regularly physically active. In fact, 25 percent of all adults are not active at all.
2. Nearly half of American youths 12 to 21 years of age are not vigorously active on a regular basis. Moreover, physical activity declines dramatically during adolescence.
3. Daily enrollment in physical education classes has declined among high school students from 42 percent in 1991 to 25 percent in 1995.

Jeez, take a hike. No, really, take a hike, a bike, or a swim. Just get up and get at it. At what? Well, first of all, the PCPFS wants you to think that size really does matter. In other words, the amount of exercise that you regularly perform is more important than the intensity of your workout. And only a moderate amount of daily activity is required. So what does the PCPFS consider "moderate" exercise?

As stated in that 1999 *Surgeon General's Report*, "a moderate amount of activity can be obtained in a 30-minute brisk walk, 30 minutes of lawn mowing or raking leaves, a 15-minute run, or 45 minutes of playing volleyball, and these activities can be varied from day to day."

That's all well and good, but what does the American College of Sports Medicine® (ACSM) have to say about this presidential program? Not surprisingly, this scientific research "health tank" concurs with the government's promotion of a leaner America. Founded in 1954, the ACSM guidelines for a healthier you include:

- exercise 3 to 5 days per week
- warm up for 5 to 10 minutes prior to your exercise
- maintain your exercise for 30 to 45 minutes
- cool down for 5 to 10 minutes following your exercise

My Heart's Beating Like a Rabbit

As soon as you start studying fitness, you'll learn that your heart is both your best friend and your worst enemy. By properly listening to your heart, you can discover your ideal fitness level. And your heart speaks to you in beats per minute.

There are two important heart rates: resting heart rate (RHR) and exercise heart rate (EHR).

RESTING HEART RATE

To determine your RHR, attach a heart rate monitor (HRM), and then lay down in a quiet location. After 20 minutes, check the recordings and record the lowest value achieved. This value is your RHR. Repeat this process for five days and record the RHR for each day. At the end of the test period, take an average of the five RHR recordings and use this average as your final RHR.

EXERCISE HEART RATE

According to the ACSM, EHR is estimated by subtracting your age from 220 (i.e., 220 – your age). This value is your estimated maximum heart rate (HRmax). Now establish your lower EHR limit by multiplying HRmax by .6 (i.e., HRmax × .6). Finally, calculate your upper EHR limit by multiplying your HRmax by .9 (i.e., MRmax × .9). Your final EHR is now a range between the lower and upper EHR limits. In normal exercise, keeping your heart rate near the lower EHR limit is the ideal fitness goal.

If you are a professional athlete or an avid fitness expert, then you probably already own an HRM. For the rest of us, an HRM is a wristwatch-like device that can measure your heart rate and track your EHR limits throughout your entire daily exercise program. In the world of HRMs, Polar® Electro reigns supreme. Models like the Polar S725X are optimized for a specific type of exercise.

This specificity helps the athlete tailor an exercise program that is exactly matched for his or her endeavor. For example, the Polar S725X is geared for cyclists and cross-training athletes. Although you can purchase optional accessories that enable the S725X to assist with running activities, this HRM is foremost for cycling.

At first glance, it might be easier to say what the S725X *can't* do, rather than listing its features:

- heart rate as beats per minute (bpm)
- heart rate as percentage of HRmax
- average heart rate per lap
- HRmax
- exercise recovery test
- lap info
- cycling cadence
- temperature
- altitude

Get Off Your Duff, Hilary

So you feel that there might be something to this exercise thing and you'd like to challenge yourself a little bit. The Presidential Champions program is for you. There are just four simple steps between you and winning the Bronze, baby.

1. Name your poison. Select an activity that makes you burn calories. And, no, I don't think those "12-ounce curls" will count.
2. Do it and do it some more. Exercise and record the event in your activity log. There is a 750-point daily "cheaters" cap to encourage you to stay active every day. A Bronze award takes 20,000 points. For example, if you run 5 miles every day, you'll "run away with the Bronze" in about 6 weeks.
3. Track it. An online activity log serves as the official registrar for all of your exercising. For those of you who live in "New York minutes," the activity log will allow you to record in increments as short as those brutal 5-minute exercise regimens when you dash to the bathroom during commercials.
4. Get it. When you reach Bronze-medal status, the activity log will remind you that it's medal time and you don't need to crack anyone's kneecap, either. Once you've tasted Bronze, you can take Silver and, even, go for the Gold.

You can sign up online at the Presidential Champions Rules (www.presidentschallenge.org/the_challenge/presidential_champion_rule.aspx).

- speedometer
- trip odometer
- data storage/transfer
- PC training analysis software
- time
- date
- stopwatch

One feature of the Polar S725X that merits further explanation is how this HRM handles your heart rate as a percentage of HRmax (i.e., HRmax-p). In this Polar HRM, HRmax-p predicts an individual HRmax value based on your exercise program as it is recorded and stored on the HRM. This method is much more accurate than HRmax derived from the age-based EHR formula.

The Beer Drinker's HRM

Shelling out $350 for an HRM can be a tough pill to swallow for the casual fit nut. An alternative HRM easily can be fabricated from a discarded portable cassette tape recorder, a pillow speaker, and a line-out patch cable.

For those of you not in the know, a pillow speaker is a thin speaker that could be stashed underneath a pillow for late-night listening. These were vogue in the late 1970s. If you're lucky, this little speaker will make a dandy microphone.

You can test the speaker as a microphone by hooking it up to a low-wattage amplifier, putting the speaker/microphone against your chest, and listening for your heart beat. If no sound is coming through the amp, you should try a more traditional microphone instead.

Now that you have an adequate input source, you can load a cassette tape into your portable deck. Attach the microphone to your bare chest (watch out for that hair, ouch!), turn on the recorder, and press the Record button. Run this contraption for about 1 minute. Disconnect the microphone, rewind the tape, and press the Play button. If you can hear your heart beat, then you are ready for the final stage in hacking an HRM from a recycled portable cassette tape recorder.

Fire up your computer and load a sound-editing application like GarageBand. Run the line-out patch cable from the cassette tape recorder into the audio line-in port of your computer, rewind your HRM tape, and press the Play

button. You should now have a rough equivalent of an electrocardiogram (ECG) waveform running along your screen. Can you imagine how hip you'll look with your retro-HRM strapped on your belt next to your iPod?

One of the nifty features of this hack is that you can record voice notes on your HRM while you are exercising. Try that with a Polar S725X. You now have visual evidence for disproving all of those people who claim that you are heartless.

HOW TO TRAIN LIKE A CHAMP WITH A HEART RATE MONITOR

1 Polar S725 cycling heart rate monitor (HRM).

2 While it looks just like a regular wristwatch, this Polar HRM is an essential link for training like a champion.

3 Just strap it on your wrist.

4 There are two more connections that you must make: to your bike and to your body.

5 The bike's connections are similar to the cyclocomputer discussed in Chapter 2.

6 The Polar transmitter sends bike data wirelessly to your Polar S725 HRM.

7 Your body connection is also a wireless transmitter that monitors your heart rate during rest and while exercising.

8 Polar Precision Performance 4 SW software assembles your training data into a meaningful output.

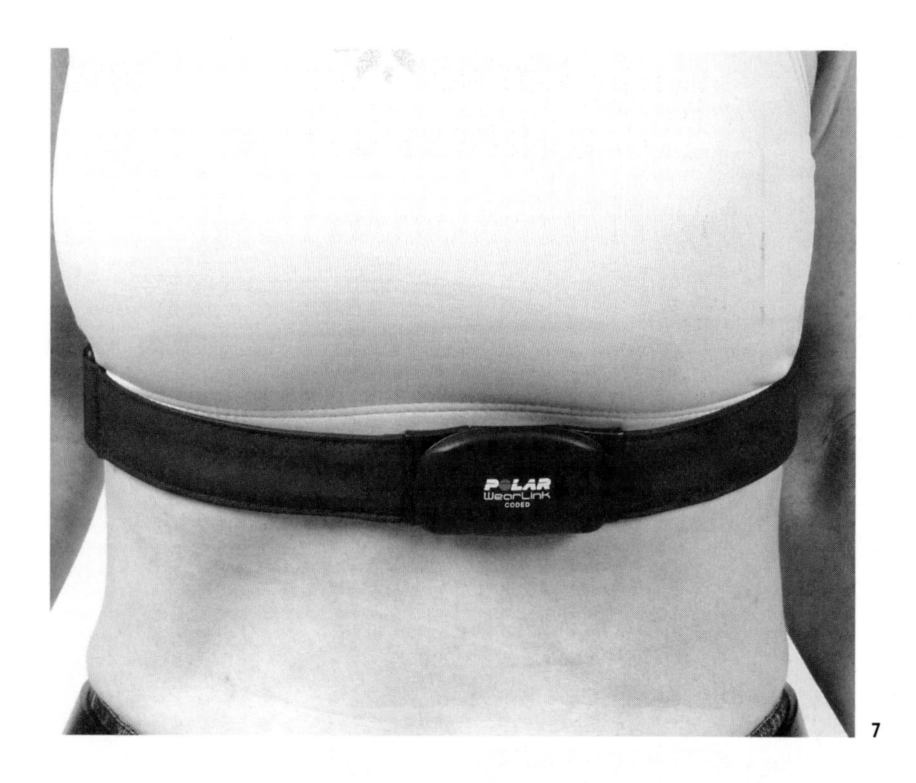
POLAR
WearLink
CODED

7

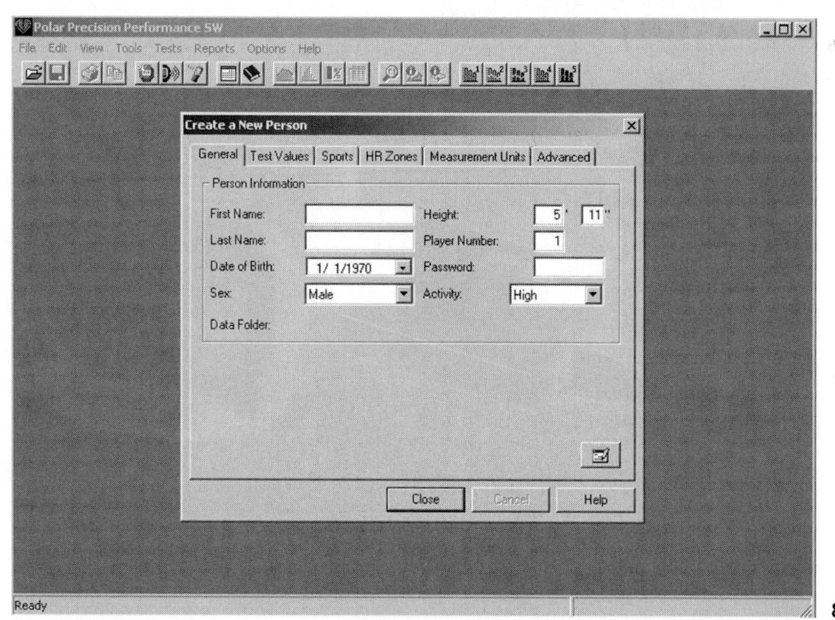
Polar Precision Performance SW
File Edit View Tools Tests Reports Options Help
Create a New Person
General
Test Values
Sports
HR Zones
Measurement Units
Advanced
Person Information
First Name:
Height:
5 '
11 "
Last Name:
Player Number:
1
Date of Birth:
1/ 1/1970
Password:
Sex:
Male
Activity:
High
Data Folder:
Close
Cancel
Help
Ready

8

9 Track your performance.

10 Create a personal training diary.

11 Graph your training.

12 Convert units.

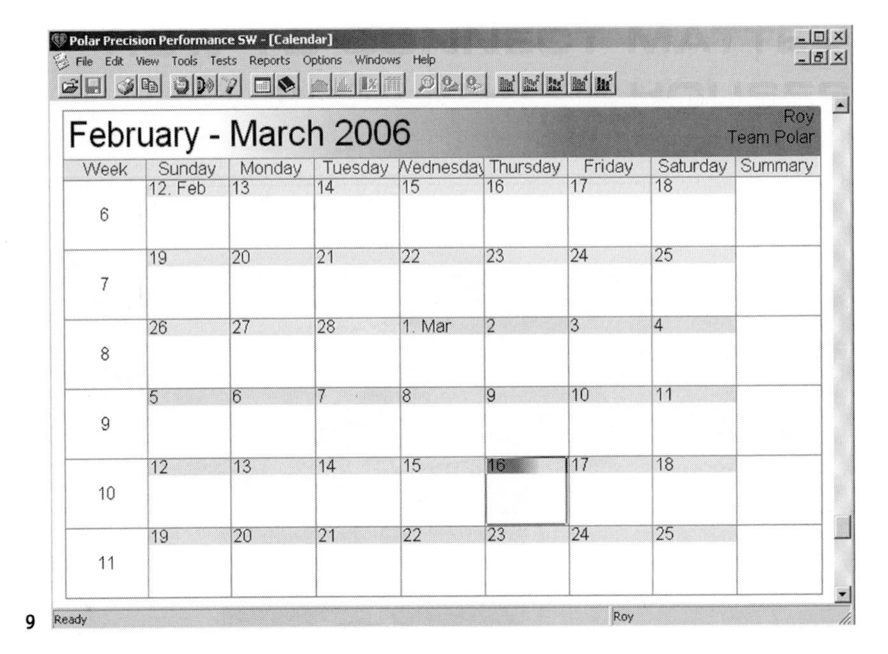

9

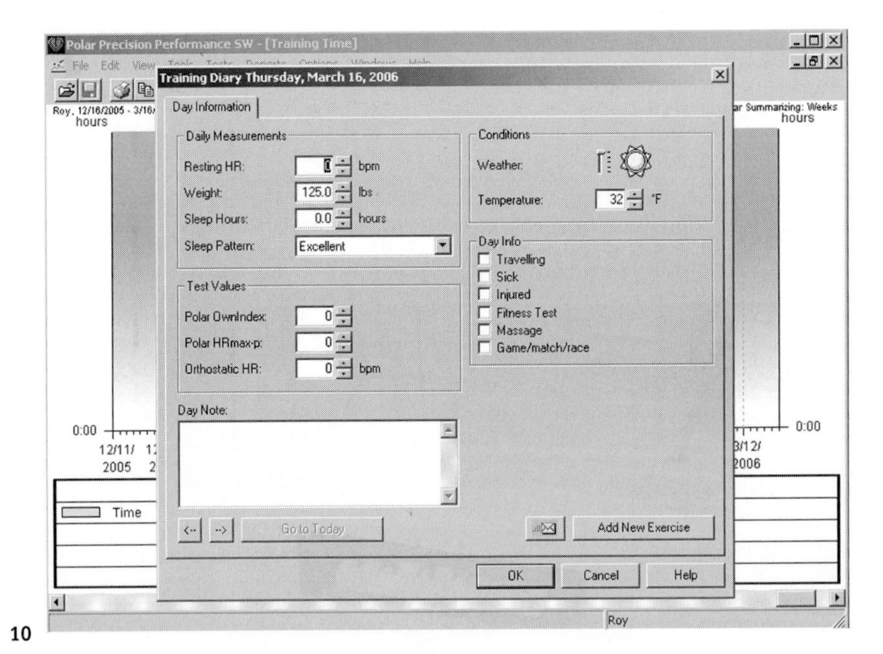

10

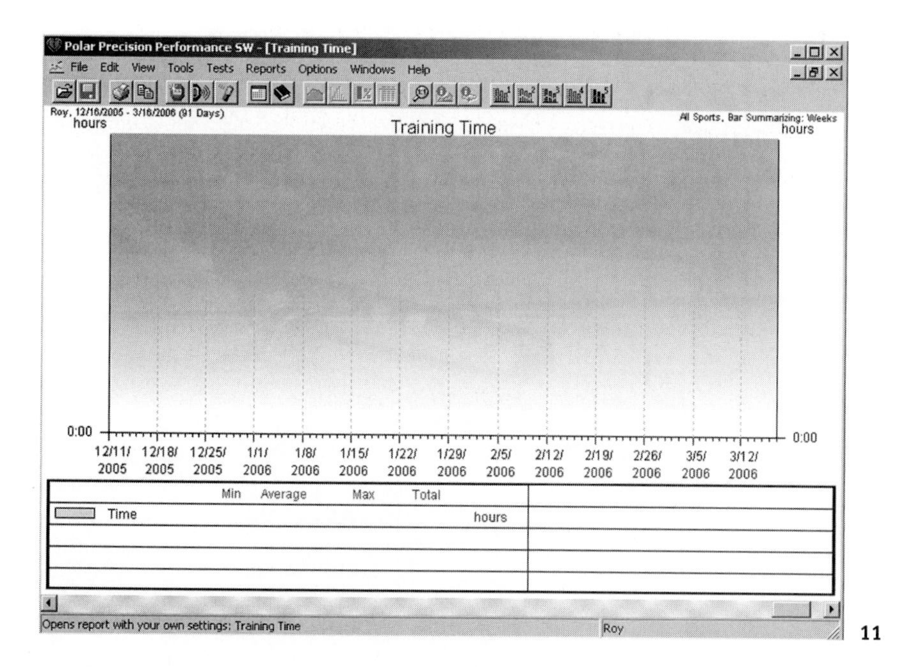
Polar Precision Performance SW - [Training Time]
File Edit View Tools Tests Reports Options Windows Help
Roy, 12/16/2005 - 3/16/2006 (91 Days)
Training Time
All Sports, Bar Summarizing: Weeks
hours
0:00
12/11/ 2005
12/18/ 2005
12/25/ 2005
1/1/ 2006
1/8/ 2006
1/15/ 2006
1/22/ 2006
1/29/ 2006
2/5/ 2006
2/12/ 2006
2/19/ 2006
2/26/ 2006
3/5/ 2006
3/12/ 2006
Min Average Max Total
Time
hours
Opens report with your own settings: Training Time
Roy

11

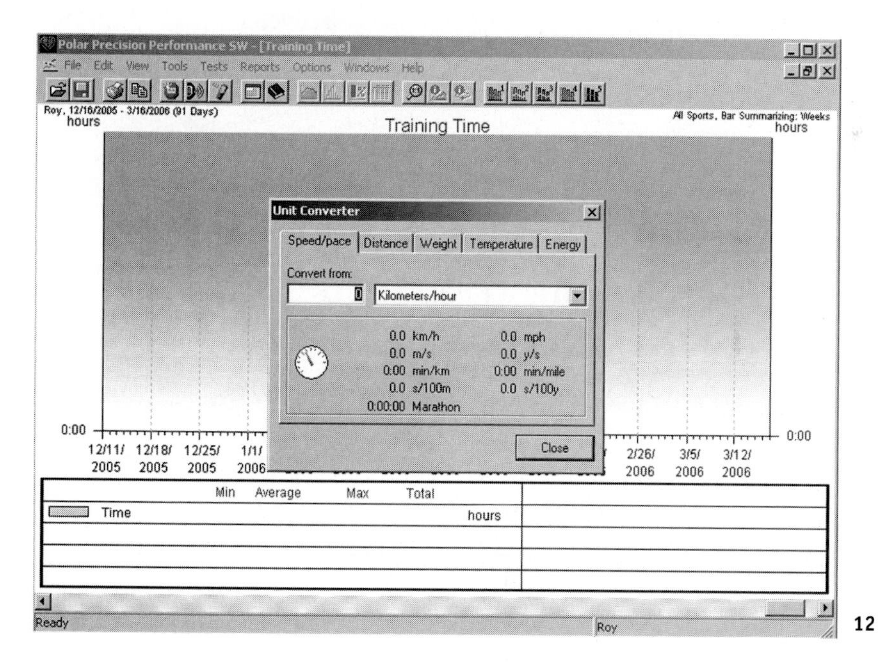
Polar Precision Performance SW - [Training Time]
File Edit View Tools Tests Reports Options Windows Help
Roy, 12/16/2005 - 3/16/2006 (91 Days)
Training Time
All Sports, Bar Summarizing: Weeks
hours
Unit Converter
Speed/pace
Distance
Weight
Temperature
Energy
Convert from:
0
Kilometers/hour
0.0 km/h
0.0 m/s
0:00 min/km
0.0 s/100m
0:00:00 Marathon
0.0 mph
0.0 y/s
0:00 min/mile
0.0 s/100y
Close
0:00
12/11/ 2005
12/18/ 2005
12/25/ 2005
1/1/ 2006
2/26/ 2006
3/5/ 2006
3/12/ 2006
Min Average Max Total
Time
hours
Ready
Roy

12

13 Send your training results via e-mail to anyone who cares about your fitness.

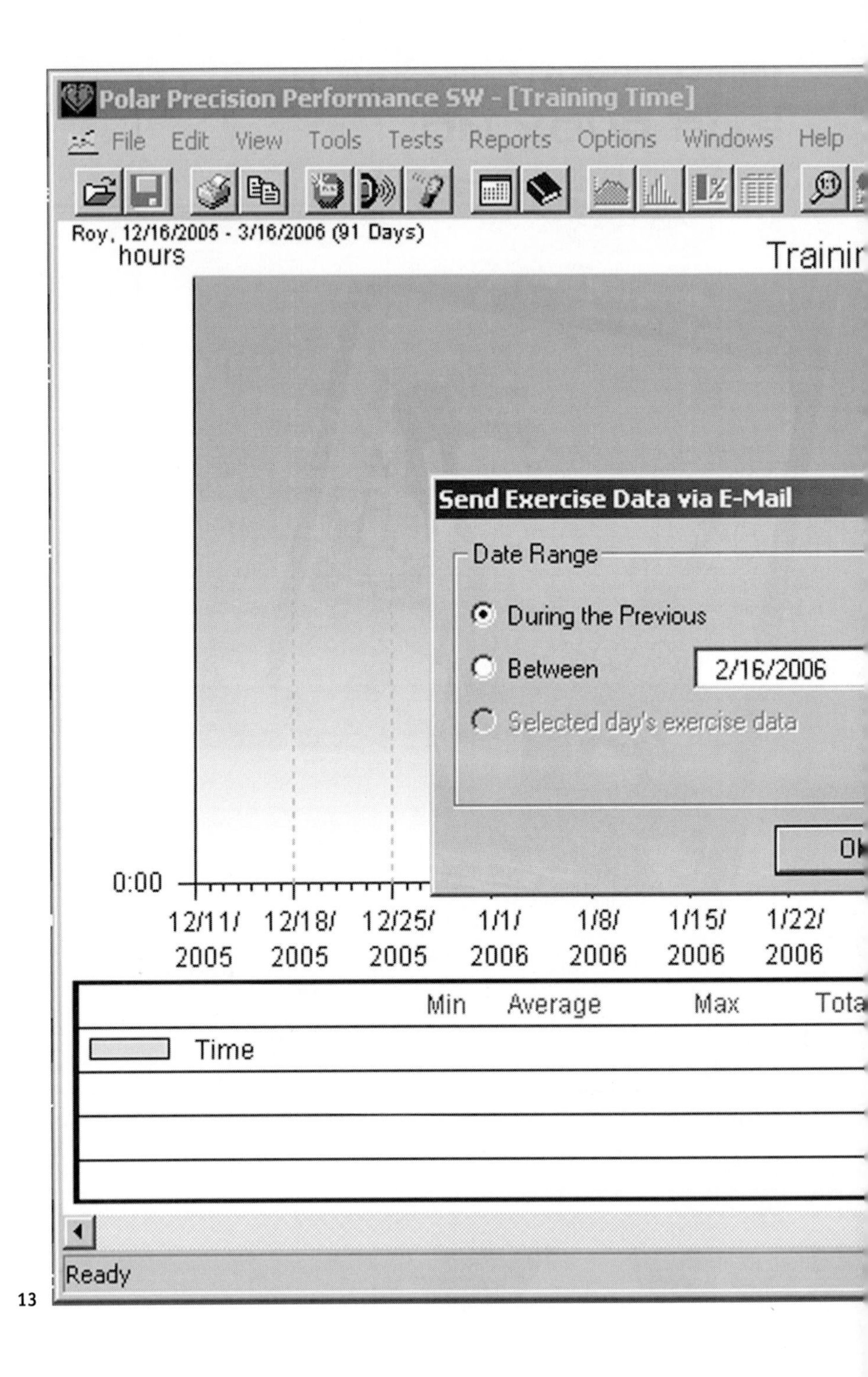

13

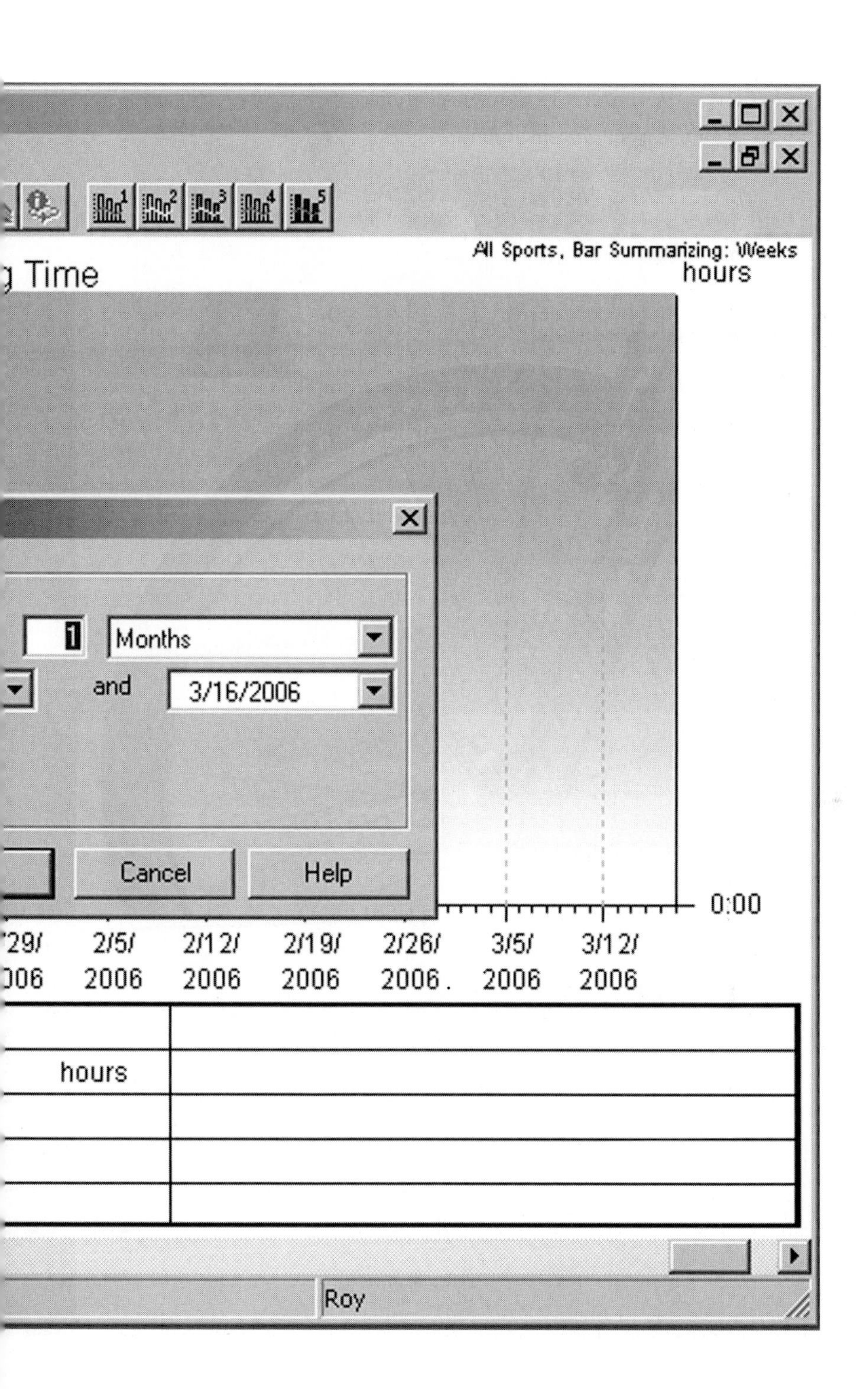
All Sports, Bar Summarizing: Weeks
hours
Time
Months
and
3/16/2006
Cancel
Help
0:00
2/5/ 2006
2/12/ 2006
2/19/ 2006
2/26/ 2006
3/5/ 2006
3/12/ 2006
hours
Roy

14 Take a walking test.

15 Upload your S725 data to your PC via an IR link.

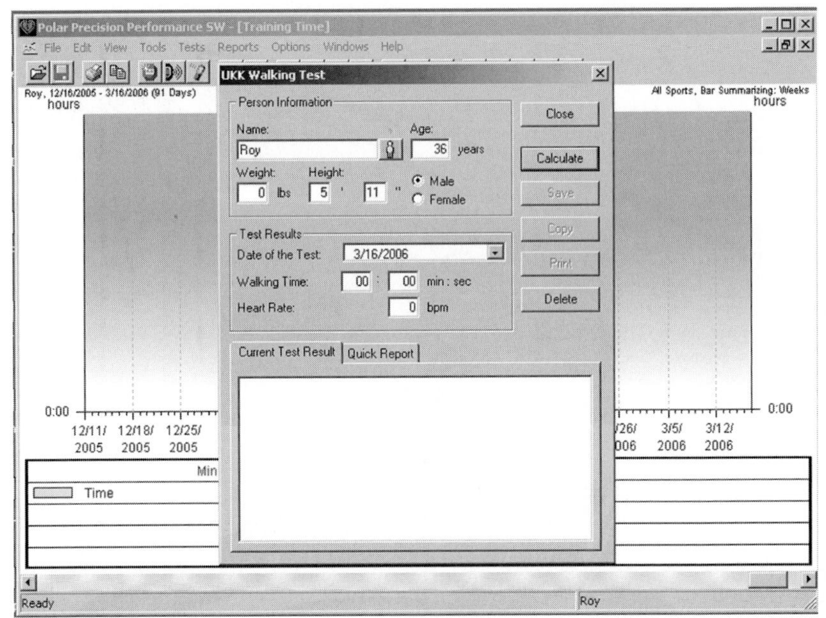

14

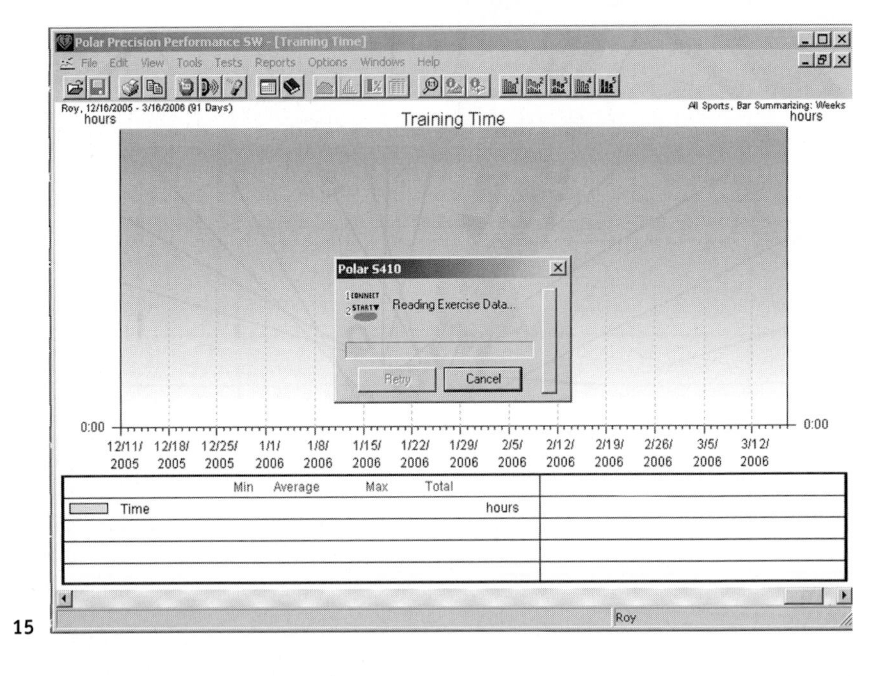

15

CHAPTER 8

You Stud Muffin

Whoever invented gypsum wallboard (commonly known as drywall) probably had honorable intentions. Ridding houses and businesses of plaster dust, lathe, and horse hair (used as a substrate or base in plaster) is a noble gesture. Thank you. Unfortunately, try to hang a photograph frame, mirror, hi-fi speaker, or towel rack on a perfectly smooth expanse of gypsum wallboard and prepare yourself for the ol' "guess 'n hit" routine.

You know the drill, you want to hang your new bookshelf at point XY on your wall. The installation instructions require that you secure the unit directly into two wall studs (see Fig. 8-1). What the heck are those? Well, more on that later, but you try to remember all of those things that your Dad told you about construction and you start gently rapping on the wall with your knuckle.

After carefully listening for a solid sound in the wall, you arm yourself with your trusty hammer and whack a nail into the

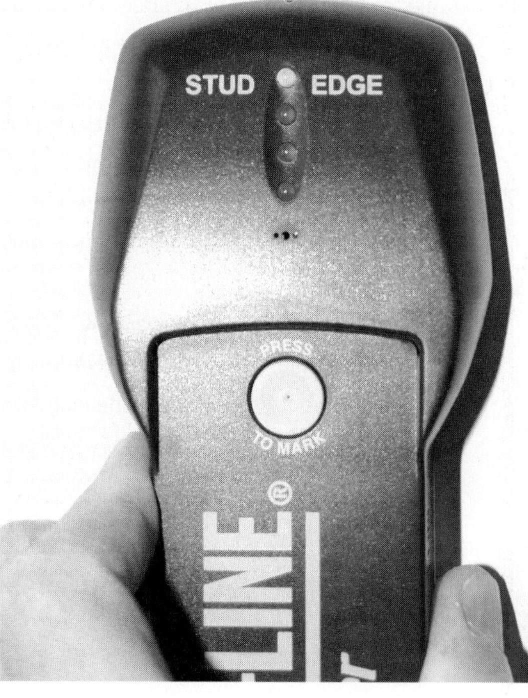

8-1 A stud finder can help you locate those pesky studs that lie hidden beneath your wall's drywall.

wall. Since you guessed wrong with your crude echolocation technique, the force of the hammer drives the nail completely through the gypsum wallboard leaving an ugly hole in the wall... in the wrong place.

Now you could try to locate a stud from this clever little "access hole" that you just punched in the wall, but you aren't thinking clearly. So, you knock, hammer, punch, try again. If you're really good, you might get lucky on your third try. But, probably not. Instead the new bookshelf is boxed back up as a "rainy day, honey do" project. Squeeze some wall joint compound into all of those holes and nobody will be any wiser. Especially, you.

There are a couple of other methods for locating studs that you should have tried:

LOOK FOR STUD DROPPINGS. Prior to painting your home's walls, most contractors have drawn stud lines (these are actually lines that have been "snapped" with a dusty tool called a chalk line) on the bare, unpainted gypsum wallboard. These lines help the workers find, you guessed it, the studs. At any rate, since these guys know where the studs are located you can try to use their screw and nail holes for (re)finding the studs. The trouble is that these fasteners can be hard to see. Try looking for them in the wooden baseboards and crown moldings or in large open areas of painted wallboard. A bright light held obliquely on these surfaces might help you spot these stud droppings.

DO THE MATH. Since most contractors frame walls with evenly spaced studs (either 16 inches or 24 inches on center), you might be able to use a tape measure for getting your stud guess in the right ballpark. Begin in a corner nearest your target wall and measure outward.

GET SOME STUD ATTRACTION. If you have a general idea where a stud might be hiding, slowly and gently run a magnet over the surface of the painted wall. You might want to wrap the magnet in a piece of cloth to prevent it from marring the wall's surface. When you feel a tug on the magnet, you might have found a screw, a nail, or a metal pipe. Be careful how you determine what tugged at your magnet's heart. Nails pounded into water pipes or drills turned into electrical conduit pipe can be hazardous to your health. Oh, and if you don't have a small magnet, try a small speaker. Speakers have tiny, powerful magnets that are perfectly suited for being attracted to studs.

You've Found the Stud of Your Dreams

Yippee, you found a stud. Now you're going to want to hang onto that divine piece of knowledge. So you'd better mark the wall. A light pencil mark is the perfect method for indicating stud placement. A level or plumb bob (just attach a weight to a piece of heavy string or cord) will make the transfer of this mark to the rest of the stud's length a snap.

As a tip to making your mark on a wall, try to hide the mark someplace inconspicuous. I like to make my mark as close to the baseboard as possible. Later, these marks are easy to locate and can be easily projected up the wall with a level.

But you need two studs. How do you find an adjacent stud? Just measure out either 16 or 24 inches from this initial stud mark. Once again, you will have to determine the type of stud spacing used in your wall. Also, beware of windows and doors in your wall. These are special architectural features that require different structural framing techniques which can throw off your expected 16-inch or 24-inch stud spacing dimension (see Fig. 8-2).

If you're looking for a high-tech alternative to that prehistoric plumb bob notion, then pull that old laser pointer out of your e-junk drawer and snap some laser lines. Fix the laser pointer to the edge of a small "torpedo" level. Hold the laser pointer light over your stud pencil mark, slowly rotate the level until the bubble is perfectly centered, and marvel at how nice your wall looks without all of those ugly nail holes in it.

I'm Lookin' for a Stud

Just what is a stud? A stud is a wooden (or, steel) 2-by-4, 2-by-6, or, if you live in Alaska, 2-by-8 piece of dimensioned framing lumber. Assembling studs together forms the inside support structure for walls. Generally, studs are evenly spaced with distances of 16 or 24 inches between the center line of each stud. Tough talkin' contractors might say something like, "Your cavity was framed with 2-by-4 studs 16 inches on center."

INSIDE YOUR WALLS

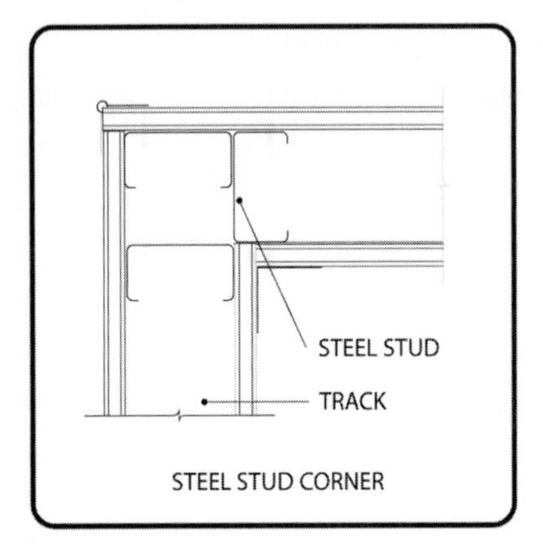

STEEL STUD CORNER

A

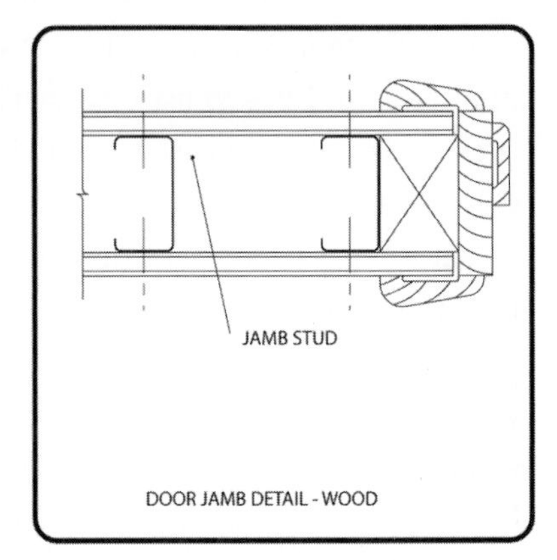

DOOR JAMB DETAIL - WOOD

B

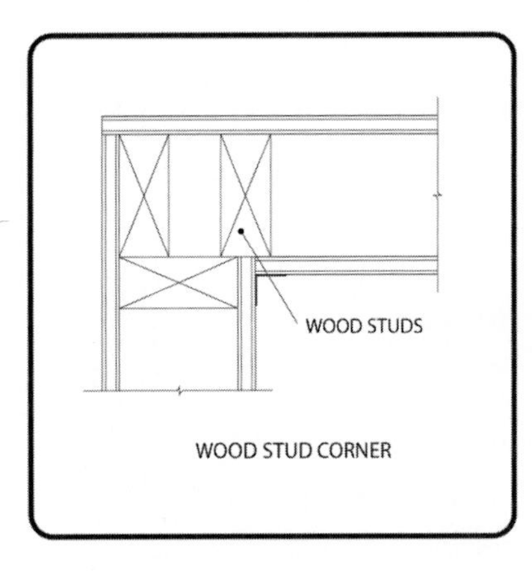

WOOD STUD CORNER

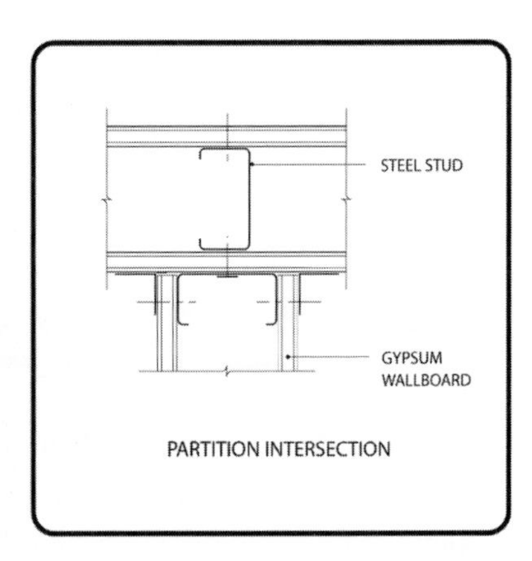

PARTITION INTERSECTION

D

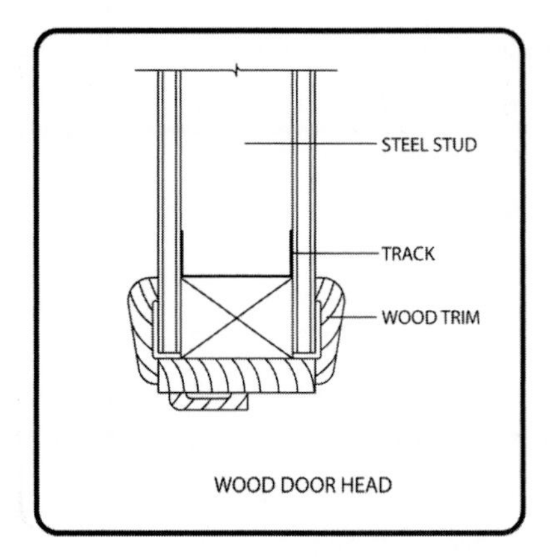

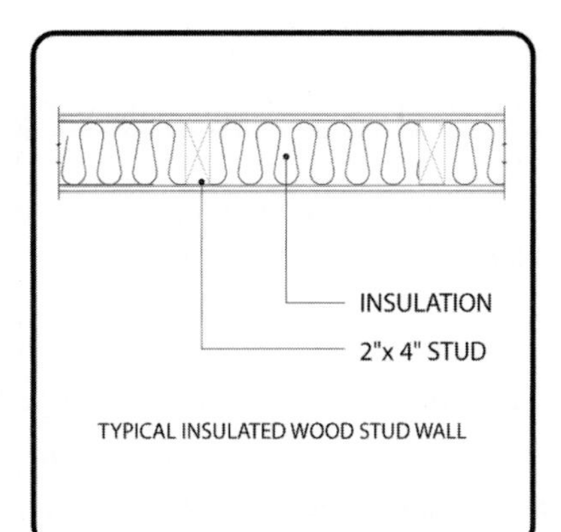

C

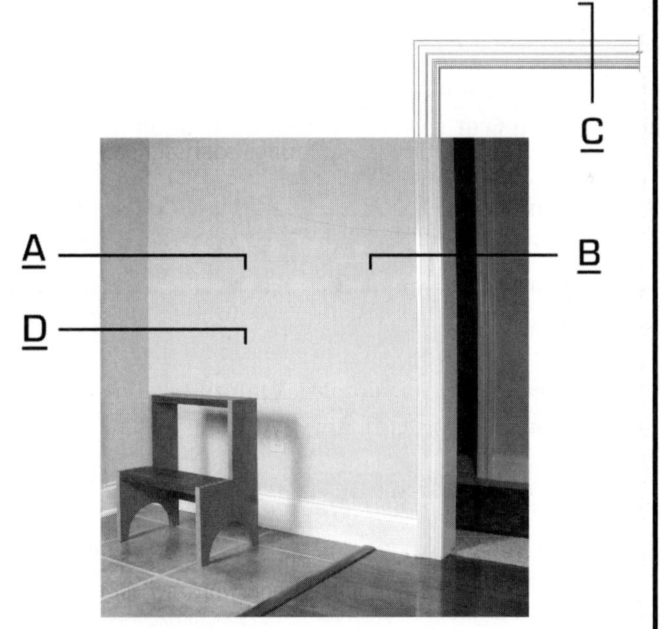

8-2 Some common building techniques used for constructing walls.

How Do Stud Finders Find Studs?

The premier brand in stud finders is Zircon Corporation. This is a well-deserved reputation. Unlike the cheapo stud finders at most home improvement stores, a Zircon StudSensor® detects the density differences between the hollows behind a wall which, in turn, reveals the studs behind a wall.

By using a unique capacitance measurement, a StudSensor can determine how much of an electrical charge wallboard and studs can absorb. Comparing the capacitance between two samples, various material densities can be referenced to each other. A higher capacitance indicates a denser material, for example, wallboard plus a stud.

There is a catch to making accurate measurements with these capacitance stud finders, however. You must calibrate the stud finder when you first turn it on. This calibration must be performed by holding the stud finder in an open area of the wall (i.e., no studs, pipes, or wiring behind the wallboard). The stud finder then pulses an electrical charge into the wall and measures the capacitance. The amount of time needed to reach a reference voltage is fixed as "zero capacitance."

Once calibrated, the stud finder is slowly moved horizontally along the wall surface. As the stud finder approaches a denser material, the time to achieve the calibrated capacitance begins to increase. When this increase reaches a factory-set threshold, the stud finder emits a beam of light which marks the edge of a supposed stud.

CHAPTER 9

High Five, Wi-Fi

It's one of those last-century marvels. Way back in 1999, Apple Computer helped establish the 802.11 wireless networking standard in personal computers and subsequent improvements resulted in the adoption by Apple of the 802.11g high-speed wireless technology. At data rates of approximately 54 megabits per second (Mbps), 802.11g can handle up to 50 users sharing a single Internet connection, as well as granting them access into an Ethernet network and allowing file sharing.

Welcome to the wonderful world of Wireless Fidelity or Wi-Fi.

So what's with all of these standards, anyway? Exactly what is 802.11 and 802.11g and Wi-Fi wireless technology? The explanation begins in 1990 with the Institute of Electrical and Electronics Engineers (IEEE) creating a wireless Ethernet (802.3). Officially named, IEEE 802.11, this wireless Ethernet used radio frequencies on the 2.4-GHz band for building wireless local area networks (WLANs). Consumers were finally able to benefit from this new standard in 1999, when Apple introduced the AirPort wireless network system, 802.11b.

Capable of delivering data up to 11 Mbps, 802.11b was replaced in 2003 by 802.11g 54 Mbps technology. Luckily for Apple, the 802.11g AirPort Extreme was backward-compatible with older 802.11b appliances. Therefore, AirPort Extreme products could integrate seamlessly with any Wi-Fi-certified network—both 802.11b and 802.11g (see Fig. 9-1).

Wi-Fi-certified is administered by the nonprofit Wi-Fi Alliance. This certification indicates that a product has been tested with other wireless products and has been found to be interoperable with other 802.11 devices. So, *any* product that is Wi-Fi-certified will

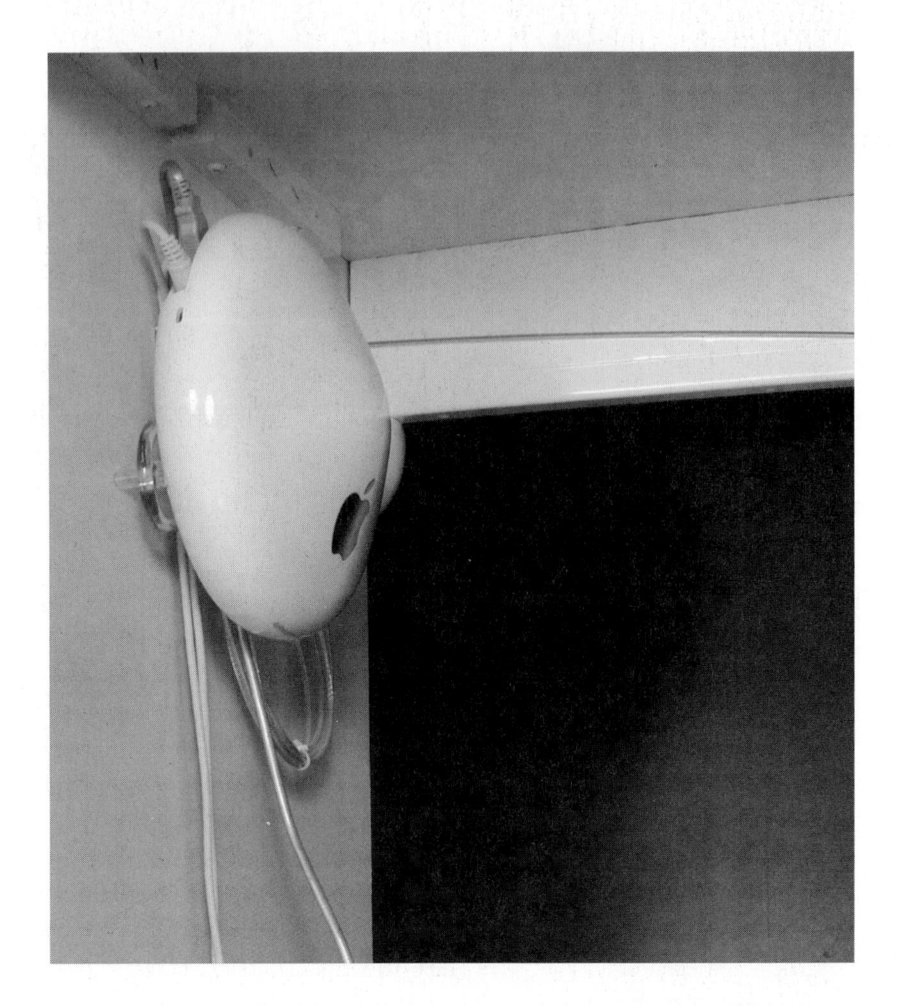

9-1 You can easily install a Wi-Fi network anywhere with the Apple Computer AirPort. This system includes both a networked Hewlett-Packard 5MP LaserJet and a range-extending external antenna. Yes, AirPort *only* enables network printing via Ethernet and USB connections, but the HP 5MP was made network-ready with a simple USB-to-Centronics converter cable.

work with *any* other Wi-Fi-certified product. All is not bliss in the wireless world, however. There is an odd man out.

Due the popularity of the 2.4-GHz band, 802.11g networks occasionally can receive interference from other devices like microwave ovens and cordless telephones. Attempting to nip this trouble in the bud, the 802.11a standard was released. Sounds odd, doesn't it? A new standard with a name preceding an older existing standard. So be it, the 802.11a standard delivers its data on the 5-GHz radio band at a rate up to 54 Mbps. Systems using 802.11a also take a power hit of 2 to 2.5 W compared to 1 to 1.5 W for 802.11b and 802.11g, respectively. Unfortunately, you guessed it, 802.11a devices aren't compatible with 802.11b networks. Furthermore, some countries haven't certified 802.11a. So this standard has been left waiting at the altar.

Oh, Get a Whiff of This Sniffer

If you're wondering how to turn your PSP into a Wi-Fi sniffer, follow these simple steps:

1. Highlight "Network Settings" from the main "Settings" menu and press the X button.
2. Highlight "Infrastructure Mode" and press the X button.
3. Determine a name for this new network connection, like "PSP-Wardriver," then press the X button. The built-in PSP keyboard will be displayed for typing in your selected name.
4. Press the right arrow button to accept the network connection name and move onto the next screen.
5. Press the up arrow key to move up to the "Scan" option and press the X button (if needed). Your wardriving, Wi-Fi-snooping PSP will now search for all access points that are within its wireless network radius.

A list of wireless access points along with their SSIDs and WEP key status will be displayed on your PSP. You'll probably be amazed at the extent of wireless communications that are going on in your own backyard.

Look Ma, I'm Gaming without Wires

Have you ever wished that you could jack your Sony PlayStation® 2 (PS2) into the net for some serious infrastructure gaming? Unfortunately, your Web connection is way over there and your PS2 is glued to your TV's side. Other than stringing a mile of Ethernet cable from A to B, how can you get your PS2 online for some hot head-to-head gaming?

The answer might be easier than you think. Simply, use a wireless Ethernet bridge like the Linksys Wireless-B Game Adapter (Model No. WGA11B).

Begin by finding a cozy spot for your bridge Game Adapter. Typically, this would be right next to your PS2. Plug the power supply in and use the supplied Ethernet cable to connect the Game Adapter to the PS2. If you have a problem "talking" between the Game Adapter and the PS2, there is a "crossover" switch located on the back of the Game Adapter (it is labeled X-II). Toggle this switch until you receive a "handshake" signal from the Game Adapter.

If you only want to play against another "local" PS2 in an ad hoc network LAN party competition, then simply match the channel selector between the two consoles. On the other hand, if you want to game online, then set the Game Adapter to the Internet channel (e.g., "In"). Game on.

There is one caveat that could prevent your jacking into the Net. The Game Adapter is set up for wireless networks that have neither Wired Equivalent Privacy (WEP) encryption settings nor a readable Service Set Identifier (SSID). This is where a Sony PSP™ can come in handy. Use the PSP in its Wi-Fi-sniffing or wardriving configuration (see below for more information on using your PSP as a Wi-Fi sniffer; see Chapter 1 for information about wardriving) to locate and identify wireless networks that match these requirements. If your network is protected from outside usage through WEP encryption, then you will have to establish an Ethernet connection between your PC and the Game Adapter for manually configuring your Wi-Fi game play. A setup "wizard" program is supplied with the Game Adapter for simplifying this stupefying process.

Well, what about that Bluetooth® thing? The official statement from the Bluetooth SIG states that: "Bluetooth wireless technology is set to revolutionize the personal connectivity market by providing freedom from wired connections—enabling links and providing connectivity between mobile computers, mobile phones, portable handheld devices and much more. Bluetooth wireless technology redefines the way we experience connectivity. The Bluetooth SIG, comprised of leaders in the telecommunications, computing, consumer electronic, network and other industries, is driving the development of the technology and bringing it to market. The Bluetooth SIG includes Promoter companies 3Com, Agere, Ericsson, IBM, Intel, Microsoft, Motorola, Nokia and Toshiba, and more than 2,000 Associate and Adopter companies."

On the technological side of the house, however, the Bluetooth specification is a low-cost, low-power radio standard that is used for connecting devices. Global acceptance of this specification has helped incorporate limited-range wireless communication into everything from automobiles to ZigBee.™ Based on a time-sharing architecture featuring frequency-hopping and tiny packet sizes, Bluetooth uses the 2.4-GHz radio band with a range of approximately 30 feet.

Once again, one of the leaders of this pack was Apple Computer with their line of PowerBook G4 portables. These notebooks were the first computers to offer Bluetooth 2.0+ Enhanced Data Rate (EDR). Other computers are stuck in the older Bluetooth 1.x support. Bluetooth 2.0+ EDR, while backwards-compatible with Bluetooth 1.x, is up to three times faster than the older standard. A maximum data rate of up to 3 Mbps is possible with Bluetooth 2.0+ EDR. This throughput plus the peripheral nature of the connectivity feature has enabled some vendors to describe Bluetooth as "wireless USB."

This wireless standard propagation stuff is far from over. There's even a whiff in the air that yet another standard is being proposed. The Intel-sponsored 802.11n standard will up the ante for data rates to 200+ Mbps. Who says that wireless communication is the future of connectivity?

When a Hotspot Comes Along You Must Zipit

If the notion of exploring this great big Wi-Fi world excites you, then head to your kid's toy box and find a Zipit Wireless Messenger. In the palm of your hand,

you are holding a terrific Wi-Fi sniffer. Or, after just a couple of hours' worth of hacking, you can transform your Zipit into a palm-sized Linux computer.

Developed by Aeronix, Inc., the Zipit Wireless Messenger is a cute pocket-sized Instant Messaging (IM) appliance that can connect to any Wi-Fi access point for on-the-go IM. Just turn on the Zipit Wireless Messenger, sniff out a wireless network access point, and start messaging. Alternatively, you can limit your IM appliance to using the wireless network access point scanner as a handheld Wi-Fi sniffer. Just take Zipit out of your pocket, turn it on, and scan for all available Wi-Fi networks.

If you'd rather use the Zipit Wireless Messenger as an IM appliance, you must already have an IM account (e.g., AOL, Yahoo, Microsoft, etc.). You will need a computer with Internet access for establishing this IM account.

Once you have an IM account, then you can freely pursue the joys of IM wherever you can locate a wireless access point. You don't need a personal computer and you don't even need Internet access. Just get an IM account and take the Zipit Wireless Messenger to your local free Wi-Fi hotspot.

There's more to a Zipit Wireless Messenger than IM and Wi-Fi sniffing, though. Somewhere hidden next to the 320 × 240 monochrome LCD screen, Agere Wi-Fi chip, stereo sound, and 16-Mb SDRAM is 2 Mb of Flash memory which holds a Linux ARM 2.4.21 Kernel. In order to fully access this Linux installation, you must hack your Zipit.

Your one-stop shop for Zipit Wireless Messenger hacking information is at the AiboHack Zipit Software Reflashing Web site. This Web site will show you how to reprogram your little IM buddy so that it can establish a wireless connection with a local network file server (NFS) and update its firmware. This firmware update will then enable you to load more versatile Linux firmware. Let there be Linux life.

CHAPTER 10

You Are Mow Wonderful, To Me

Ask any homeowner what is the most dreaded and disliked chore of home ownership and he or she will probably say, "mowing the lawn." Whether you have to maintain a 5,000-square foot postage stamp or manicure a 5-acre estate, mowing the lawn is an arduous ordeal that is universally despised.

You know what? This pervasive disdain might not be a sign of laziness on the part of homeowners. Rather, there might be some sound health and environmental issues that mitigate our aversion to lawn care.

In a research report by Roger Westerholm, Ph.D., assistant professor in the Department of Analytical Chemistry at Stockholm University, and funded by the Swedish Environmental Protection Agency ("Grass Cutting Beats Driving in Making Air Pollution," *ENS*, May 31, 2001), lawn mowers are fingered as major polluters of polycyclic aromatic hydrocarbons (PAHs). Furthermore, this study suggests that lawn mowers contribute about one-tenth of the U.S. total mobile source hydrocarbon emissions.

Westerholm also claims that one old gas-powered lawn mower running for an hour emits as much pollution as driving 650 miles in a 1992 model automobile. During his tests, the Swedish research group used regular unleaded fuel in a typical four-stroke, four-horsepower lawn mower engine and found, after one hour, that the PAH emissions were equivalent to a modern gasoline-powered car driving about 150 kilometers (93 miles).

Similar claims were echoed two years later in the United States. Again, California was in the environmental pollution hot seat—this time for reducing lawn mower pollution.

A Chocolate Frog Cluster

It doesn't take a rocket scientist to see that the world's amphibian populations are declining. Just take a late spring stroll along the Platte River in Nebraska and you will be hard pressed to see very many frogs plopping into the water. This reduction in amphibian populations is even more pronounced in Central and South America.

According to J. Alan Pounds, resident scientist at the Tropical Science Center's Monteverde Cloud Forest Preserve in Costa Rica, his team of researchers might have discovered a link between rising tropical temperatures and the rise in the chytrid fungus. His claim is reported in an article from THE WASHINGTON POST ("Warming Tied To Extinction Of Frog Species" by Juliet Eilperin, January 12, 2006).

This killer fungus, the chytrid fungus, is now thriving in Costa Rica. It is deadly to frogs because it releases a toxin as it grows on the skin and attacks the epidermis and teeth. How does this happen? Pounds' study suggests that global warming has increased water vapor which forms a denser cloud cover resulting in cooler days and warmer nights. And these conditions are an ideal breeding ground for the fungus.

Not everyone is convinced by this study, however. Climatologists, like Stanford University's Stephen Schneider, think Pounds' work is a "step in the right direction" but not conclusive in its attributions to global warming.

According to Pounds, "Disease is the bullet killing frogs, but climate change is pulling the trigger. Global warming is wreaking havoc on amphibians and will cause staggering losses of biodiversity if we don't do something first."

According to the article, "Cutting Back On Lawn-Mower Pollution" by Davide Castelvecchi in the *Santa Cruz Sentinel* (November 29, 2003), lawn mowers emit at least 10 times the amount of smog-causing gases as an average automobile. Citing Jerry Martin of the California Air Resources Board, "During one hour of (mowing), you could have been driving a minimum of 10 cars," Martin said, "and if you have an old mower, up to 30 or 40."

All of the environmental impact indicators aren't that clear, either. On one hand, you have the EPA recommending "frequent mowing" in a 1992 publication (e.g., "Healthy Lawn, Healthy Environment: Caring for Your Lawn in an Environmentally Friendly Way"). Four years later, the same EPA (e.g., "Your Yard and Clean Air"), now informs you that "small engines are big polluters" and "use low-maintenance turf grasses" that "require less mowing."

In your own backyard, it doesn't take much head-scratching to figure out that lawn mowers are a big polluter. For me, every time I push that darn mower around in the South's delightful 100/100 (that's degrees/humidity) summer weather, I curse at the exhaust choking my breath, the high cost of refueling, and the strenuous labor that I have to expend. Sure call me a wimp, but I still think that gas-powered lawn mowers are polluters that have seen their last day in the sun.

Testosterone, Chlorophyll, and Electricity

OK, smarty pants, how would you mow your lawn? First of all, I think that the 1992 EPA suggestion of mowing a lawn frequently is a good one. In that same publication, the government recommends that you should "mow often enough that you never cut more than one-third of the height of the grass blades." I like that recommendation.

Following on, in that same publication, the EPA also contends that you should leave "short clippings on the grass—where they recycle nitrogen." I like that idea, too. Bagging clippings for hauling away to a landfill is silly and wasteful.

Finally, in that previously cited EPA fact sheet, "Your Yard and Clean Air," the first hint at alternative lawn care products is mentioned. In this fact sheet, the EPA says that, "electric equipment is cleaner than equipment powered by gasoline engines. Electrically powered lawn and garden tools produce essentially no pollution from exhaust emissions or through fuel evaporation. However, generating the power to run electric equipment does produce pollution."

So is an electric lawn mower solely the answer to pollution? The short answer is no, but eliminating gas-powered lawn mowers does go a long way toward the elimination of PAHs.

The EPA is not alone in recommending electric lawn mowers. According to a spokesperson for Atco-Qualcast, one of Britain's largest lawn mower manufacturers, in Britain, more than 70 percent of lawn mowers sold annually are electric.

Returning to the Castelvecchi article, a strong case for electric lawn mowers is made in California, as well. This article claims that the California Air Resources Board funds rebates to homeowners who replace older gas-powered lawn mowers with an electric version. This offer is valid only in "smoggy regions" of the state such as San Francisco.

For me, it's a done deal, a no brainer—retire that old gas-guzzling lawn mower and replace it with an electric lawn mower. My electric lawn mower, however, won't be purchased from some lawn care store. My mower will be assembled from recycled pieces and parts. Specifically, a Traxxas E-Maxx RC truck body, a Black & Decker electric line trimmer, and a solar battery–charging panel.

Better yet, salvage the brain from a Roomba floor vacuuming system and add an external Mind Control programmable control stick from Element Products. After some programming gaffs and experimental tests, my lawn mowing robot is ready to save the world's environment. Moo-ha-ha; err, mow-ha-ha.

HOW TO BUILD A ROBOT LAWN MOWER

1

2

1 The robot lawn mower begins life as a radio-controlled (RC) car.

2 Not any RC vehicle will work as a mower. Make sure that the model you use has an enclosed chassis; big, wide, knobby tires; and a flat surface for mounting the line trimmer.

3 The cutting of the robot lawn mower is performed with a battery-powered line trimmer.

4 Remove the trimmer's head, wiring, and battery.

5 Assemble the parts that you'll need for attaching the trimmer's head to your RC car.

3

4

5

6 Use PVC pipe for holding the trimmer's head and wiring.

7 You will need some couplers for fixing the trimmer's head to the PVC pipe.

8 Fix the trimmer's shield to the PVC pipe.

9 Wire up the trimmer's head to the RC car. Test your robot lawn mower with the RC system *before* you install a robot brain.

6

7

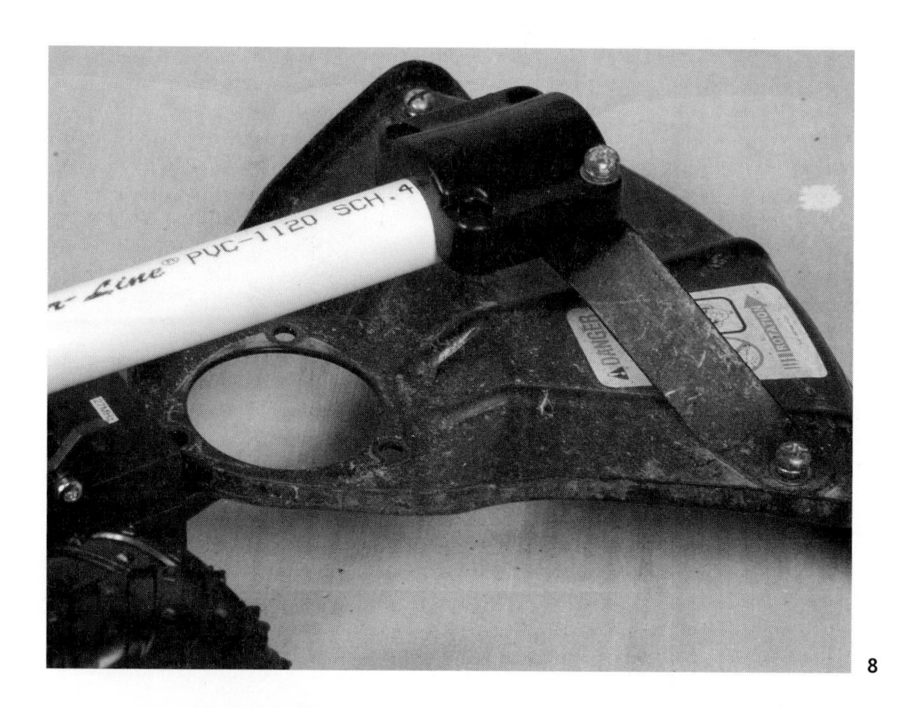

8

9

10 Disconnect the RC circuit and install your preferred brain. The Roomba PCB might be a good brain to try (see Case Study 1).

11 Grab a beer and let your robot mow the lawn.

10

11

CHAPTER 11

Wanna Play House, Doll?

If you're interested in toys (and who isn't?), two of the most significant events that you should attend are Nürnberg Toy Fair and the International Toy Fair. While both of these annual events are open to "members of the trade" only, let me loan you my "press" pass and give you an insider's look at next year's greatest toys.

According to Spielwarenmesse EG, the International Toy Fair Nürnberg is the world's foremost exhibition of "creative, leisure" products and activities. Held in Nürnberg, Germany, the show annually attracts over 77,000 members of the toy trade from over 100 countries and features more than 2,700 exhibitors representing over 65 countries.

From these exhibitors, over 1 million products are displayed, including more than 60,000 new product lines which will be commercially available in the upcoming year. All of these toys, gadgets, and gizmos are displayed in over 155,000 square meters of floor space. Now if only they could tie this exhibition in with Oktoberfest, then they would have something really BIG.

The Spielwarenmesse International Toy Fair Nürnberg products and activities include: model construction, hobbies, model railways and accessories, mechanical and electronic toys, dolls, plush toys and accessories, games, books, learning and experimenting, multimedia, carnival, festive and joke articles, fireworks, wooden articles, arts and crafts, handicrafts, painting, creative design, outdoors, and leisure. And let me tell you, it's really hard to pass up a review of the "festive and joke articles" category.

Whereas Spielwarenmesse International Toy Fair Nürnberg is the world's largest collection of creative and leisure products and activities, the American International TOY

FAIR,® or "TOY FAIR" is the largest toy trade show in the Western Hemisphere. Staged in two distinct locations, the 325,000-square-foot Javits' Convention Center and at various corporate venues scattered around the "toy district" on Fifth Avenue and in the West 23rd Street vicinity in New York City.

Attending the TOY FAIR are over 20,000 toy buyers from over 94 countries. These are the folks who make sure that your local toy store has the hottest toys for the next holiday season. Supplying these goodies are over 1,500 manufacturers, distributors, importers, and sales agents from over 30 countries who display a vast assortment of toys. TOY FAIR regularly showcases toys from many various categories:

- action figures and dolls
- games and puzzles
- bicycles, tricycles, and ride-ons
- radio-controlled vehicles
- infant and preschool toys
- cars, trucks, and trains
- puppets and plush animals
- audio and video cassettes
- computer software and video games
- playground and sporting equipment
- seasonal gifts
- books
- stationery
- party supplies

The TOY FAIR also parcels unique and landmark toys into special "feature" sections. From the TOY FAIR 2006, for example, these special sections included:

- Collectibles—One-of-a-kind doll artists, handmade teddy bears, collectible dolls, finely crafted miniatures
- e@play—Electronic, educational, *edutainment* toys, games, and software
- Game Zone—Board games, card games, puzzles, and brainteasers
- HobbyTech—Model and hobby products, die-cast and radio-controlled vehicles, model trains, boats, planes, and cars
- International Pavilions—Exhibitors from China, Hong Kong, Spain, and Thailand

- Reading, Writing, and Rhythm—Children's books, music, instruments, and recordings
- Specialty Source—Unique product lines that promote healthy play and educational content. Supported by the American Specialty Retailing Association (ASTRA).
- Sweet Spot—Candy-themed products; candy-related novelty items
- Urban Bazaar—Indie art toys and collectibles with a pop-culture/designer spin

Three Toys to Hack

Do you want to know what the coolest toy for next year will be? Tomorrow's hot new toy will actually be a design that you hacked from last year's hot toy. Just rummage around inside your kid's toy box and pull out the toy that's sitting on the bottom of the box. This will be the subject for your futuristic toy hack.

When I scavenged through my kid's three toy boxes, I unearthed these discarded gems:

HOW TO CONTROL IR TOYS

1 These IR text communicators are ideal candidates for connecting to IRTrans.

2 IRTrans connects to your computer (either Mac or PC) via a USB port. IRTrans can both receive and transmit coded IR signals.

3 Special software enables IRTrans to read IR signals, combine them into über-IR signals, and then transmit them around your house.

1

2

3

HOW TO CONNECT MATTEL® PIXEL CHIX™ HOUSES

1 Mattel Pixel Chix houses have fantastic LCD screens that are suspended in a household-like environment. The results are very realistic and highly interactive.

2 Pixel Chix houses are meant to be connected.

3 If you wish to extend the range of your Pixel Chix "neighborhood," use micro clips and wiring to connect each house to one another.

4 Amelia now loves talking to Penelope and Katherine in their big Pixel Chix neighborhood.

1

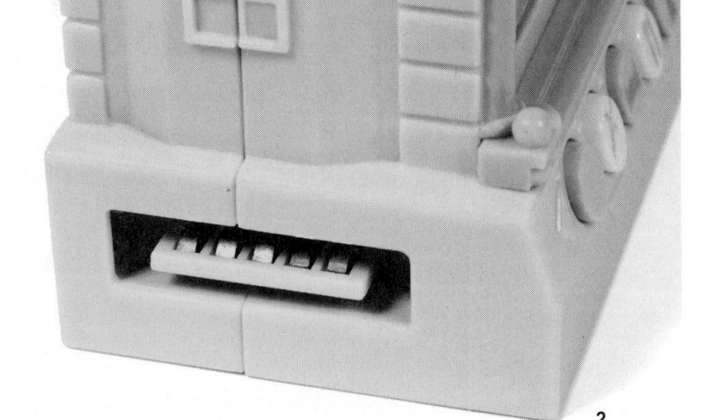

2

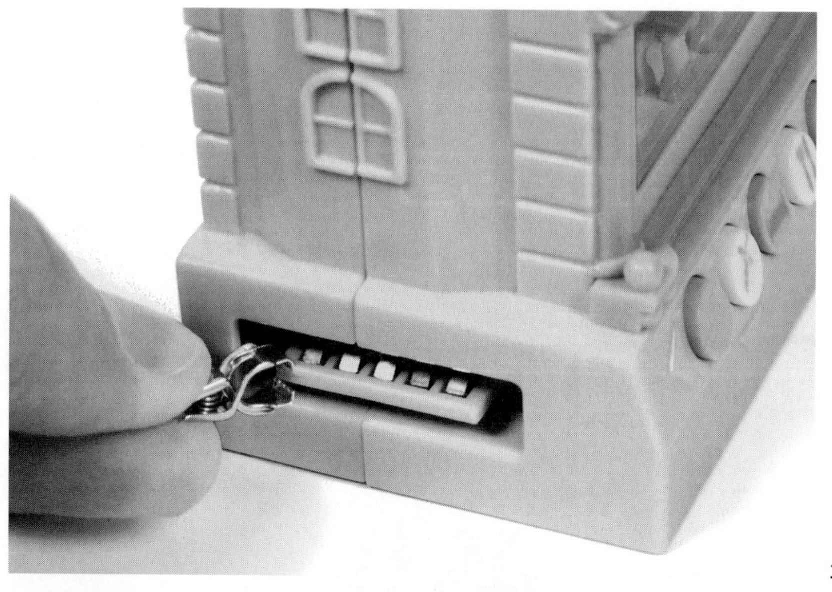

3

4

WRITE THE NIGHT FANTASTIC ON YOUR BIKE

1 Spy Gear™ Spy Nightwriter.™

2 You can create words that float in the air with Spy Nightwriter. This technique is sometimes called Persistence of Vision (POV).

3 A series of buttons enable you to display one of five different messages.

4 You can also key in your own message.

5 Attaching Spy Nightwriter to your bike can make night riding extremely fun.

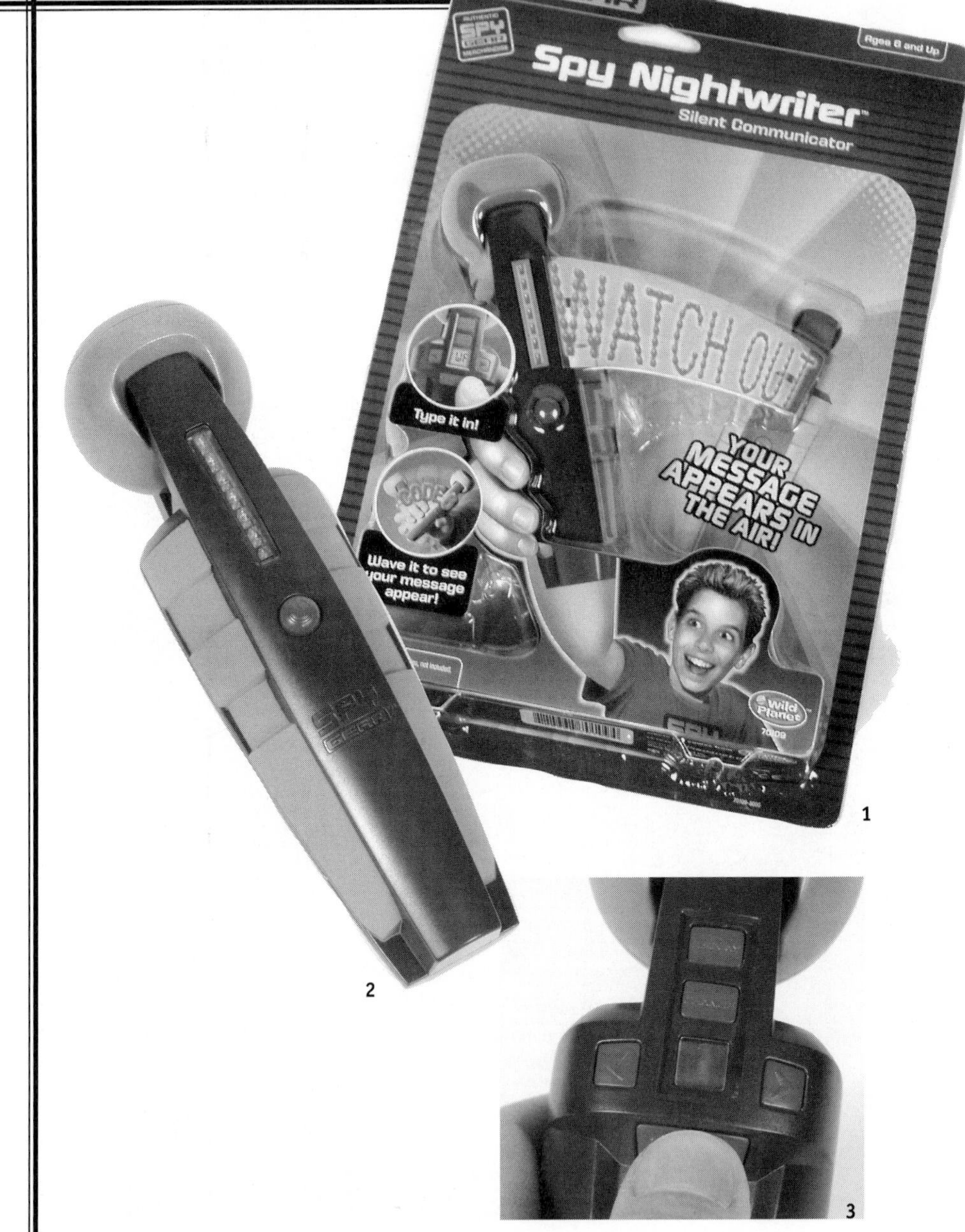

1

2

3

4

5

6 Use cable connectors for fastening Spy Nightwriter to your bike's front fork.

7 Key in your message, turn on the LED flashlight, and market your new bike billboard for sale.

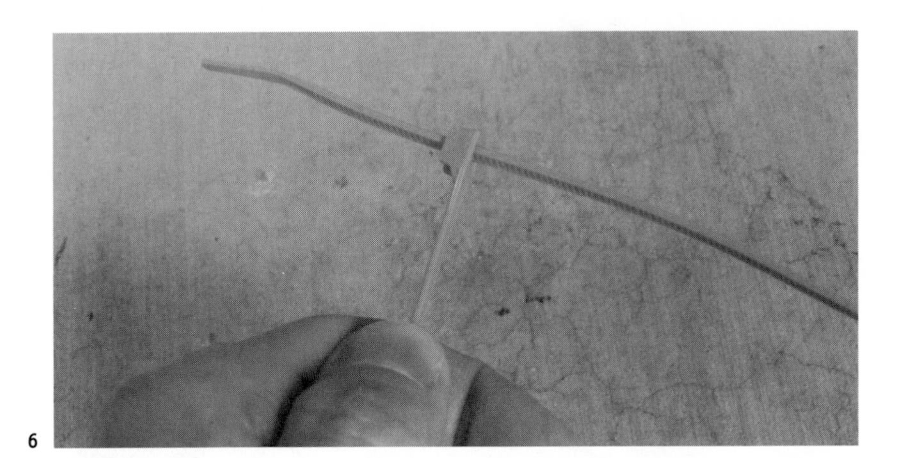
6

7

CHAPTER 12

Recycle, Reuse, Redux

We all have something that was once our pride and joy, but now it has been relegated to the junk drawer. Or worse, we succumb to greed and throw away our former treasures on eBay or at "analog" garage sales (see Fig. 12-1). Typically, these disposals result in a rate of return amounting to only a couple of cents on the original dollar cost (see Figs. 12-2 through 12-4).

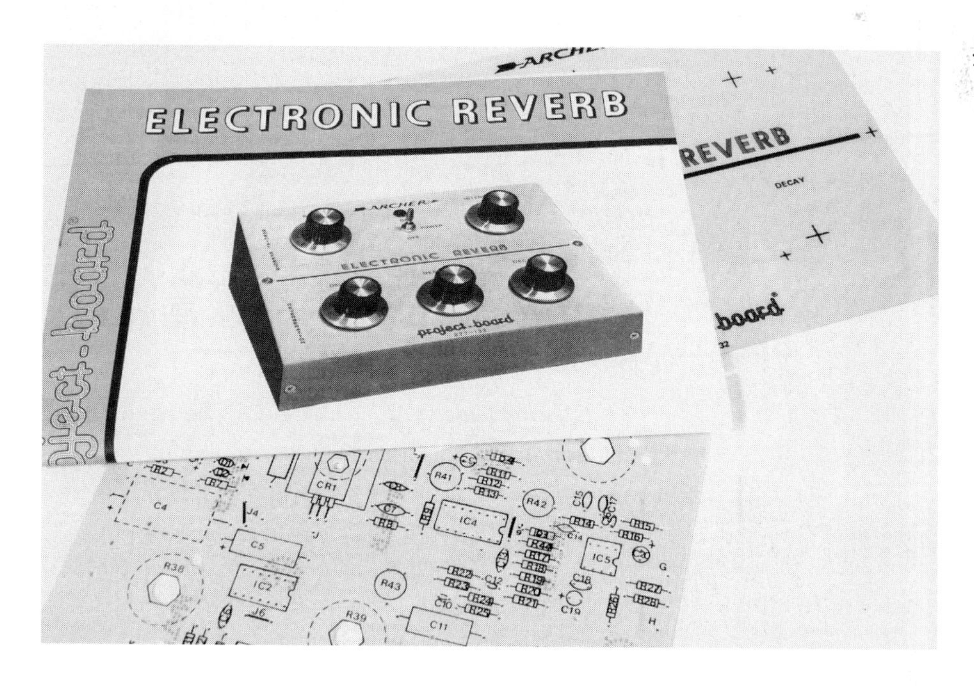

12-1 Old electronics can be found and recycled in a number of odd ways. I found this old Archer (now RadioShack®) kit at a garage sale. Ah, I can almost smell the solder.

12-2 Notebook computer cases, like this SportFolio Sleeve from Marware can breathe new life into older electronics.

12-3 Depending on your notebook computer model, Marware makes a good case for recycling.

12-4 Screen protectors and keyboard pads from Marware can also make an older computer useable.

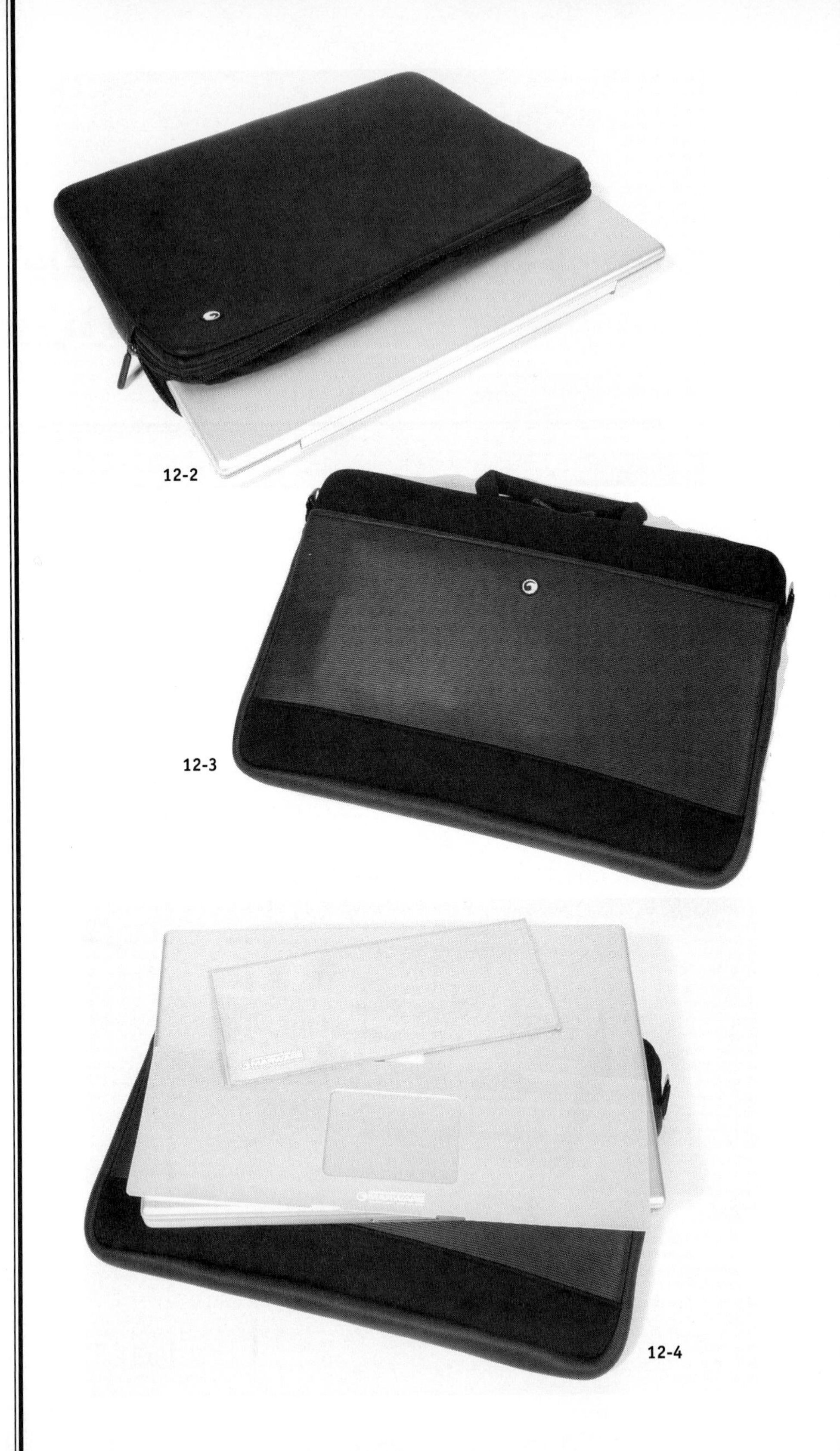

The message couldn't have been clearer—hackers rule. Specifically, hackers who recycle old spacesuits.

In 2006, an unlikely team of an astronaut and a cosmonaut aboard the International Space Station (ISS) launched the first hacked satellite to orbit the Earth. Embodied inside a discarded Russian Orlon® spacesuit were three batteries, a radio transmitter, and an array of sensors for reading temperature and battery strength.

Dubbed SuitSat (Spacesuit plus Satellite), this orbiting spacesuit along with its hacked transmitter circled the Earth and relayed its status to ground stations monitoring its telemetry. Remarkably, these ground stations were not the usual multimillion dollar government facilities that typically monitor NASA missions; rather these ground stations were housed in, well, houses. And these ground stations were manned by, well, anyone. Anyone, that is, who had access to a police-band scanner, handheld radio scanner, or ham radio.

The brainchild of Sergey Samburov, RV3DR, ARISS-Russia team, SuitSat was proposed at the joint AMSAT Symposium/ARISS International Partner meeting in October 2004. After approval, project manager A. P. Alexandrov and Deputy Project Manager A. Poleshuk from RSC Energia, located in Korolev (Moscow area) guided the development of a joint team of Russian and U.S. engineers.

The U.S. contribution was toward the development of the hardware. This development was led by AMSAT member Lou McFadin, W5DID. Inside the recycled Russian Orlon spacesuit (decommissioned August 2004) were two boxes holding the radio transmitter and the digital recorded message playback system. A battery power supply was nestled inside the spacesuit as well, for powering this limited-lifespan transmission satellite. Rounding out this

I Want My SSTV

How do you send a picture when all you can transmit are audio tones? Why, you use slow-scan television (SSTV). Capable of providing low-quality photographs over ham radios, an SSTV image was embedded in SuitSat's digital recorded message playback system. The data transmission format that SuitSat used for its SSTV image was called Robot 36.

hardware installation was the external mounting of an antenna and interface control box.

Broadcasting on 145.990-MHz FM, SuitSat could be received by anyone on the ground with the suitable radio equipment. Lacking any kind of propulsion system, SuitSat shadowed the flight path of the ISS. Therefore, in the United States, SuitSat passed overhead once or twice each day—typically between 12 AM and 4 AM. Depending on your receiving conditions (terrain around your radio, length of antenna, and atmospheric conditions), these overhead passes lasted approximately 5 to 10 minutes.

The Next Voice You Hear

But what did SuitSat broadcast? The transmission was cyclic—30 seconds of data, pause for 30 seconds, and then repeat. Each transmission cycle began with:

"This is SuitSat-1, RS0RS" [SuitSat Voice ID—5 seconds in length]

The remainder of the 25-second transmission consisted of:

- International voice messages [10 seconds in length]

  Russian message
  Europe student messages (Spanish and German)
  Bauman Institute Message (Russian)
  Canada student message (French)

The Hottest CD in Space

One of the oddest items that was bundled inside SuitSat was a CD disc loaded with school photographs, artwork, poems, and student signatures. Known as the "School Spacewalk" CD, this disc was onboard SuitSat when it reentered the Earth's atmosphere and burned up. Another, duplicate copy of this disc was retained by the crew for viewing during those long, cold, dark, and lonely nights in space.

Mr. Alexandrov message (English)
Japan student message (Japanese)
United States student message (English)

- Mission telemetry [5 seconds in length]

  Mission time
  Suit temperature
  Battery voltage, where 28 V is the nominal voltage

- SSTV image [10 seconds in length]

- 30-second pause

On February 3, 2006 at 11:27 PM EST, SuitSat-1 was launched into orbit during a 5-hour, 43-minute extravehicular activity (EVA) by ISS Expedition 12 Commander Bill McArthur and Flight Engineer Valery Tokarev. Launched into a "retrograde" orbit (an orbit that slows down and drops below the ISS), SuitSat was placed into orbit at the beginning of the second spacewalk of Expedition 12.

During the preparation of SuitSat for the EVA launch, the ISS crew had to:

1. Connect cabling between the transmitter and the controller box.
2. Insert both boxes inside a fabric container inside the spacesuit.
3. Mount the antenna and the interface control box to the outside of the spacesuit helmet.
4. Connect all external cables (antenna, interface control box) and the battery to the two internal boxes.

SuitSat was now ready for launch. Once outside the ISS, the EVA members flicked three switches on the external interface control box and "jettisoned" the satellite into its retrograde orbit. This jettison technique was similar to the method that McArthur had used during the launch of the Floating Potential Probe experiment during a November 2005 spacewalk.

In order to achieve the proper retrograde orbit, McArthur had to push SuitSat away at about a 30-degree angle upward and about 10 degrees to the left of the back of the station. SuitSat was now operational and transmitting its messages back to listeners on Earth. Unfortunately, SuitSat only worked for about two orbits of the Earth. It then died. Or, did it?

The official NASA report was contradicted by members of the Connecticut-based American Radio Relay League (ARRL). An ARRL member, Allen Pitts, reported on February 5, 2006, that his group had indeed received "weak, cold, and really hard to copy" transmissions from SuitSat.

About two weeks later the silent SuitSat reentered Earth's atmosphere and burned up—an ignoble end to an ingenious blend of recycling, hacking, and international cooperation.

I think that CNN.com said it best, "One small step for trash is giant leap for ham-kind."

THE INTERNATIONAL SPACE STATION GOES DIY

1

1 The Expedition 12 patch represents both mankind's permanent presence in space and future dreams of exploration. The International Space Station (ISS), featured prominently in the center, will continue to grow in its capability as a world-class laboratory and test bed for exploration. The vision of exploration is depicted by the moon and Mars. The star symbolizes mankind's destiny in space and is a tribute to the space explorers who have been lost in its pursuit. The Roman numeral XII in the background signifies the twelfth expeditionary mission to the ISS. (*Illustration courtesy of NASA.*)

2 **Exploded view of International Space Station (ISS).** (*Drawing courtesy of NASA.*)

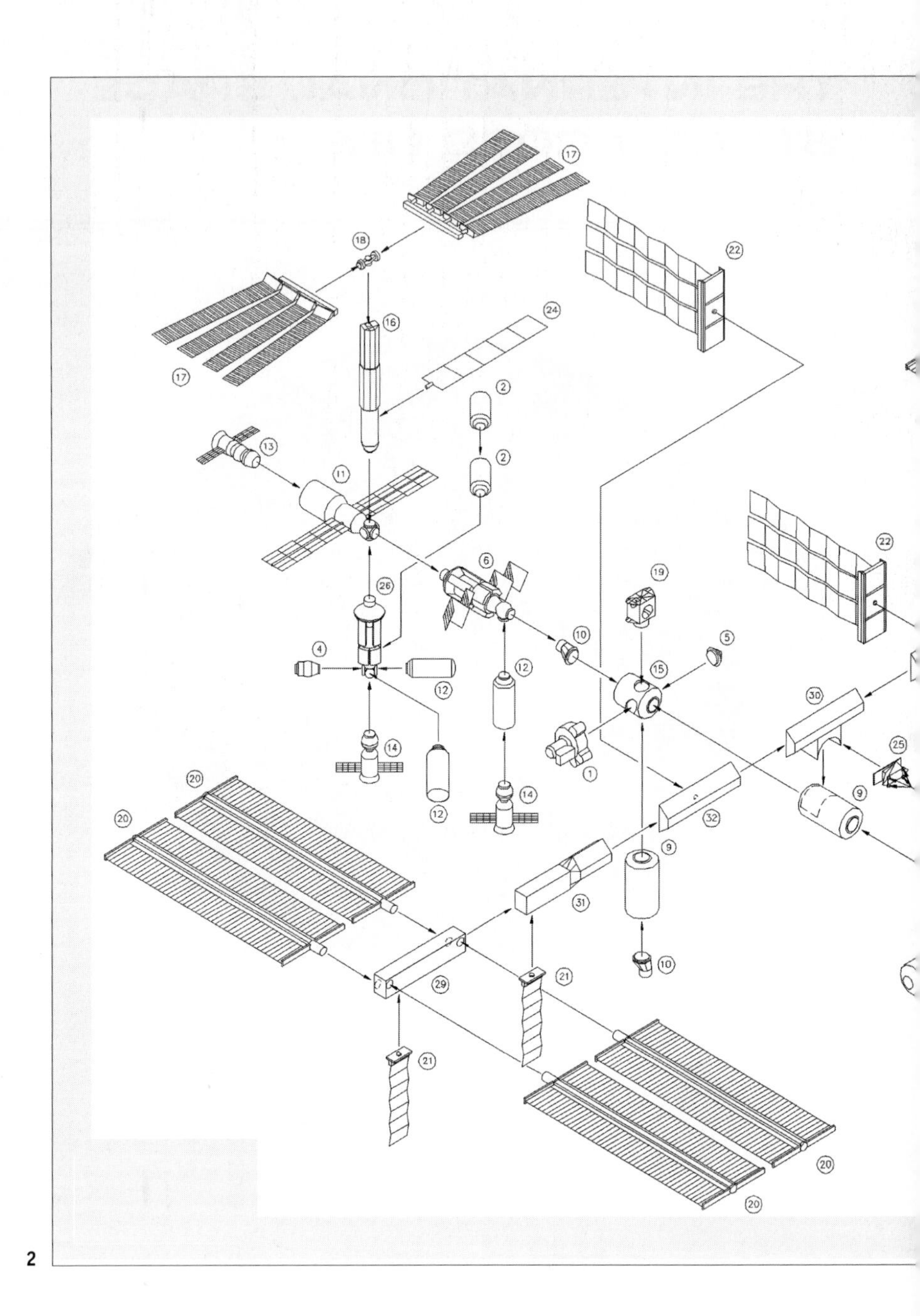

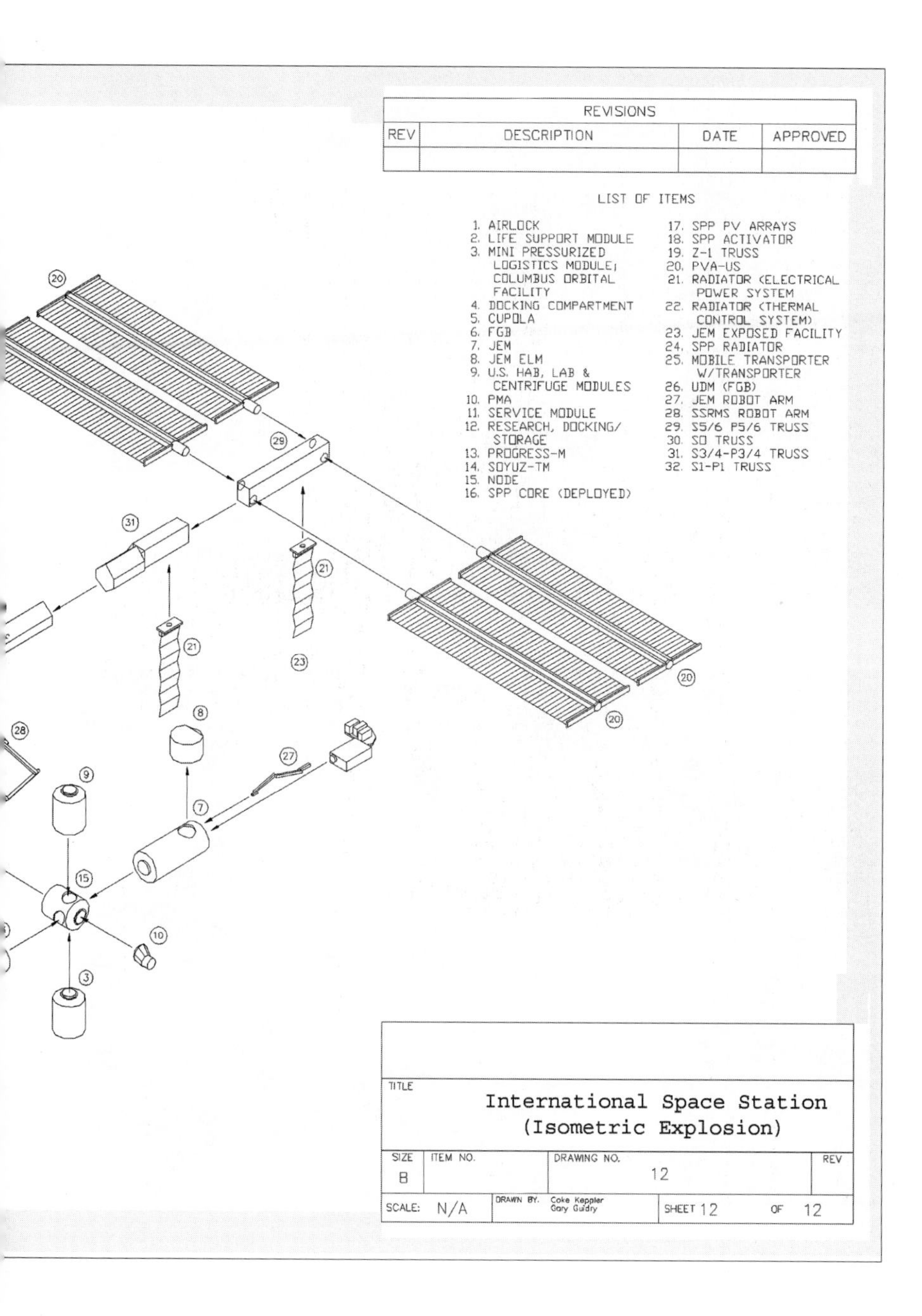

REVISIONS
REV
DESCRIPTION
DATE
APPROVED
LIST OF ITEMS
1. AIRLOCK
2. LIFE SUPPORT MODULE
3. MINI PRESSURIZED LOGISTICS MODULE; COLUMBUS ORBITAL FACILITY
4. DOCKING COMPARTMENT
5. CUPOLA
6. FGB
7. JEM
8. JEM ELM
9. U.S. HAB, LAB & CENTRIFUGE MODULES
10. PMA
11. SERVICE MODULE
12. RESEARCH, DOCKING/ STORAGE
13. PROGRESS-M
14. SOYUZ-TM
15. NODE
16. SPP CORE (DEPLOYED)
17. SPP PV ARRAYS
18. SPP ACTIVATOR
19. Z-1 TRUSS
20. PVA-US
21. RADIATOR (ELECTRICAL POWER SYSTEM
22. RADIATOR (THERMAL CONTROL SYSTEM)
23. JEM EXPOSED FACILITY
24. SPP RADIATOR
25. MOBILE TRANSPORTER W/TRANSPORTER
26. UDM (FGB)
27. JEM ROBOT ARM
28. SSRMS ROBOT ARM
29. S5/6 P5/6 TRUSS
30. SO TRUSS
31. S3/4-P3/4 TRUSS
32. S1-P1 TRUSS
TITLE
International Space Station (Isometric Explosion)
SIZE B
ITEM NO.
DRAWING NO. 12
REV
SCALE: N/A
DRAWN BY. Coke Keppler Gary Guidry
SHEET 12 OF 12

3 This image of Hurricane Wilma was taken at 8:23 AM CDT Wednesday, Oct. 19, by the crew aboard NASA's international space station as the complex flew 222 miles above the storm. At the time, Wilma was the strongest Atlantic hurricane in history, with winds near 175 miles per hour. The storm was located in the Caribbean Sea, 340 miles southeast of Cozumel, Mexico. (*Photograph courtesy of NASA.*)

4 An old Russian Orlan spacesuit is photographed in the Unity node of the International Space Station, which was released by hand from the space station during a spacewalk Feb. 3, 2006. Outfitted with a special radio transmitter and other gear, the spacesuit comprises a Russian experiment called SuitSat. It will fly free from the station as a satellite in orbit for several weeks of scientific research and radio tracking, including communications by amateur radio operators. Eventually, it will enter the atmosphere and be destroyed. (*Photograph courtesy of NASA.*)

CHAPTER 13

Knit Wit

The press release couldn't have been less surprising in its subject—Levi's® announced the first pair of iPod-compatible denim jeans. Broadcast from their San Francisco office on January 10, 2006, Levi Strauss declared that it would be launching a "wearable technology revolution" with the Fall 2006 release of the new Levi's® RedWire™ DLX Jeans. Yawn.

According to Robert Hanson, Levi's U.S. brand president, these jeans, for both men and women, are "the latest extension of the Levi's brand leadership position by merging fashion and technology that provides consumers with the most innovative way to enhance their portable, digital music lifestyle." Innovative? Really, I just slip my iPod into my back pocket. And I did this "merging of fashion and technology" back in 2004.

Yes, this 1873 company, makers of the "original, authentic jeans" was finally jumping onto the digital lifestyle bandwagon. Albeit, about four years too late. Most of us have already learned how to integrate our technology into our daily lives. There has to be more to this story than just stitching an iPod pocket onto a pair of jeans, right?

Yup. In his announcement of these new jeans, Hanson declared, that, "in designing the jeans we considered both function and fashion—the result is a uniquely functional, yet stylish, great fitting jean." Specifically, these design features were:

> **EASY POCKET STORAGE**—an iPod docking cradle is secreted away in a side pocket. This pocket was specially designed to minimize any iPod "bump" that would be externally visible. Additionally, a "red conductive ribbon" (hence "RedWire") enables users to pull the iPod out of this pocket for referring to its screen without having to disconnect your music player.

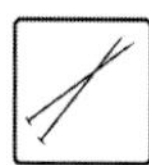

"HIP" CONTROLS—this isn't what you might be thinking (e.g., body movement control—mixing while dancing), rather a joystick remote control has been embedded in that little watch pocket that most of us never use. Just stick your finger into this pocket and you can play/pause, track forward, track backward, and adjust the volume. I can see it now—this feature will bring a new definition to "playing pocket pool."

HANDY WIRE RETRACTOR —a retractable headphone wiring manager for keeping your iPod cables tangle-free.

Let me see: iPod nano or a pair of pants? Hmm, which should I buy? Oh, and Levi Strauss did add this caution to their press release, "the jean is washable once the iPod is removed." Duh.

Knee Deep in Kenpo

Rising a little higher up on your body, the Kenpo® Jacket for iPod was declared a "pinnacle of form, function and personal style" by the vice president of Kenpo, Inc., Joel Bernstein. Maybe not a household brand name, this Los Angeles, California–based manufacturer of general consumer apparel has created a niche market specializing in outerwear, denim, and fleece.

In the case of the Kenpo Jacket for iPod, a clever swatch of "smart fabric" has been stitched into the sleeve. Known as ElekTex,® this patented fabric control system is a set of five electronic buttons that enable the wearer to control an iPod without removal from its special "padded interior pocket."

By pushing these touch-sensitive buttons, you can control the iPod's play/pause, track forward, track backward, and adjust the volume. Remarkably, the jacket is devoid of any wires and can be washed or dry-cleaned following the removal of the iPod and the internal iPod connector. Power for this all-fabric control system is derived from the iPod's batteries.

Armed with a built-in key lock feature, pressing the play/pause button for 3 seconds unlocks the fabric control system. The system is then relocked after 7 seconds of inactivity. Available in four men's styles, the Kenpo Jacket for iPod has a retail price of $275. It is available for purchase from Eclipse Solar Gear at www.eclipsesolargear.com.

It's Your Bag

One of the winners of the 2006 Innovations Design and Engineering award at the International Consumer Electronics Show in Las Vegas during the first week in January 2006, was a little messenger bag from Innovus Design, Inc. The "Fusion" Solar Messenger Bag with optional PowerPac was the winner in the Best of Innovations in the Portable Power category. What made this bag really neat was the integration of a photovoltaic panel capable of trickle-charging a set of batteries while you strut your stuff around town.

Based on a simple design interface of a 12-V cigarette lighter adapter, "Fusion" Solar Messenger Bag is capable of helping give a power boost to cell phones, iPods, and radios, as well as keeping your removable, rechargeable batteries topped off.

Constructed of a durable nylon exterior, this messenger bag is very similar to another popular Innovus Design product —the Solar "Flare" Messenger Bag. Able to lug around your notebook computer, cell phone, and PDA, the "Flare" is also able to accommodate your outdated equipment like paper, binders, and, gasp, books. Likewise, all of the usual suspects like pens, pencils, CDs, and your house keys tethered by an elastic lanyard can be safely tucked away, somewhere inside this messenger bag. Oh, and if you're looking for that proverbial "kitchen sink" adage; yes, you can even stick a water bottle inside an internal pocket.

All of this cargo-carrying capability aside, the "Flare" (and "Fusion," for that matter) really comes alive during its patented solar-powered recharging mode. Coupled to an ingenious 12-V, 2.5-W automotive-type cigarette lighter adapter, you can easily attach any compatible personal electronics device to the bag for a drink of some solar electric juice. Be forewarned, however, the "Flare" is incapable of recharging your notebook computer's battery.

You could try to make your own "power" bag. Make sure that you begin with a rock-solid case that will be able to handle your hack (see Figs. 13-1 through 13-3).

13-1 The CEO Milano from Marware is a great hackers' case.

13-1

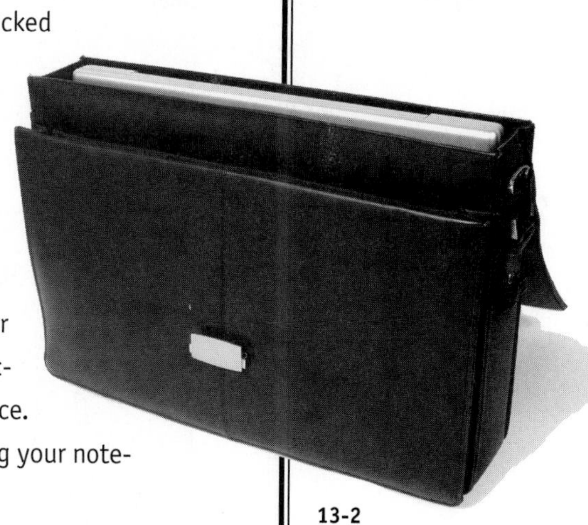

13-2 Able to hold all of your electronic computing gizmos, this CEO Milano case by Marware can also hold your Sony PSP.

13-2

13-3 Tuck an Apple Computer PowerBook 17-inch inside the CEO Milano case from Marware and you can still hold a Wacom Bluetooth tablet, cell phone, and Sony PSP.

13-4 Learn to knit and you can purl, too.

13-5 American Buffalo Products knit.

13-3

Tall Yarn Tails

Move over cashmere, there's a new animal in that sweater. Built from a heritage as old as America, American Buffalo Products is the manufacturer of yarn made from American buffalo. Painstakingly designed by Ruth Huffman, this patented process takes the brutish bad-boy looks of the bison and transforms it into a creamy chocolate brown yarn that rivals the finest cashmere (see Fig. 13-4).

13-4

But this isn't your grandma's yarn. This buffalo yarn retains many of the same traits found in the hoofed animal version. Namely, garments made from this yarn can be washed and they will not shrink. Additionally, it is both lightweight and insulating. So, go ahead, run out in the rain or snow with just a sweater; just make sure that it's a buffalo sweater.

Looking at a buffalo, it's tough to foresee a delicate yarn emerging from its fur (see Fig. 13-5). Even Huffman wasn't originally convinced that her idea would work. It took over two years' worth of research and development to get her technique just right. Then, on a lark, she purchased 1,000 pounds of buffalo hair. Her concept became a reality and a new fashion statement was ready to be made. Oh, and in case you're keeping score at home, that 1,000 pounds of hair translated into 377 pounds of yarn.

So how do you get buffalo hair off a buffalo? Well, very carefully. Seriously, either by stealing or shearing. In stealing, you wander around aimlessly picking up bits of loose hair from trees and bushes. If you'd rather have a life, you can shear the hair from each animal. In shearing, you get about 4 to 5 pounds of hair from a single buffalo.

According to Huffman, the American buffalo has five different kinds of hair. Through her patented process, she was able to sift through all of the rough, coarse outer hairs and extract just the downy soft undercoat fleece. It is this

13-5

fleece that she then spins into yarn. The result is a rich, soft cashmere-like yarn with a deep earthy brown color.

For the high-tech consumer, all of this research is great news. Now you can add your own iPod pocket, for example, without paying an arm and a leg for a pair of blue jeans. Luckily, Huffman has plenty of yarn for meeting this new consumer demand. In 2006, she spun 10,000 pounds of buffalo yarn. Even with that sizeable output, this modern-day Rumpelstiltskin only used about one-third of the total amount of buffalo hair sheared in 2006. Now that's a miller's tale.

Dr. Doolittle Does More

As if making yarn from American buffalo hair was hard to believe (never underestimate the ingenuity of hackers and makers), then the story of Linda Niemeyer might be even tougher for you to swallow. After watching an educational

I'm Not Tryin' to Buffalo Ya

American Buffalo Products have some great buffalo "fast facts":

400,000 American Buffalo in the United States

About 250,000 American Buffalo in Canada

Bison are molting animals that shed their coats in the Spring of each year.

They have dense, woolly winter coats they shed in the Spring and Summer.

They have 330-degree field of vision and hearing and sense of smell are quite exceptional, with their hearing range similar to that of humans.

A raised tail is a sure sign of trouble. The higher the tail, the bigger the trouble.

13-6

13-6 Alpaca knit.

television program on alpacas and their fibers (sounds like a great title right there, eh?), Niemeyer discovered the following fuzzy facts about alpacas:

- Alpacas are small.
- Alpacas are a fiber-producing animal that is sheared once a year.
- Alpaca fiber is warmer and softer than wool.
- Alpaca fiber comes in 22 natural colors.
- Alpacas have been domesticated by the South American Andean cultures for over 6,000 years.

Now this former graphic designer had an idea. What if she raised her own alpacas; yes, a herd, a fiber empire, a yarn dynasty? So she bought a pregnant alpaca and had an instant herd. Gathering the fiber and spinning the yarn, Niemeyer was well on her way to knitting a yarn dynasty.

Blue Sky Alpacas is now a thriving business with an exclusive selection of fine yarns (see Fig. 13-6). High-tech knitters can choose from eight different alpaca yarns and blends:

- 100 percent Alpaca—the luxury classic
- Alpaca and Silk—a 50/50 premium blend
- Bulky—a big, chunky favorite
- Bulky Hand Dyes—kettle-dyed 50/50 alpaca/wool blend

- Worsted Hand Dyes—a superior 50/50 alpaca/merino blend
- Organic Cotton—all natural cotton
- Dyed Cotton—organic cotton enriched with dyed colors
- Duotones—a big, chunky yarn in tone-on-tone dyes

If you thought that knitting was a hobby restricted to the XX-chromosomal sex, think again. Knitting is quickly gaining a resurgence among the under-35 crowd. And that crowd is populated by both men and women.

In a February 2005 CBS News report ("Men & Boys Knitting Up A Storm"), a survey by The Craft Yarn Council of America was cited for reporting an increase in the number of women knitters aged 25 to 34 from 13 percent in 2002 to 33 percent in 2004. Unfortunately, this same survey did not study the number of male knitters.

Several knitting supply stores that were mentioned in this article, however, did indicate that they had seen a marked increase in the number of men frequenting their stores. How many of these men are shopping and how many are stalking? You think that's funny? One shop owner did indicate that her son uses knitting as bait for meeting girls. She sheepishly admitted that her son uses "knitting as a babe magnet." Read that as: knit one, and purl the girls.

Headware

Some common knitting abbreviations follow:

BO = bind off	K = knit
CO = cast on	P = purl
Dec = decrease	pm = place marker
Dpn = double-pointed needles	Rnd = round
Inc = increase	sts = stitches
K2tog = knit 2 together	tog = together

The following pattern is for a hat that should fit a small adult head—like that of a pinheaded teenager.

NOTE: Always make a test-knitting or gauge before starting your project. This gauge will tell you how many stitches per inch you are capable of making with your particular yarn. This project is for beginner knitters. Finished head circumference is 21 inches.

MATERIALS

- Two (2) skeins of 110 yards, 50 grams. I chose Classic Alpaca Wall St. Flannel, color 403 from The Alpaca Yarn Company. The gauge is 5 1/2 stitches per inch.
- U.S.: Six (6) 4-mm double-pointed needles

1. **BRIM.** Cast on 128 stitches (sts) evenly on 3 double-pointed needles (dpn). Place marker (pm). Work in a knit (K), purl (P) ribbed stitch for about 3/4 inch.
2. **CROWN.** Change to a straight stitch and work in rounds until piece measures around 5 inches. Decrease (Dec) round (Rnd). *K2, knit 2 together (k2tog), repeat to the end of the round. K 1 round*. Repeat step between ** until only 8 sts remain.
3. **FINISHING.** Cut yarn and thread tail between remaining sts, pull tight, and fasten to inside.
4. **INSIDE CASING.** Cast on 15 sts. K1,P1 in a ribbed stitch for about 2 1/2 inches. Bind off (BO). The completed patch should measure 1 1/2 × 2 1/2 inches. Make 1 more patch in the same manner. Sew these patches to the inside of the hat on opposite sides. Cast on 20 sts. K1,P1 in a ribbed stitch for about 1 inch, BO. Sew this patch to the inside back of the hat. Now you are ready to insert your ear buds.

HOW TO ADD IPOD WIRING INSIDE YOUR HAT

1 Katherine's hat holds three alpaca knit pockets inside for holding the ear buds and the iPod's Y-connector.

2 Use Alpaca kit to add comfy inserts inside any knit hat for holding your iPod wiring.

3 Looking good and sounding great.

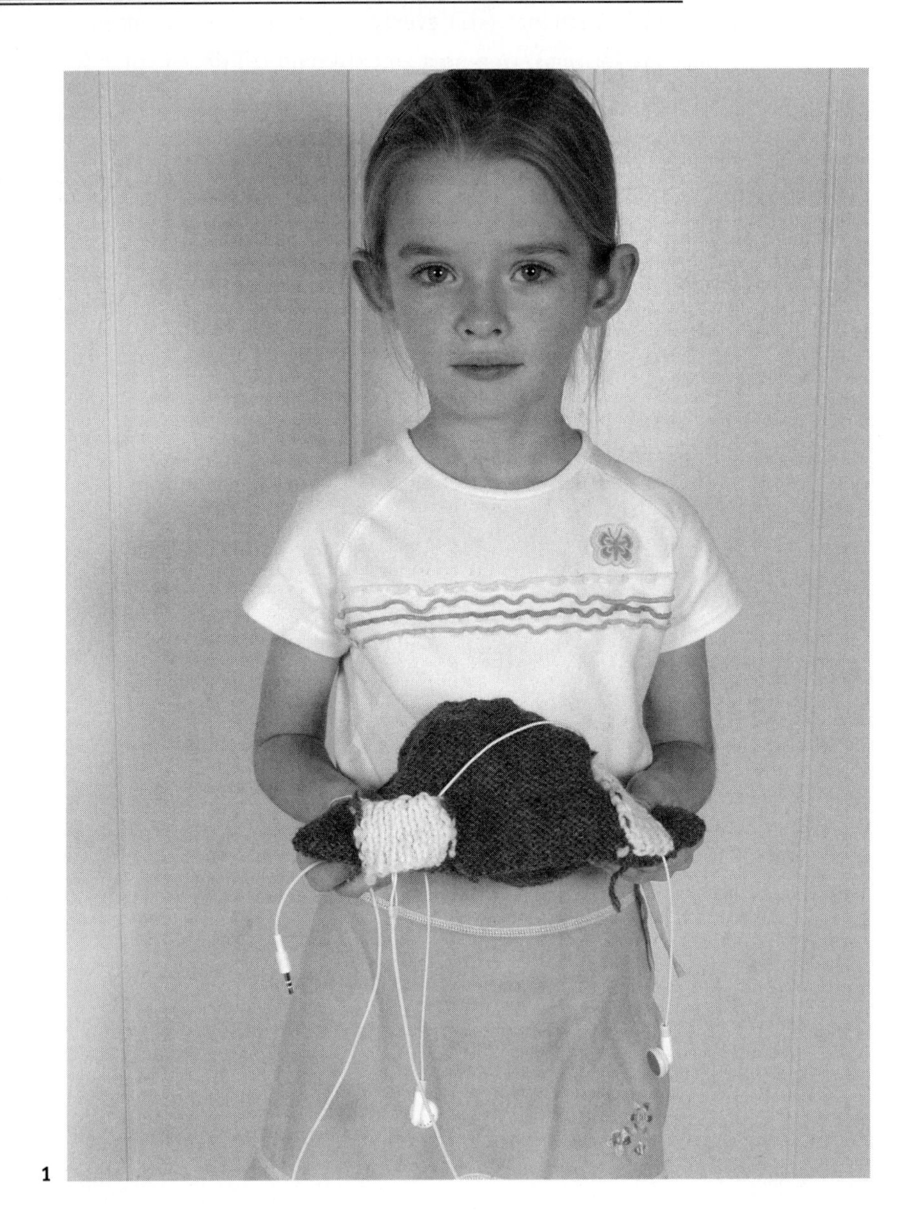

1

2

3

HOW TO WIRE YOUR PANTS—PSP POCKET

1 Don't wait for Levis, convert your own jeans into a multimedia console.

1

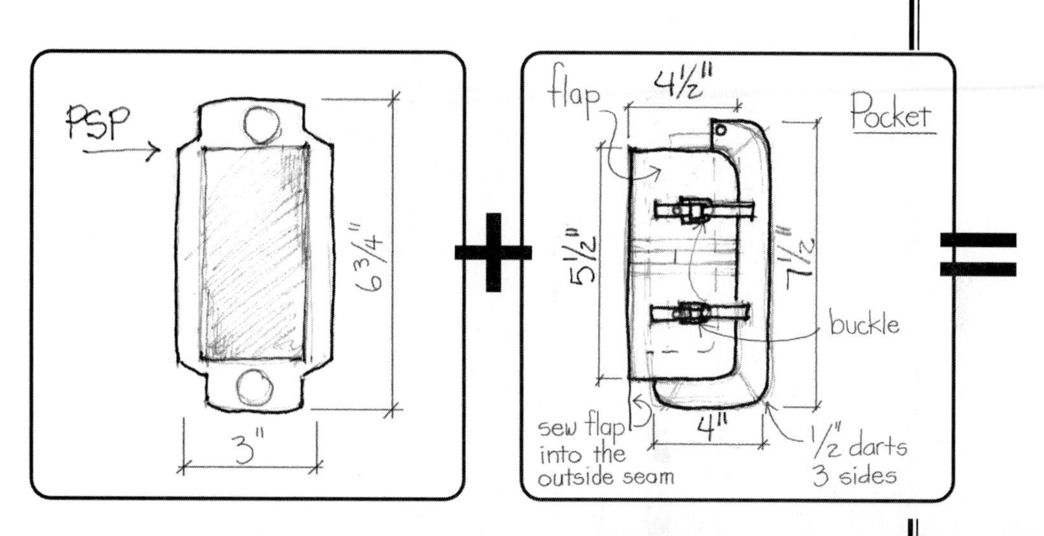
PSP
3"
6 3/4"
flap
4 1/2"
Pocket
5 1/2"
7 1/2"
buckle
sew flap into the outside seam
4"
1/2" darts 3 sides

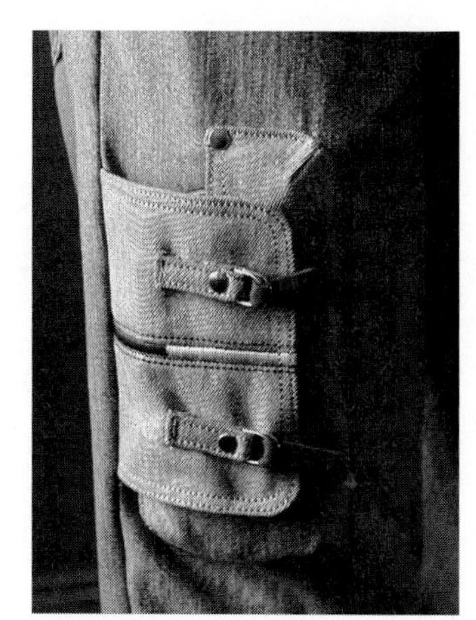

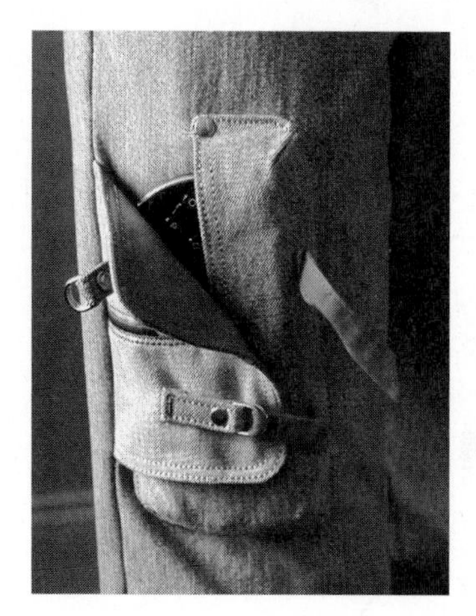

2 **Assemble this piece with the next piece for making your PSP Pocket.**

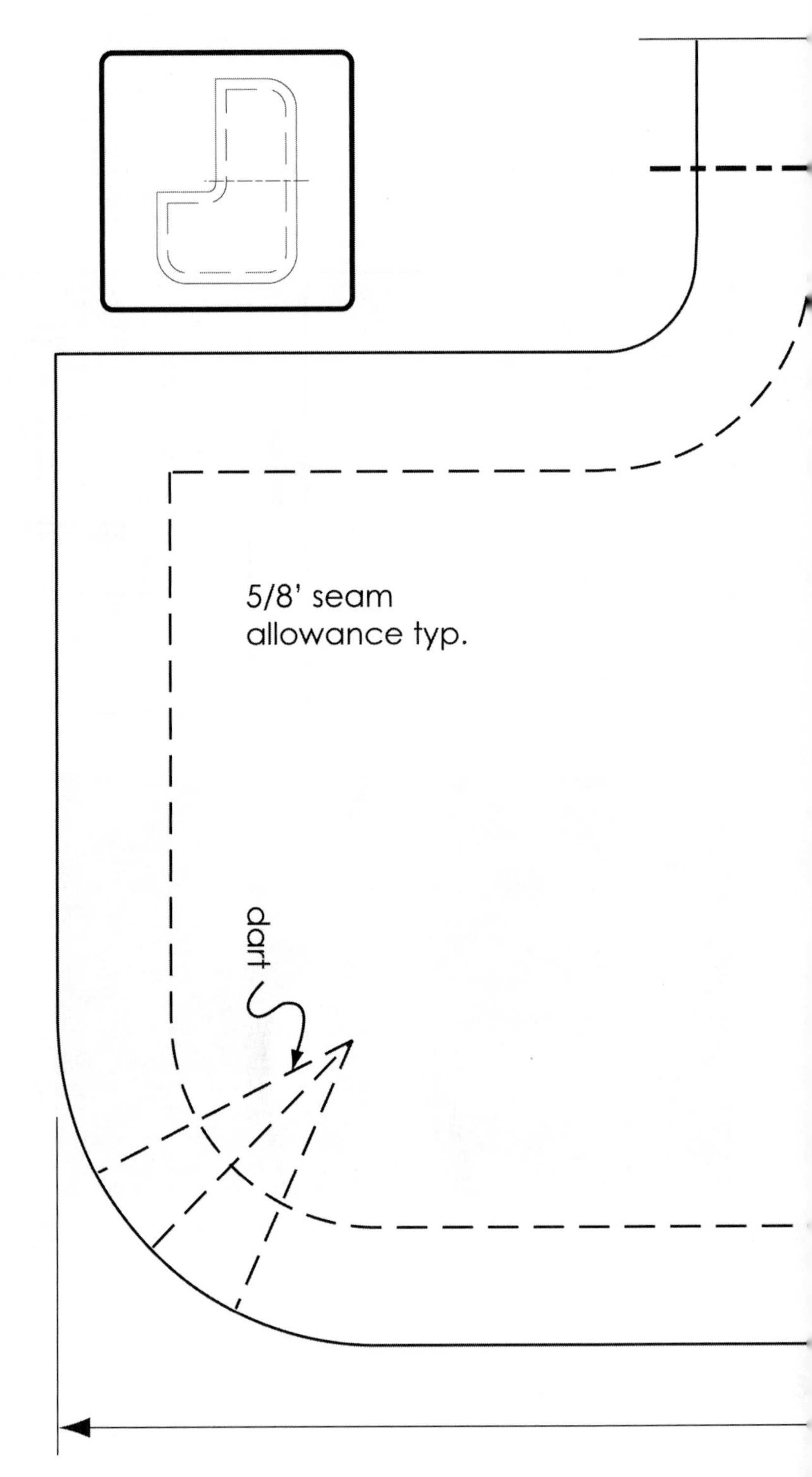

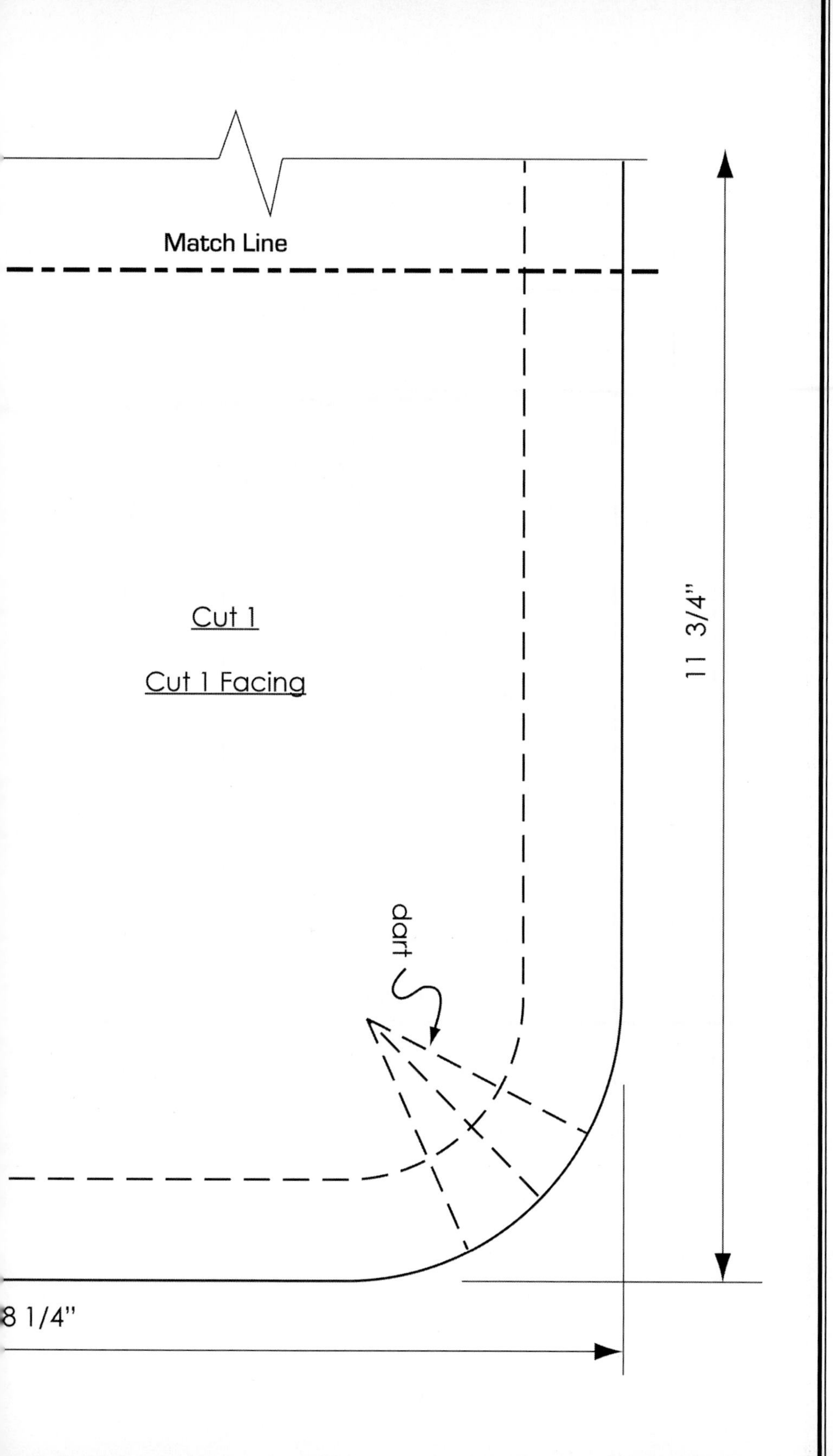

Knit
Wit

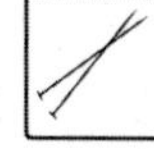

3 This piece connects to the previous piece for making the PSP pocket.

4 This piece makes the PSP pocket flap.

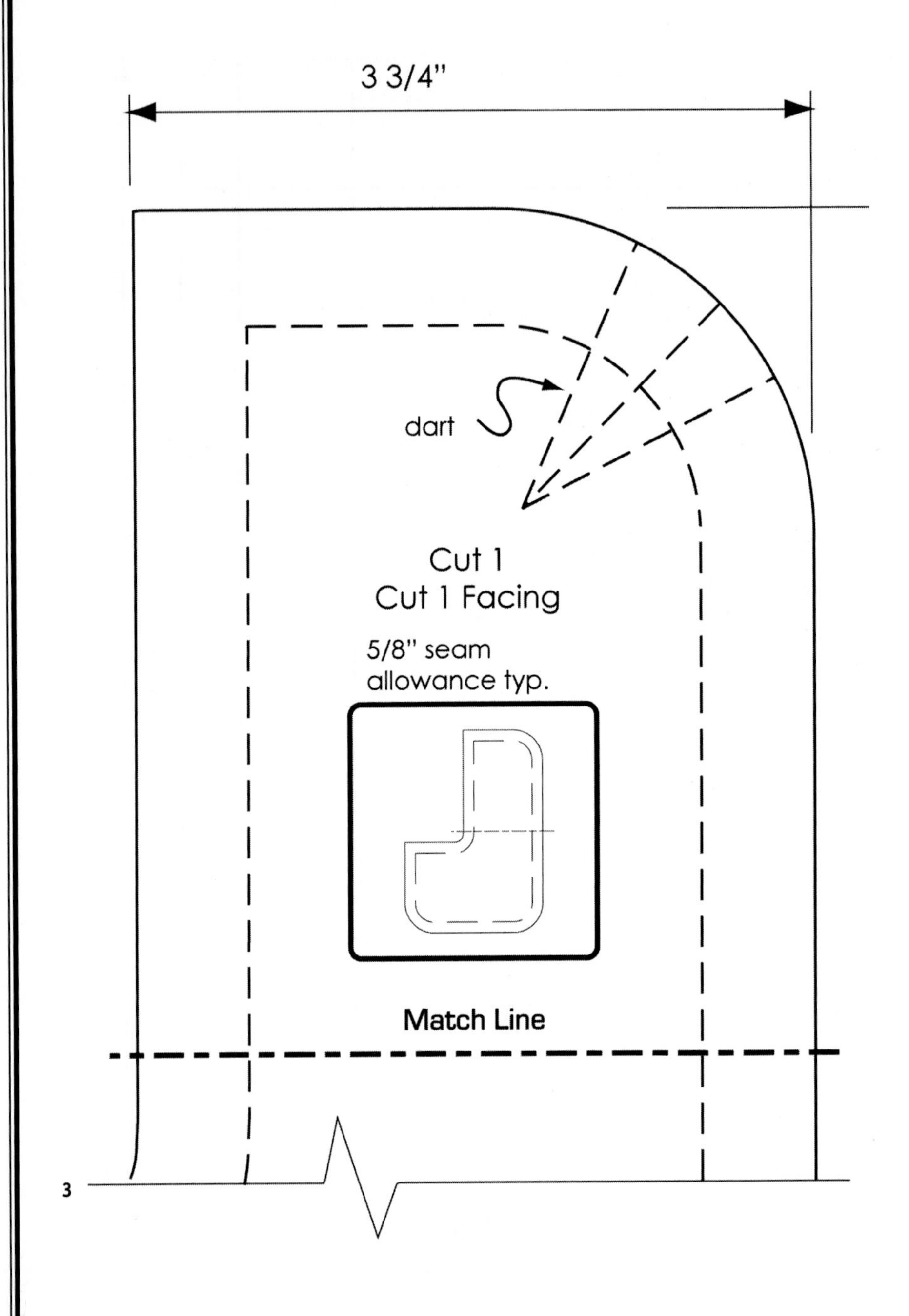

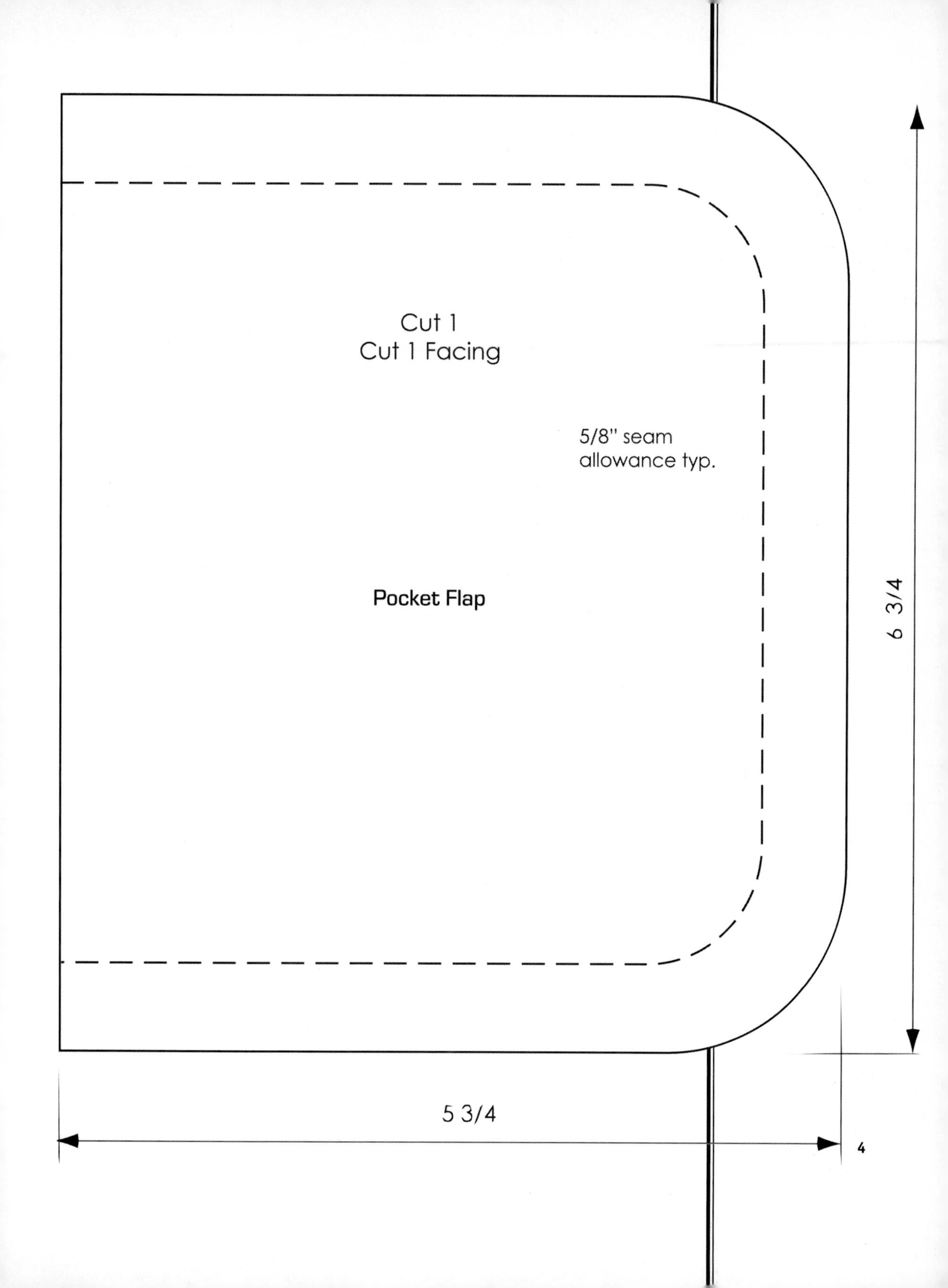
Cut 1
Cut 1 Facing
5/8" seam
allowance typ.
Pocket Flap
6 3/4
5 3/4

CHAPTER 14

Google(plex)

Who can argue that the successful venture started in 1998 by Larry Page and Sergey Brin isn't one of the greatest startup stories in the neurotic dot-com industry? The project began as BackRub, and was initially funded by Andy Bechtolsheim for the sweet sum of $100,000. It has remained an iconic "beta" search engine for over one year, climaxing in 2000, when Google became the world's largest search engine. For most companies, that might be the end of the story. For Google, it was only the beginning (see Fig. 14-1).

Like some mathematical transcendental expression, Google just keeps going and going. Or, should that be growing and growing? Today that modest "beta" search engine has matured into a robust search engine that is capable of more than just spitting out a list of links onto your monitor. Today's Google can:

- blog your brains out
- cache Web pages
- create mailing lists
- download source code
- edit photographs
- give driving directions
- provide Instant Messaging
- send e-mail
- shop for prices
- translate text
- oh, and search for a bunch of stuff, too

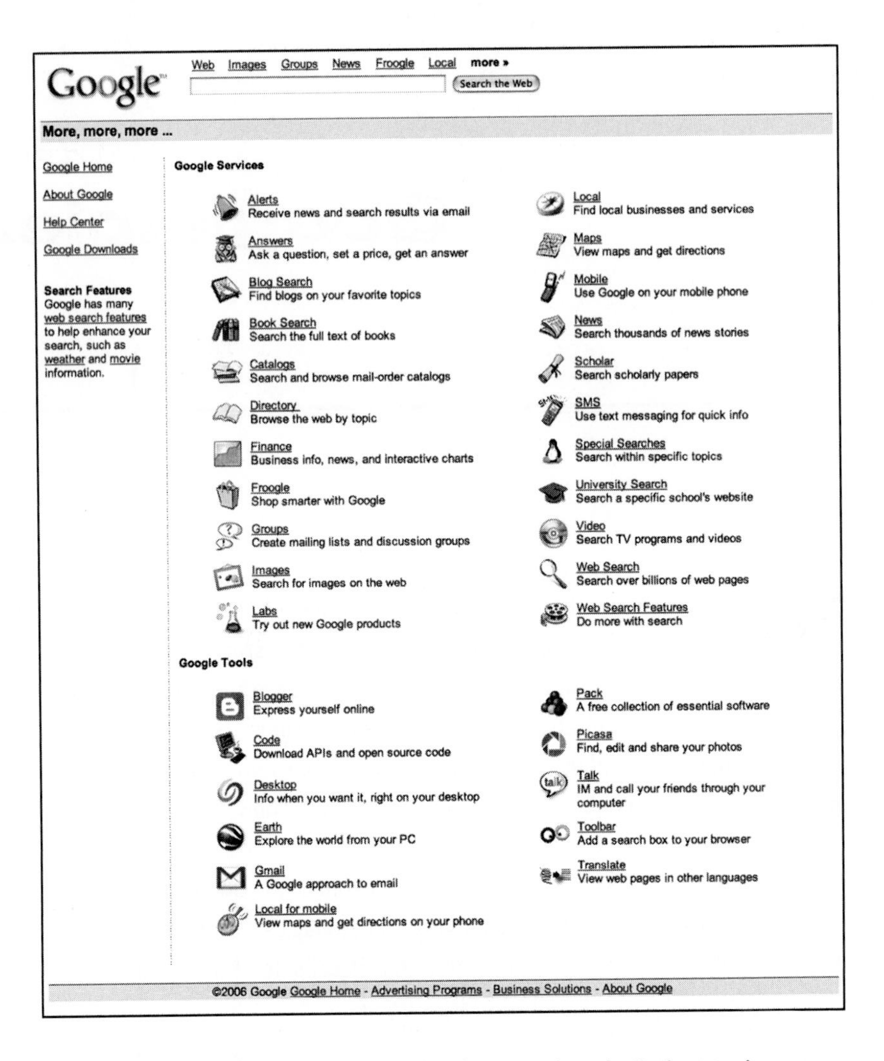

14-1 Remember the good ole' days of the Internet when you couldn't find anything? Google changed that.

Don't pooh-pooh that search engine feature, though. By just typing a special keyword inside any Google search text box, you can unleash a hacker's delight in hidden or little-known capabilities. In the following examples, if you type the sample phrase you will implement the labeled feature. For example, to conduct a book search, just type the special keyword "books" along with your search criteria.

- Books—<u>books</u> by prochnow
- Calculator—1 × 10^ 100
- Currency—20 euros <u>in</u> USD
- Definitions—*define* googol or googol *definition*

- Facts—*what is* the best book on hacking and recycling
- Search by filetype—sony psp filetype:pbp
- Flights—pib weather or northwest 777
- Images—hurricane pics or pics of hurricanes or hurricane *pictures*
- Links—*link*:books.mcgraw-hill.com
- Maps—street address, city, state, ZIP
- Movies—movie:gumby
- Numbers—UPS, FedEx, USPS, VIN, UPC, area code, patent, FAA, FCC
- Phonebook—name, city, state or name, ZIP or name, city or name, state
- Sites—*site:*amazon.com psp books
- Stocks—aapl
- Weather—*weather hattiesburg, ms*

Big Enough to Hate

Not everything Google touches turns to gold, however. In 2005, one of their planned conquests (digitizing all English books in and out of print) ran afoul of the Association of American Publishers.

Google had originally stated an intention in 2004 to scan or digitize the library collections of books from Harvard, Stanford, University of Michigan, Oxford

Is it Googol or Google?

"Googol" is the fanciful mathematical term for a number one (1) followed by 100 zeros. The term was coined by a nine-year-old named Milton Sirotta. Sirotta was the nephew of Edward Kasner, who was an American mathematician. Kasner unleashed googol on the world in his 1940 book MATHEMATICS AND THE IMAGINATION by Kasner and James Roy Newman (London: Penguin, 1940). Computer folklore attributes the difference in spelling between googol and google as a mistake by Larry Page. Ironically, it's Page who gets the last laugh, since most people today incorrectly spell Kasner's mathematical toy as "google."

University, and The New York Public Library. They also added the caveat that these scans would be restricted to books in the public domain from the collections of Oxford University and The New York Public Library. And all publishers, scholars, and researchers cheered, "hurrah." Then the other shoe dropped.

Google Print Library Project was now going to be called Google Book Library Project and the goal was to scan all public domain books and all copyrighted books currently (and formerly) in print. And the Association of American Publishers (AAP) along with the big names in publishing cried "foul." Then another shoe dropped.

As if this complete disrespect for copyright law was not enough, Google then did the dirty deed: They unilaterally mandated that authors (like myself with 26 published books) and publishers would have to send a detailed list of books that were not eligible for Google to digitize. In other words, we would have to "opt out" of the Google Book Library Project. And the AAP and major publishers sought legal action and filed a lawsuit against Google.

On October 19, 2005, Pat Schroeder, a 12-term member of Congress from Colorado who served as the Ranking Member on the House Judiciary Subcommittee on Courts and Intellectual Property, submitted a detailed press release on behalf of AAP that clearly defined the lawsuit against Google. Due to the large volume of misinformation that has surrounded this lawsuit, I present here an excerpt of the text of Ms. Schroeder's press statement:

PUBLISHERS SUE GOOGLE OVER PLANS TO DIGITIZE BOOKS— Google Print Library Violates Publishers' and Authors' Rights

The Association of American Publishers (AAP) today announced the filing of a lawsuit against Google over its plans to digitally copy and distribute copyrighted works without permission of the copyright owners. The lawsuit was filed only after lengthy discussions broke down between AAP and Google's top management regarding the copyright infringement implications of the Google Print Library Project.

The suit, which seeks a declaration by the court that Google commits infringement when it scans entire books covered by copyright and a court order preventing it from doing so without permission of the copyright owner, was filed on behalf of five major publisher members of AAP: The McGraw-Hill Companies, Pearson Education, Penguin Group (USA), Simon & Schuster and John Wiley & Sons.

The suit, which is being coordinated and funded by AAP, has the strong backing of the publishing industry and was filed following an overwhelming vote of support by the 20-member AAP Board which is elected by, and represents, the Association's more than 300 member publishing houses.

Over the objections voiced by the publishers and in the face of a lawsuit filed earlier by the Authors Guild on behalf of its 8,000 members, Google has indicated its intention to go forward with the unauthorized copying of copyrighted works beginning on November 1.

As a way of accomplishing the legal use of copyrighted works in the Print Library Project, AAP proposed to Google that they utilize the well-known ISBN numbering system to identify works under copyright and secure permission from publishers and authors to scan these works. Since the inception of the ISBN system in 1967, a unique ISBN number has been placed on every book, identifying each book and linking it to a specific publisher. Google flatly rejected this reasonable proposal.

Noting the existence of new online search initiatives that respect the rights of creators, such as the "Open Content Alliance" involving Yahoo, Hewlett-Packard, Adobe and the Internet Archive, Mrs. Schroeder said: "If Google can scan every book in the English language, surely they can utilize ISBNs. By rejecting the reasonable ISBN solution, Google left our members no choice but to file this suit."

Mrs. Schroeder noted that while "Google Print Library could help many authors get more exposure and maybe even sell more books, authors and publishers should not be asked to waive their long-held rights so that Google can profit from this venture."

Is it a Noun or a Verb?

Has anyone ever said to you something like, "Just google for that information"? It's doubtful that you've ever heard, "Go ahead and Black & Decker your lawn." Or, "Anheuser-Busch me." OK, maybe I have heard that last one, Bud.

CHAPTER 15

Some Heavy Petting

Are you a cat or a dog person? Believe it or not, that was an actual interview question during my application with a graphic design studio. Even though I indicated that I had no preference, I still got the job. Go figure.

Well, as I found out, there are a lot of cat fanciers and dog lovers. And if you're keeping score at home, the office was evenly divided 50 percent cat and 50 percent dog. Oddly enough, there was no relationship between pet choice and sex, either. Although I'm sure that cat fanciers have sex as often as dog lovers.

This 91.67 percent pet ownership statistic with the graphic design studio (remember, I was the odd man out: pet-less) is much higher than the national average. According to a 2005/2006 survey by the American Pet Products Manufacturers Association (APPMA), 63 percent of all U.S. households own at least one pet. Even more remarkable, 45 percent of the households in the United States have more than one pet.

So what do these statistics translate into in terms of total pet numbers? Surprisingly, the largest pet population is freshwater fish, where 139 million are floating around in tanks in the United States. The second highest population is cats with 90.5 million felines stalking American homes. Dogs fetch the third most populated spot at 73.9 million.

These pets aren't free to keep, either. This same APPMA survey states that, in 2005, the United States spent about $35.9 billion on its pets. Specifically, $14.5 billion on food, $8.6 billion on veterinarian care, with the remainder divided among grooming, boarding, medicine, supplies, and live pet purchases. Let's put these expenses into perspective.

According to the U.S. Department of Homeland Security in a summary of government reaction to Hurricane Katrina, FEMA had distributed about $4.4 billion in federal aid to about 1.4 million households. This distribution is in addition to the $11.5 billion of federal money that was given to five Gulf Coast states for rebuilding and redevelopment.

Looking for a little more perspective in analyzing your pet expenses? At the other extreme of spending, the cost of war in Iraq and Afghanistan ranges between $170 billion (based on two government budget requests) to over $240 billion, as reflected on the National Priorities Project "Cost of War" Web site.

Doggone It, Bored

Just like that age-old question, "Does the light stay on when you close the refrigerator door," try asking yourself what your pet does when you're gone from home. PetPlace®.com thinks that your dog might be bored or stressed with separation anxiety when you're out of the house. How can you tell? Just ask yourself these questions:

- Does Fido follow you around the house?
- Is your pup anxious when you're getting ready to leave the house?
- Does Rover bark enough to raise the dead after you close the front door?

If you are truly worried by this line of questioning and you feel that your "best friend" might have a problem, you can find a much more detailed set of psychoanalytic lines of inquiry on the PetPlace.com Web site.

My Dog Stinks

The surface area of a human's nasal epithelium is approximately 10 square centimeters. A dog's nose, however, has a nasal epithelium with a surface area that is approximately 170 square centimeters, studded with over 100 times more olfactory receptor cells per square centimeter than we smell-challenged people.

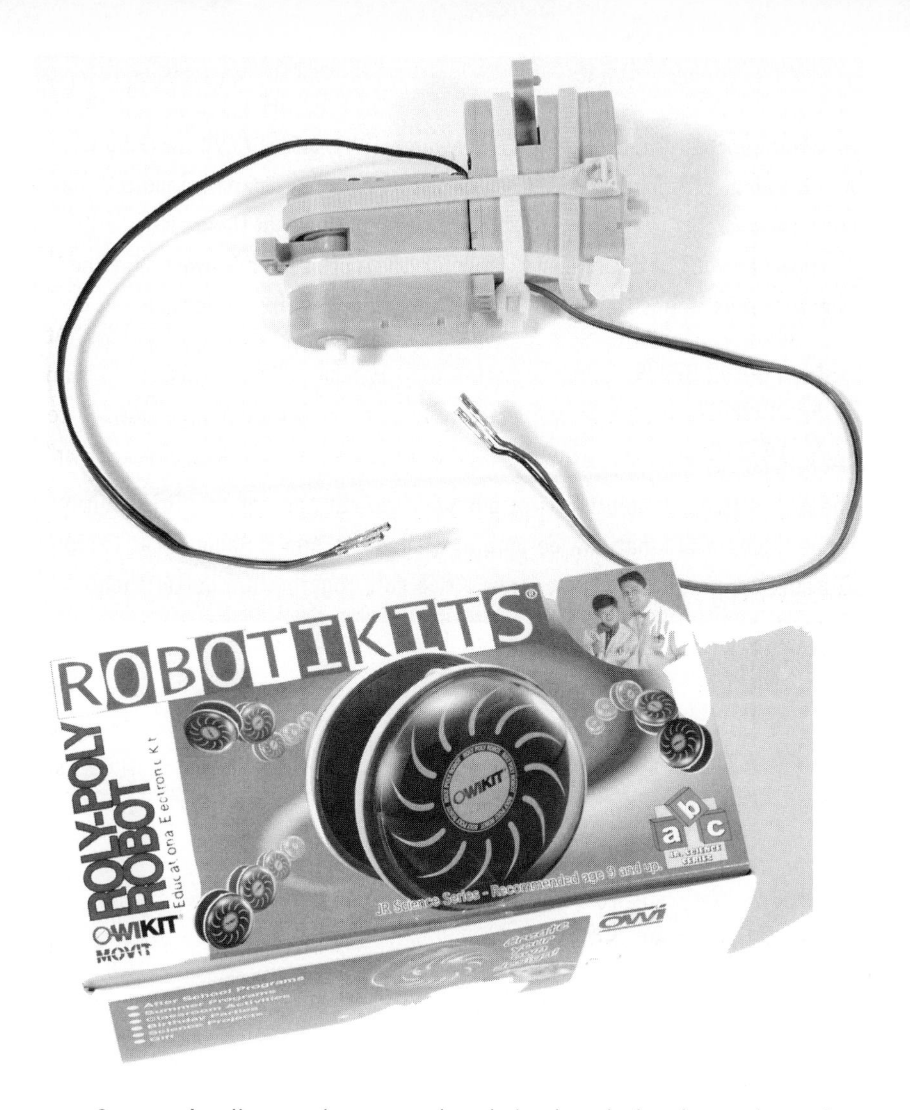

15-1 You can make your own pet toy robot from a couple of servo motors and a robot brain.

15-2 Low-cost robots can be easily adapted into pet sitters.

Once you've diagnosed your pooch as being bored when home alone, then you can return to PetPlace.com for some solutions (see Figs. 15-1 and 15-2). The key to these answers is what the Web site calls, "environmental enrichment." The top three things that you can do to help your dog are:

1. Get a doggie partner.
2. Hire a professional dog walker.
3. Send your dog to a doggie day care.

The Purr-fect Answer

Just because you own a cat don't think that your pet is immune to loneliness. A January 2005 issue of *New York Tails: A Magazine for the People and Pets of*

New York reported that Dr. Stefanie Schwartz, DVM, MSc, DACVB, working with Arm & Hammer,® found that 76 percent of surveyed veterinarians reported that most cat owners are unaware of the signs of loneliness in their cats.

How can you tell if your pussy is lonely? In a reprint of Dr. Schwartz' findings, *New York Tails* claims that Tabby is meowing in misery if you notice:

- Over-grooming
- Spraying
- Vocalizing

The Arm & Hammer solution is to buy your sad cat another cat. This solution is the linchpin of a nationwide program called the Arm & Hammer Cat-Panion Crusade. Or, in more succinct terms, "two cats together are better than one home alone."

Cat's Eyes

All kittens are born with blue eyes. In about 6 to 7 weeks, after the eyes open, they slowly change color to the final adult cat eye color.

MODIFY AN RC CAR INTO A PET TOY

1 Air Hogs® Zero Gravity.™

2 This RC toy will drive your pet crazy.

3 It looks like a rather conventional Hummer.

4 The real secret lies in this toy's underside.

5 An incredible vacuum system holds Zero Gravity against your walls and ceilings while wheels drive it around.

6 Remove the RC circuitry and install a timer circuit and relay.

3

4

5

6

7 Position Zero Gravity near a wall and activate the timer.

8 At the prescribed time, the relay will be activated and Zero Gravity will climb up on your wall.

9 Beware, when the batteries expire in Zero Gravity it will fall like a rock and then Fido will get his revenge.

7

8

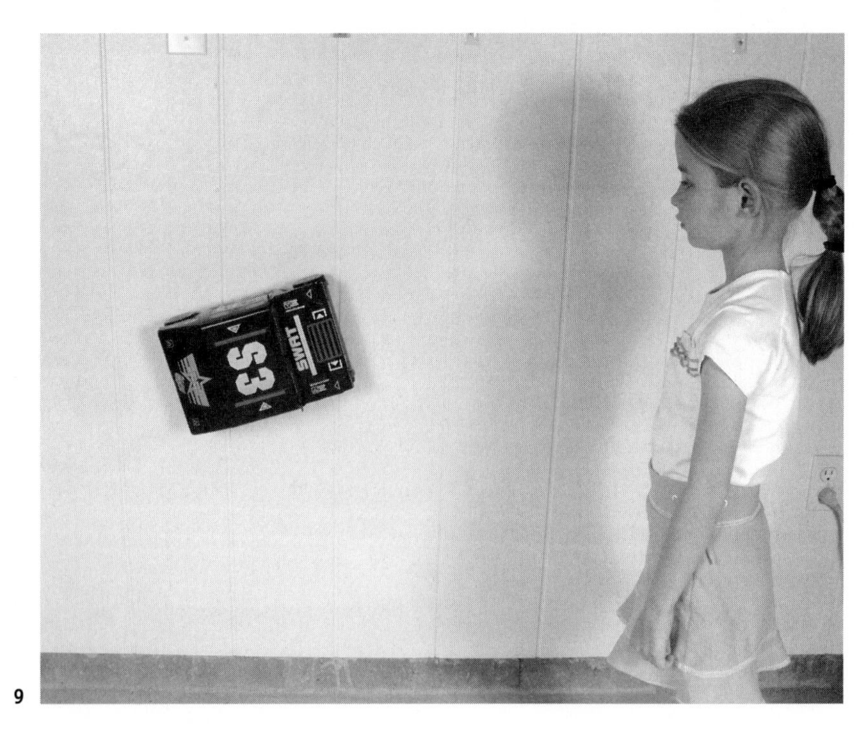

9

CHAPTER 16

Harem Scare 'em

It seems like just about everyone is putting Linux into something. Devices run under Linux, operating systems run on Linux, and cell phones run around the world with Linux. How many of these Linux-based devices do you own?

- Aeronix Zipit Instant Messenger
- Archos PMA400 Pocket Multimedia Assistant
- Buffalo Technologies Kuro Box
- iRiver PMP-120 and PMP-140
- Linksys Network Storage Link for USB 2.0 Disk Drives (NSLU2)
- Mattel Juice Box
- Motorola A780
- Motorola Rokr E2 Music Phone

Castrating Operating Systems

In the late 1960s, three AT&T Bell Labs employees, Ken Thompson, Dennis Ritchie, and Douglas McIlroy entered their names into computer lore for initiating the creation of Unix. Originally called Unics, Unix, or UNIX, it was officially coined as the name for this new operation system in 1970. Three years later a decision was made to rewrite Unix in C and the legend was born. Linux is a derivative of Unix.

- Nokia 770 Internet Tablet
- Pepper Computing Pepper Pad 2
- Sharp Zaurus SL-5x00
- Sony PlayStation® 3
- TiVO Personal Video Recorder (PVR)

OK, if Linux is so popular and it's free, how do I get it onto my computer? Unless you are both technically gifted and a masochist, you should purchase a Linux distribution. Whoa, what do you mean *purchase*? Isn't Linux free? Yes, and you can "freely" download a Linux distribution, which is also known as a *distro*, from hundreds of different Web sites. That is, you can download it if you have a broadband connection. Most Linux distributions are chunky downloads weighing in between 320 megabytes to 3 gigabytes worth of data. Ouch!

It's much better to purchase a Linux distribution CD/DVD. These are generally reasonably priced ($1 to $5) and shipped directly to your mailbox. Typically these distros contain the Linux kernel, the GNU tool chain, and an incredible sampling of free and open-source apps. Believe me, these distros represent one of the best buys in the computer world today. Buy a couple of them and sample software the way it could be; that is, if we could all just get along.

Just how many distros are out there? At last count, over 94. Here is an abbreviated shopping list that represents a great sampling of various distros for you to purchase (or, download) and evaluate:

Free is Good

In 1991, Linus Torvalds had way too much time on his hands and he created Linux. His initial effort was refined with the assistance of a loose-knit group of developers from around the world. Derived from Unix, Linux is a free, open-source operating system. The Linux operating system is called the Linux kernel and is the basis for all Linux distributions.

Size Does Matter

There is a special Linux distro that is made specifically for use with microcontrollers and embedded systems lacking a memory management unit. This version of Linux is called uClinux or microcontroller Linux. The "u" is a printing simplification for the micro symbol (μ) which is representative of microcontrollers (e.g., μC). This microcontroller distro is based on the 1996 version (e.g., Version 2.0) of the Linux kernel.

- Ark Linux (a claimed "4 click" installation)
- ASPLinux (a multipurpose Red Hat distro)
- BlackRhino Linux (for PlayStation® 2 development)
- Debian GNU/Linux (a family favorite)
- easyLinux (a GUI installer)
- Fedora Core (a general Red Hat distro)
- Gentoo Linux (a power user distro)
- KNOPPIX (boots from distro CD)
- Linux/Mac68k (a 68xxx Mac distro)
- Mandriva (a wide user base)
- MkLinux (an early Apple attempt)
- Puppy (a small distro)
- Red Hat Enterprise Linux (for the serious user)
- Slackware Linux (a multiprocessor speed demon)
- SUSE Linux (now Novell)
- Ubuntu Linux (an international flavor)
- Yellow Dog Linux (a PPC distro)
- Zen Linux (can be run directly from CD)

HOW TO CHANGE THE PEPPER PAD BATTERY

1 Pepper Pad is an Internet appliance that can deliver a knockout punch to Microsoft's Origami project. Why? Pepper Pad is a rugged, portable touch-screen computer that uses a Linux OS and costs less than $850. Oh, and did I say that it's splash-resistant, too? Finally, pool-side computing is a reality.

2 All of the ports on Pepper Pad are covered with flexible rubber plugs.

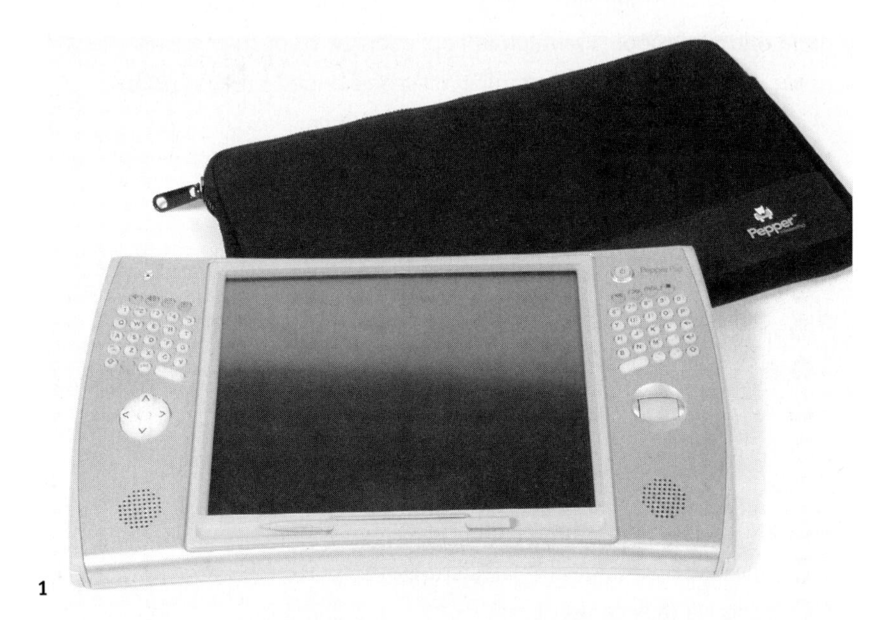

1

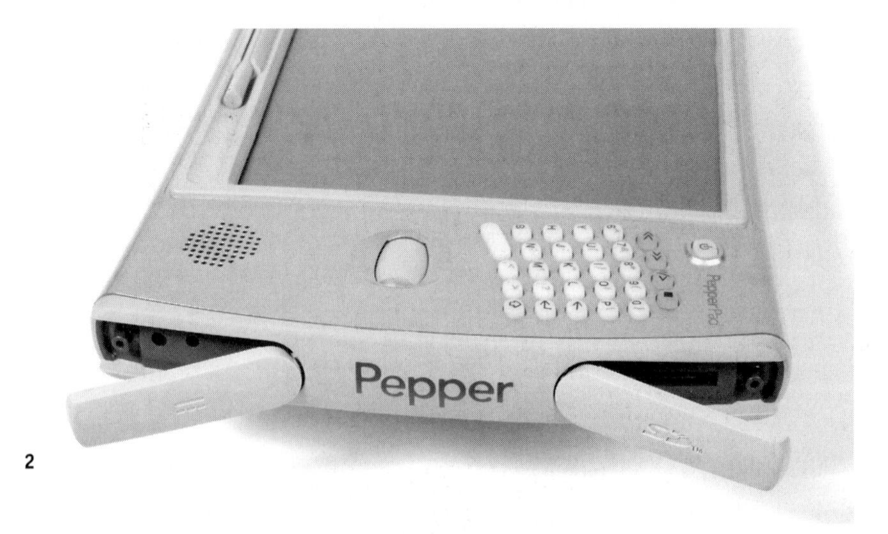

2

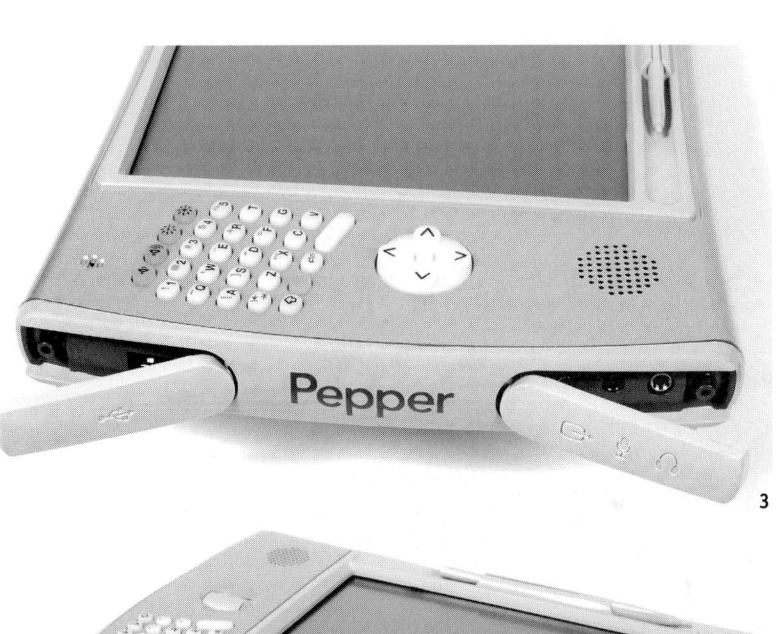

3

4

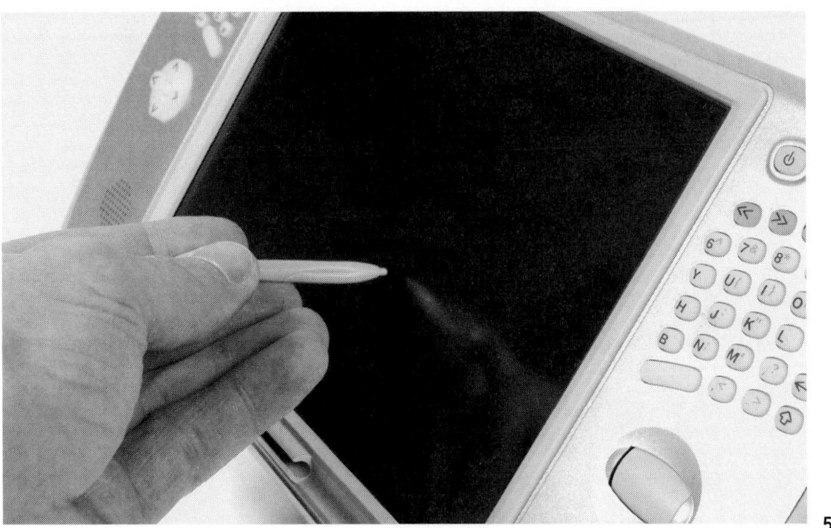
5

3 USB, audio in, audio out, and an external monitor port make Pepper Pad a portable that is equally at home on your desk, in the kitchen, or on a kid's bed.

4 An IR transmitter inside Pepper Pad can be configured to operate some popular TVs.

5 A built-in stylus controls Pepper Pad. Alternatively, for the touch screen–challenged, you can use a scroll wheel or arrow keys for input.

6 Pepper Pad uses tamperproof fasteners.

7 You will need a Torx 8 bit for opening up your Pepper Pad.

8 If the batteries on Pepper Pad fail, just open 'er up and replace 'em.

9 These plug-n-use batteries can be ordered online.

6

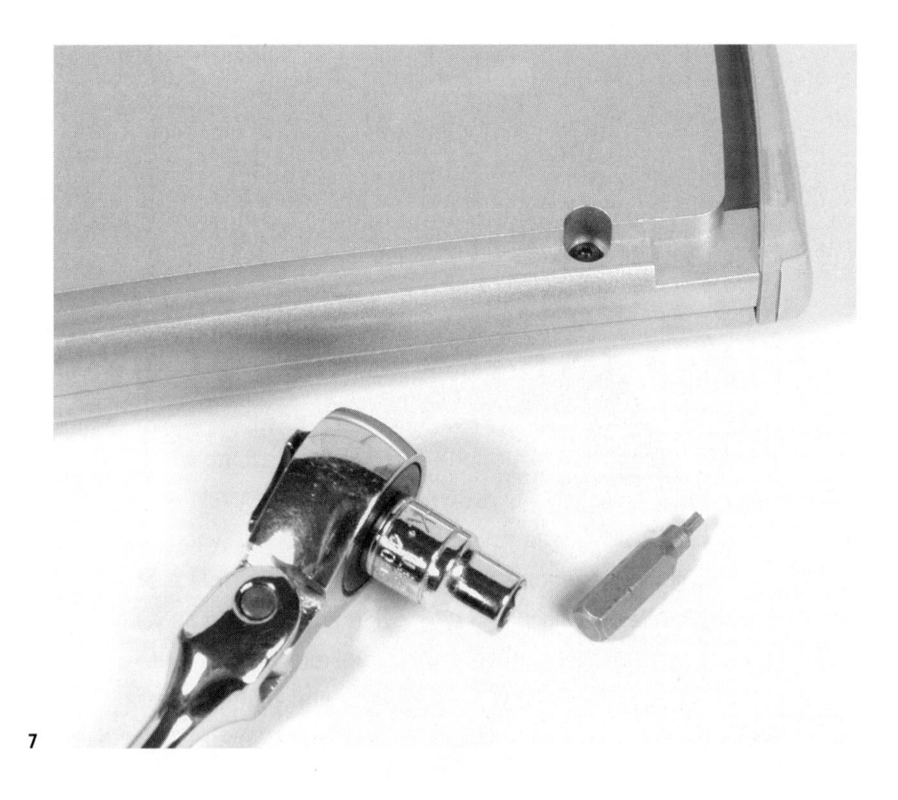

7

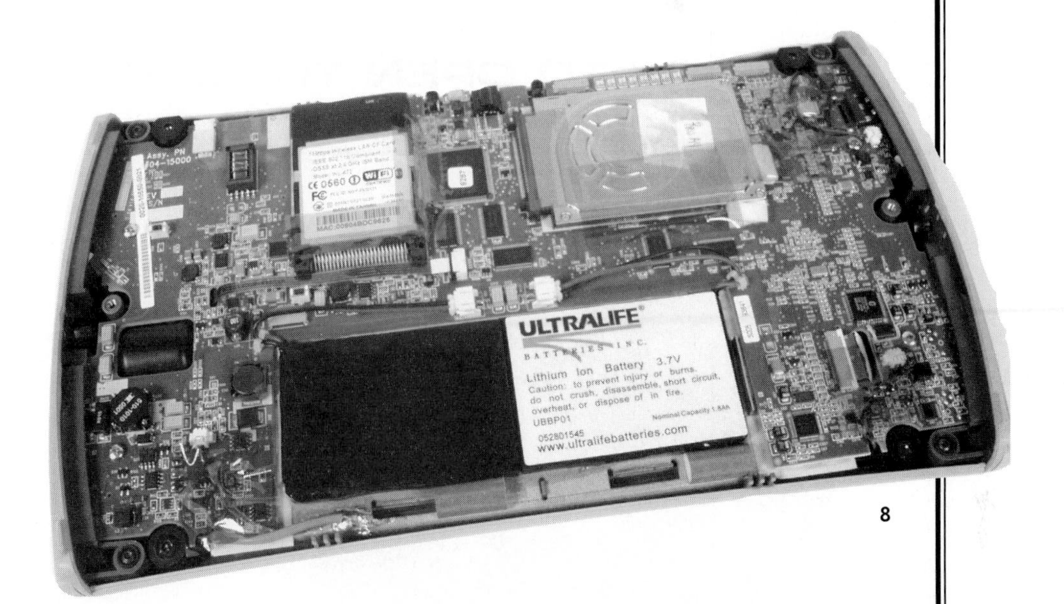
ULTRALIFE®
BATTERIES INC.
Lithium Ion Battery 3.7V
Caution: to prevent injury or burns.
do not crush, disassemble, short circuit,
overheat, or dispose of in fire.
UBBP01
052801545
www.ultralifebatteries.com

8

Gemtek
DC9625
ULTRALIFE®
BATTERIES INC.
Lithium Ion Battery 3.7V
Caution: to prevent injury or burns.
do not crush, disassemble, short circu
overheat, or dispose of in fire.
UBBP01
Nominal Capacity
052801545
www.ultralifebatteries.com

9

HOW TO OPEN THE ZIPIT™

1 The Zipit Wireless Messenger uses a Linux kernel and has a great Wi-Fi interface.

2 With a useable rubber "chiclet"-style keyboard, Zipit can be used as a Wi-Fi "sniffer" right out of the box.

3 A headphone jack is the only port on Zipit.

1

Prev
Next
Mute
P1
P3
P2
Q W E R T Y U I O P
A S D F G H J K L BS
Z X C V B N M Enter
Alt
Ctrl
Space
Menu

2

3

4 Zipit uses a passive tamperproof system—the screws are hidden underneath rubber feet.

5 Remove the battery before you attempt to open Zipit.

6 The Zipit keyboard is a rubber membrane that can be lifted off the keypad contacts.

7 There are some useful Zipit hardware hacks online for adding a backlight and a removable memory card.

4

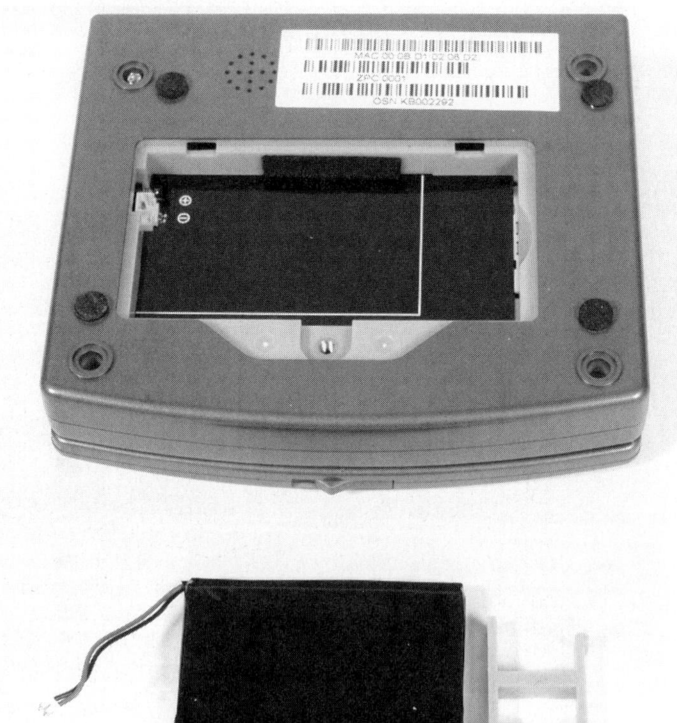

5

Prev
Next
Mute
P1
P3
P2
Q W E R T Y U I O P
A S D F G H J K L
BS
Z X C V B N M
Enter
Alt
Ctrl
Space
Menu

6

PCBA AX100344-
S/N-
ESIGNED IN THE USA
WL-100BA51/A1
AX100343-007
BTI ML-1
94V-0
KA901902
MADE IN CHINA
agere
BAT
AERONIX

7

HOW TO OPEN THE JUICE BOX™

1 The Mattel® Juice Box™ was intended to be the multi-media player designed "just for kids." Too bad kids liked the Apple Computer iPod, instead. There's a lesson here: Don't dumb down stuff for kids.

2 Removable cartridges called Juiceware™ enable you to add your own music and images to Juice Box.

3 Remove the batteries, then remove the screws.

1

2

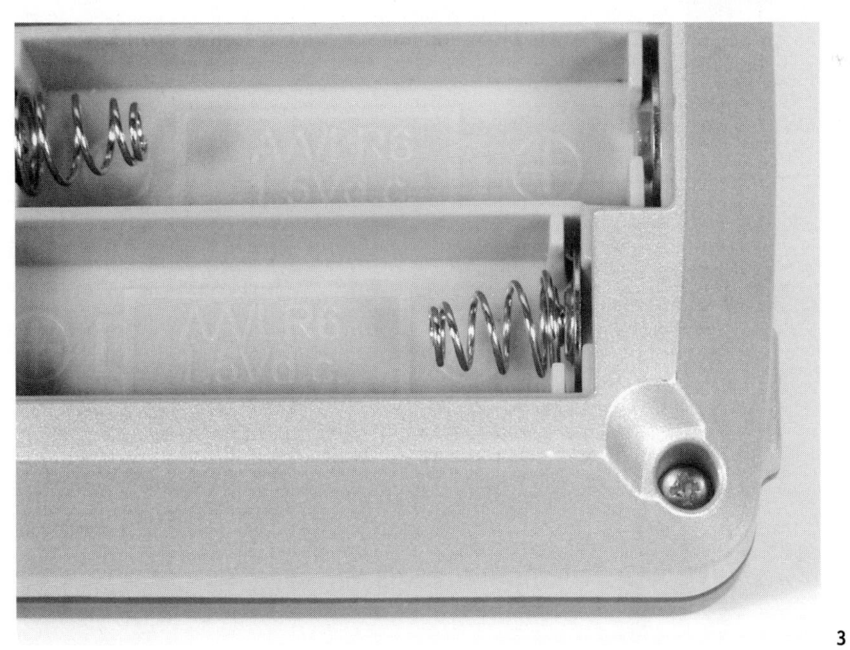
3

4 There's not much to see inside the Juice Box, just like outside the Juice Box.

5 You can find a Juice Box hardware hack online for swapping the Juiceware cartridge with a more conventional SD media card.

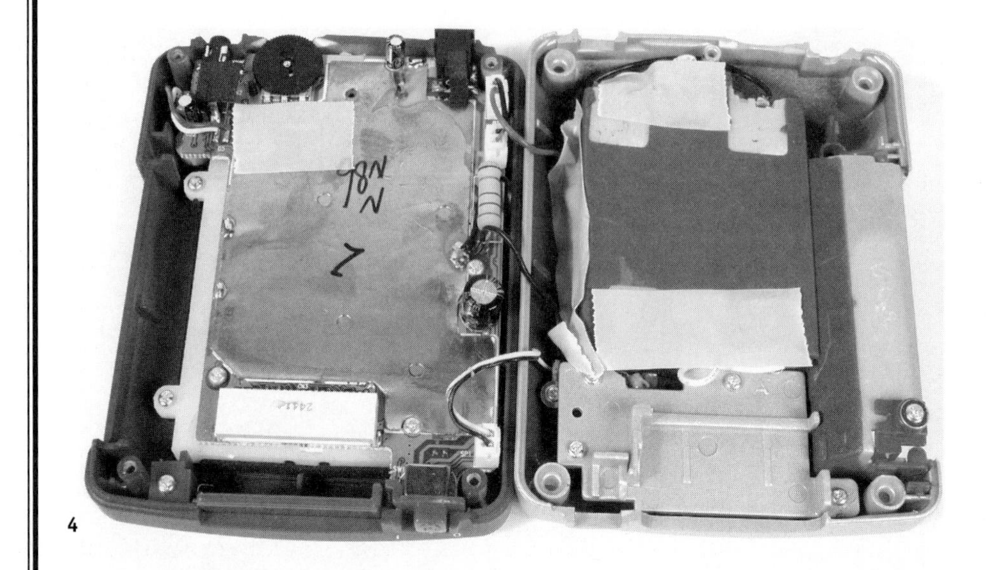

4

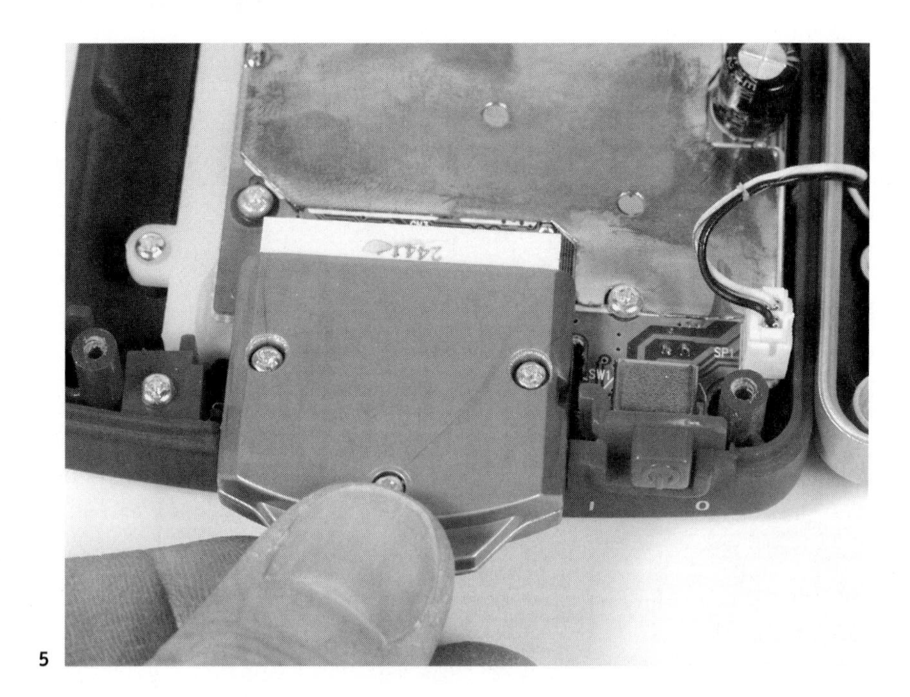

5

CHAPTER 17

Beat Your Beat

"Music hath charms to soothe the savage breast, to soften rocks, or bend a knotted oak." This statement is attributed to William Congreve, 1670–1729, from *The Mourning Bride* (Act I, Scene 1) by John Bartlett (1820–1905) in the tenth edition of his compendium, *Familiar Quotations* (1919). You don't think that this citation is correct? Just like with politics and the weather, everyone has their own opinion, interpretation, and misrepresentation of this familiar quotation.

You've probably heard so-called attributions like "music soothes the savage breast" or, my favorite misquote, "music soothes the savage beast." All it takes is a little sleuthing at your local library (preferably a major university academic library) and you can enlighten yourself. Even so, most people believe that they know the real quote—even though it is incorrect. Such is the case with the *theremin*.

The what? While you might not know the name, you will certainly recognize the sound of this ethereal electronic musical instrument. Just listen to a recording of "Good Vibrations" by The Beach Boys (1966) or "listen" to the science fiction movie, *The Day the Earth Stood Still* (1951). In both cases you will hear the eerie melodic strains of the theremin reverberating through the air. Now that you have the music echoing through your mind, who do you think invented the theremin? If you thought Bob Moog, that's OK. It's great that you're thinking, but you're wrong. The theremin was invented by Russian Lev Sergeivitch Termen (later changed to Léon Théremin or Leon Theremin).

In 1919, Theremin was the Director of the Technical Laboratory at the Physics and Technical Institute in Russia. During a conference in 1920, he presented an electronic device capable of delivering a vibrato out of seemingly thin air. This device was originally

called the Aetherphone. Other devices were designed, Termenvoksa, Heterophone, and Thereminvox and were eventually collectively called the theremin.

One of Theremin's earliest admirers of his unusual music was Lenin. No, it was not one of the founding members of The Beatles, John Lennon but, rather, one of the founding members of Russian Communism—Vladimir Illich Lenin. In 1929, after Theremin had brought his invention to the United States, Radio Corporation of America (RCA) licensed his electronic musical device. Based on this licensing deal, RCA built 500 theremins.

Can't Touch This

If you're looking for a way to get your hands on the vibrating world of theremin music, you can buy a DIY kit from Moog Music and PAiA.

Moog™ Music Etherwave® Theremin Kit

This build-it-yourself kit has an unfinished wood cabinet and prebuilt circuit board. Several wiring points require soldering. $349.00

PAiA Theremax Kit

Electronic parts only with knobs, wire, circuit board, 12-VDC power supply. No case. $94.75

Theremax Wooden Lectern Case Kit

All wooden pieces are precut and predrilled from white pine with antennas, hardware, and printed metal front panel. $83.25

NOTE: The good folks at PAiA have the complete construction plans (including plans for the wooden case) for building the Theremax available online as a FREE download. Go get 'em, read 'em, marvel at the complexity of the design, then buy the kit.

Fast forward to 1991 and music genius Bob Moog designs a series of commercially available theremins that are sold through his company, Big Briar. Like Theremin, Moog is an electronic musical instrument inventor. He is best known for his line of Moog synthesizers like the incredible, Minimoog.® In 1996, *Electronic Musician* published an article by Moog that showed readers how to build their own theremin.

Experiments in theremin designs didn't stop with Bob Moog, though. One of America's other brilliant electronic musical instrument designers, John Stayton Simonton, Jr., created the Theremax. This was a do-it-yourself (DIY) kit theremin design that was one of the last major projects that Simonton worked on prior to his death.

HOW TO KNIT YOUR OWN MUSICAL FABRIC

1 Take an ordinary tape recorder, some old audio tape, and a loom, and you can weave your own musical fabric.

2 Any knitting loom will work for this project.

3 Locate an audio cassette tape that you want to weave into your fabric.

1

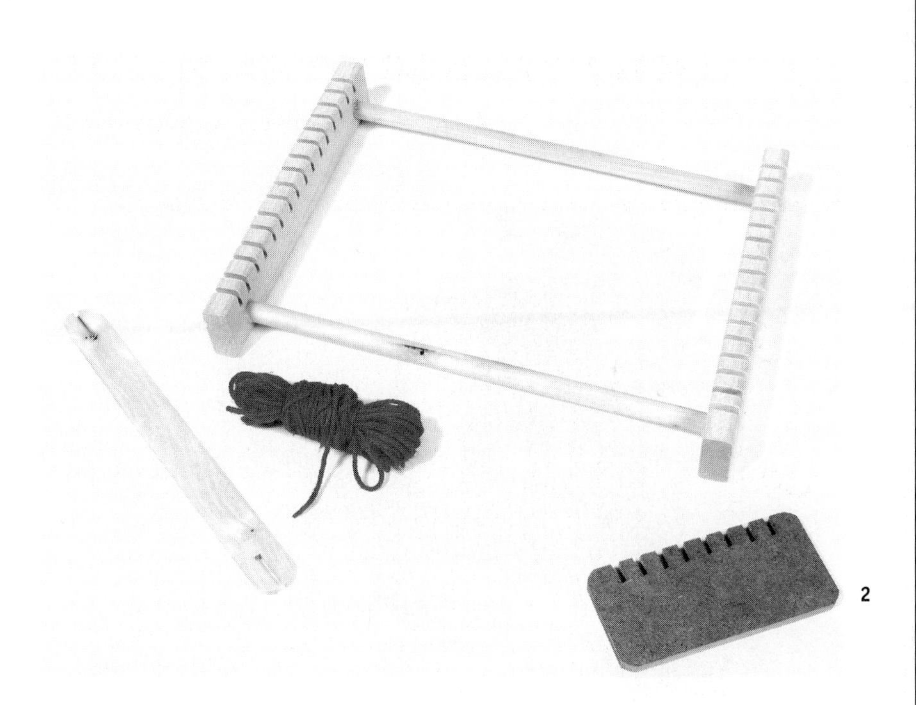

2

3

4. Pull out the leader and discard it. Remove enough audio tape for your project. A test sample will only require a couple of feet of tape.
5. Follow the instructions with your loom for attaching the tape.
6. Wrap the audio tape on the loom (this step is called the warp). Make sure that the recorded side of the tape is *always* up.
7. Temporarily tie off the audio tape so that it won't come loose during your weaving steps.

4

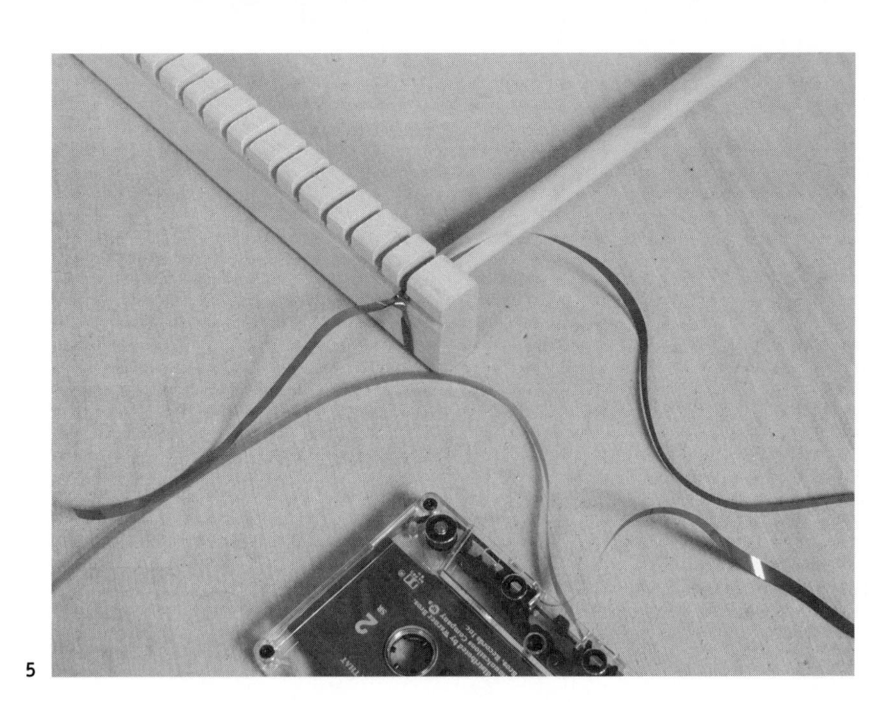
5

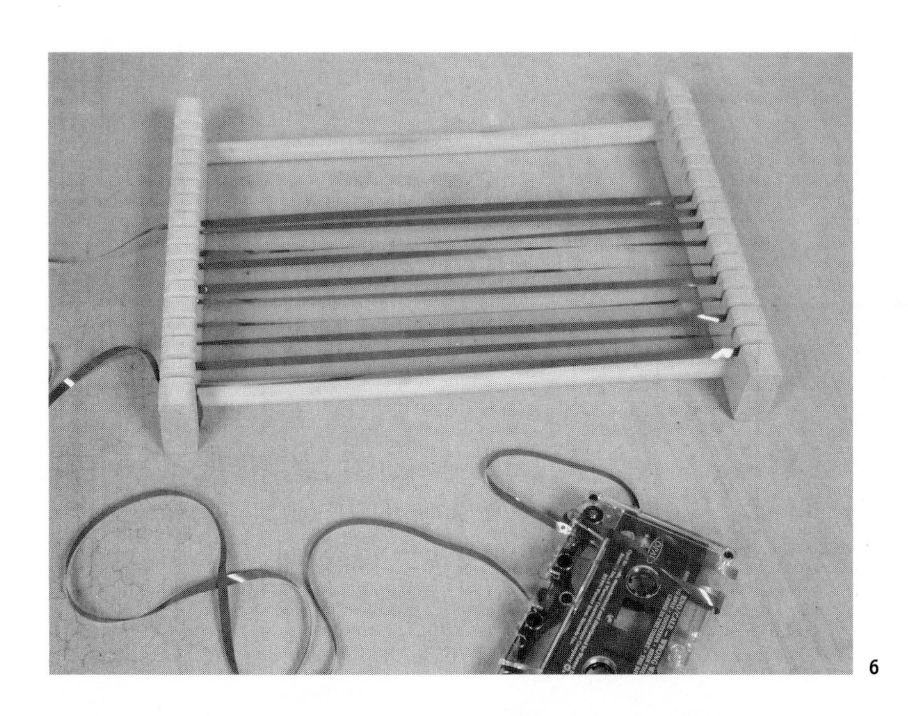

6

7

8 Insert the shuttle stick *over* and *under* the audio tape warp.

9 Twist the shuttle stick to make a tunnel through the warp and run the yarn through this tunnel. Remove the shuttle stick and insert it *under* and *over* the warp. Adept weavers will actually use two shuttle sticks for this step; raising then lowering each shuttle stick.

10 The completed musical fabric. I call it SonKnit.

11 You can "play" your SonKnit with an old tape recorder. Remove the play head from the recorder.

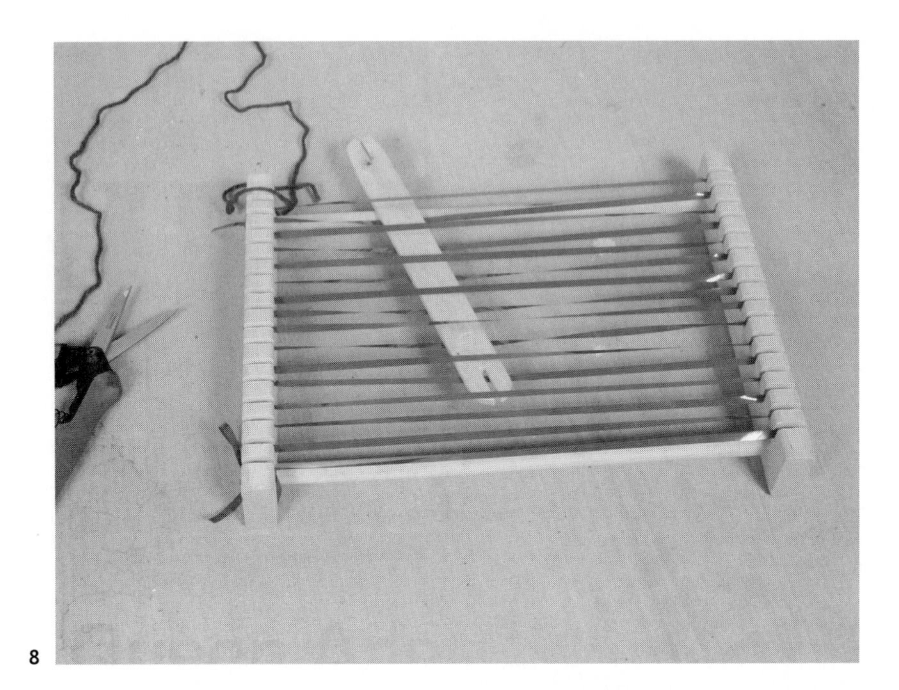

8

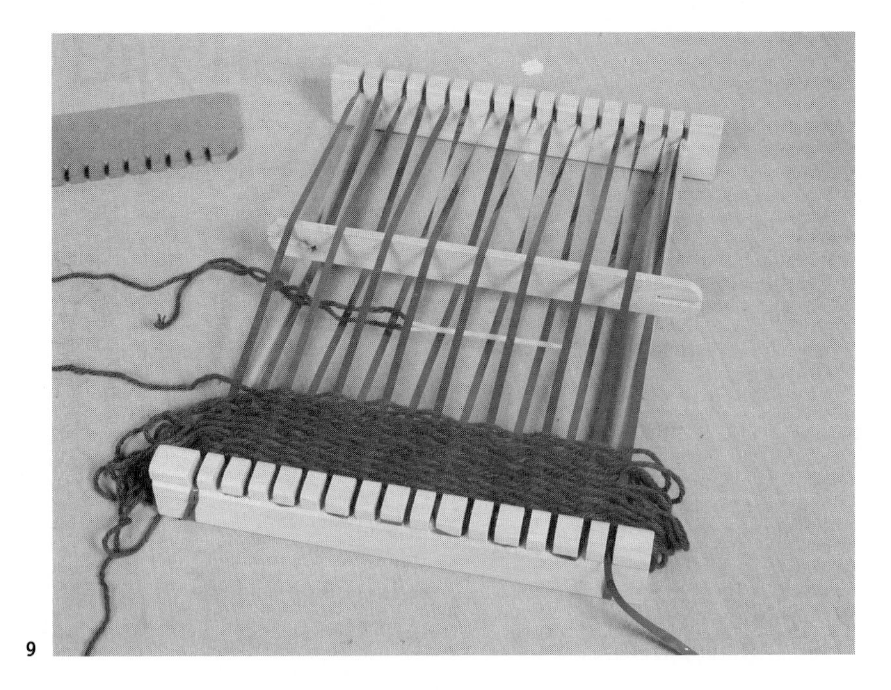

9

10

11

12 Press the Play button on the recorder and run the play head over the SonKnit.

12

CHAPTER 18

One Way, Feng Shui

According to Kartar Diamond, Senior Feng Shui Consultant, the five elements of Feng Shui and their colors are:

- Earth—orange, yellow, brown
- Fire—red, maroon, burgundy, hot pink
- Metal—white, silver, gray, gold
- Water—all shades of blue or black
- Wood—all shades of green

While you might think that Feng Shui is a new way of decorating your home or office, it is more properly a life-enhancement philosophy that can be most visibly applied to our daily lives through interior design. Pronounced like "fong shway," Feng Shui can trace its roots back over 3,500 years in China. Described in a *Book of Burial* as when the life energy and spiritual life force or "chi" (also "qi") disperses on the wind and then collects on the boundaries of water, Feng Shui means "Wind and Water."

Feng Shui attempts to harmonize your life by accenting the positive flow of your chi, while minimizing all negative effects. One way of achieving this harmony is through the design of our architecture. Granted, most of us can't control or dictate the orientation, construction, and layout of our dwellings, but we can control the interior design of our living spaces. In this capacity, Feng Shui is the art (or, science) of choosing, placing, and arranging all of the objects within our homes, offices, and yards for achieving an optimal flow of chi. Be forewarned, this interior application of design is a modern (post-

nineteenth-century Spiritualist movement; also called Black Sect) interpretation of Feng Shui that is in conflict with traditional Feng Shui. Traditional Feng Shui stresses the relationship between the position of the actual structure or building and its surrounding environment.

It's Elemental, My Dear Reader

So what about these Feng Shui elements: earth, fire, metal, water, and wood? Well, think of these elements as life energies or forces that are combined and interrelated to each other and help to shape your life.

EARTH—stability, real estate, and legacy; color yellow; direction center. Earth is patient, just, honest, and methodical, representing puberty.

FIRE—energy and enthusiasm, danger, and leadership; color red; season Summer; direction south. Fire also can represent prepuberty. If there is too much fire the results can be destructive.

METAL—harvest, business, and success; color white and gold; season Autumn; direction west. Metal represents the adult years.

WATER—travel, communication, and learning; color black; season Winter; direction north. Water represents old age.

WOOD—creative and innovative, sociable, and communal; color green; season Spring; direction east. Wood also can represent birth and early childhood.

HOW TO MAKE FENG SHUI FLOWER POTS

1

1 Choose from a variety of colored LEDs to add a suitable influence to your outdoor patio.

2 IKEA makes flower pots that are ideal for adapting to Feng Shui design principles.

3 Drop your LED circuit into the bottom of the flower pot. Make sure that you don't cover the drainage holes in the bottom.

4 Insert your plant into the pot. I prefer to keep my plants inside separate liners so that I can interchange them according to the season.

5 The glowing LEDs reflect off the inside of the flower pot and create a soft, colored glow.

2

3

4

5

HOW TO GET RID OF ROOM CLUTTER

1 Katherine's room needs some help with this clutter.

2 The *Hensvik* bookcase by IKEA is a perfect solution.

1

2

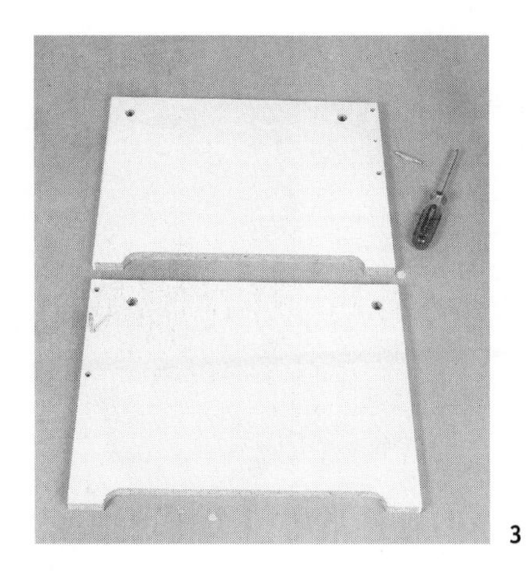

3

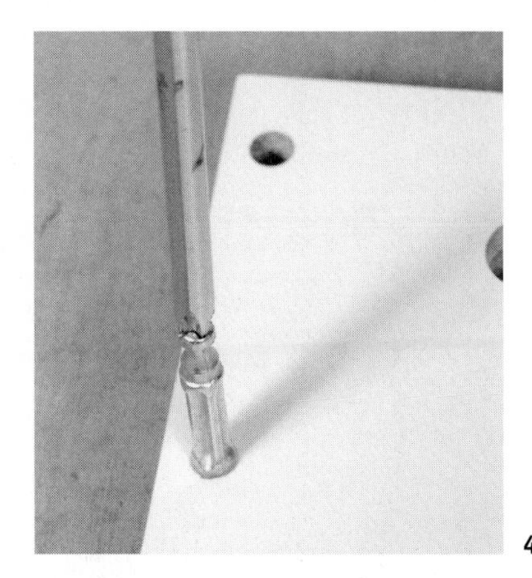

4

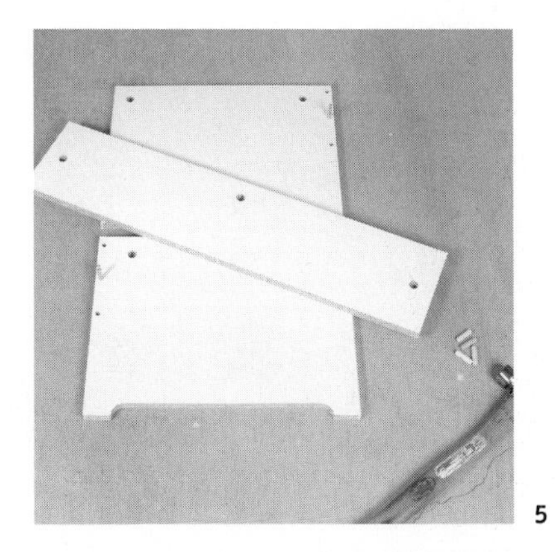

5

6

7

8

3 Layout the base end pieces.

4 Attach the mounting posts.

5 Find the base back skirt.

6 Gently hammer in the wooden pins.

7 Fit the skirt into the base end pieces.

8 Line up the pins and posts on the skirt and one base end piece.

9 Lock the skirt into place.

10 Line up the other base end piece.

11 Lock the skirt into place.

12 Lay this assembly aside.

13 Locate the bottom shelf pieces.

14 Gently hammer the wooden posts into the middle end pieces.

15 Slide the bottom shelf onto the middle end pieces.

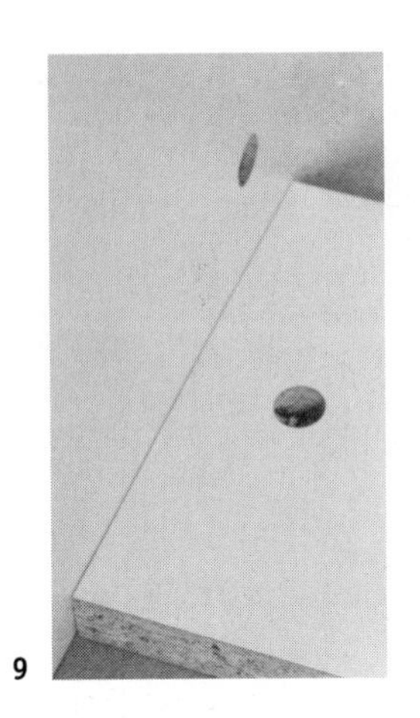

9

10

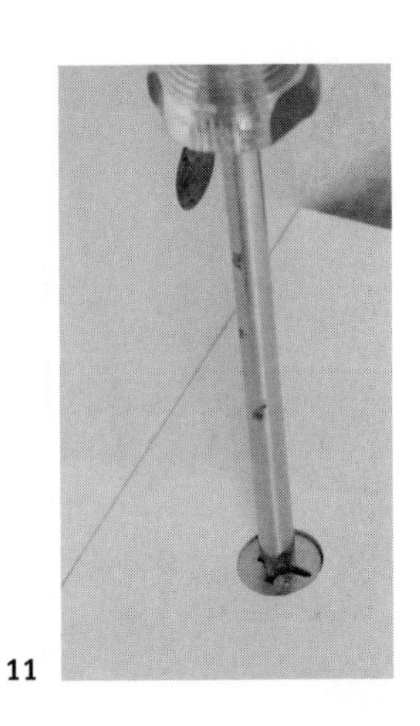

11

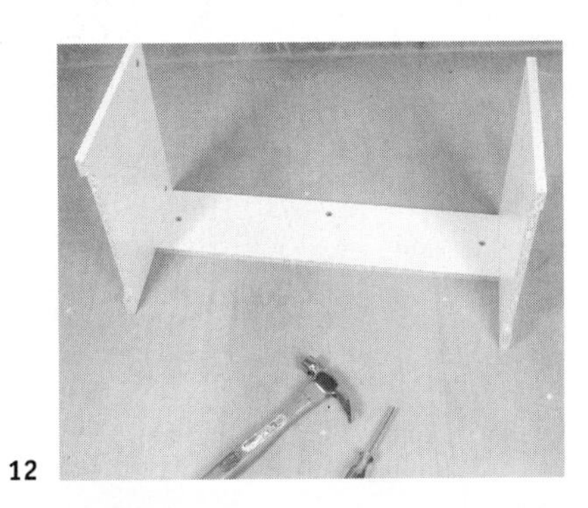

12

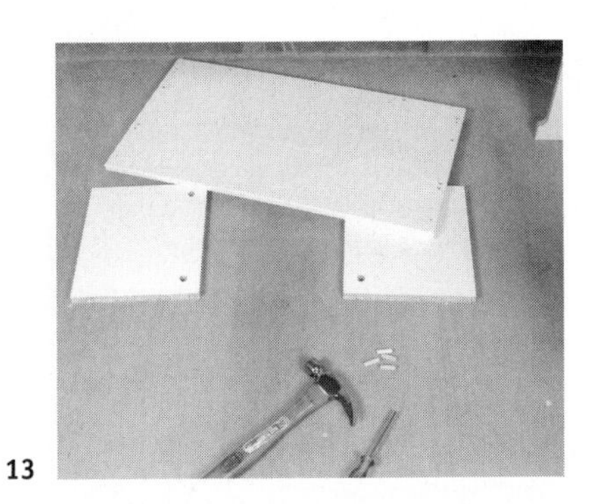

13

14

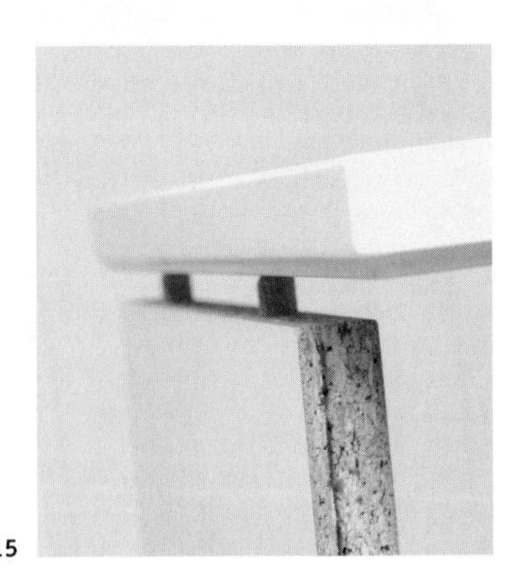

15

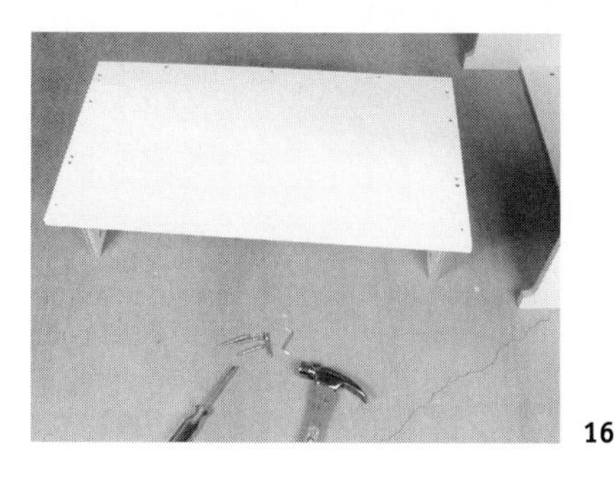
16

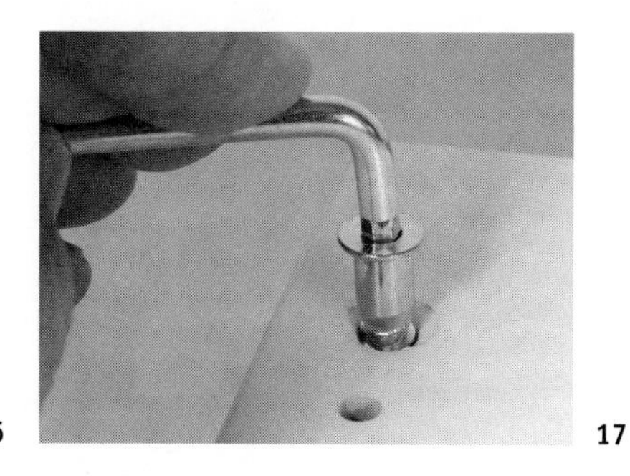
17

18

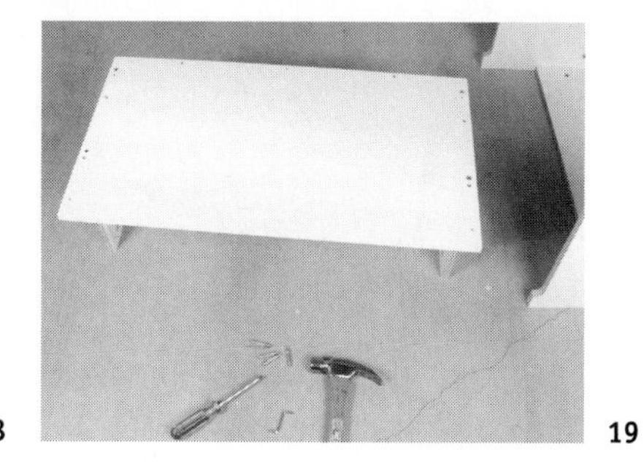
19

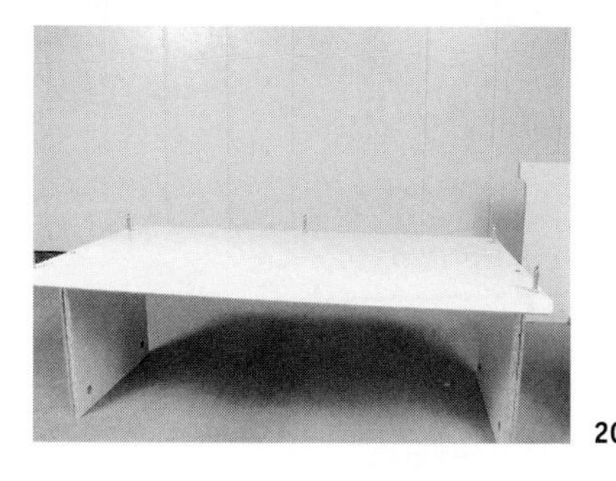
20

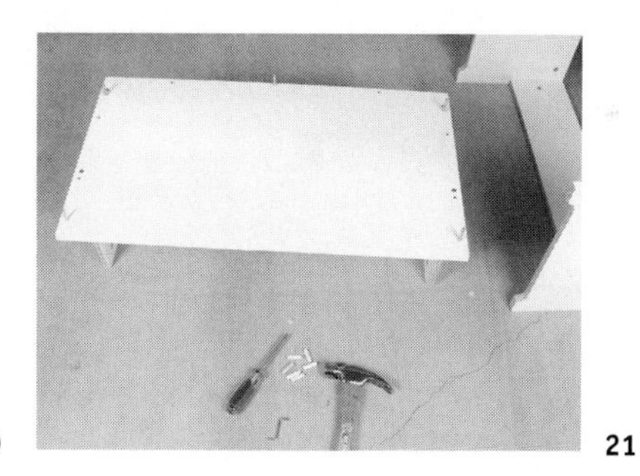
21

22

23

16 Set the bottom shelf assembly on its two end pieces.

17 Insert screws into the bottom shelf.

18 Fasten these screws with the provided tool.

19 Keep the bottom shelf assembly upside down.

20 Attach the mounting posts.

21 Insert the wooden pins into the bottom shelf assembly.

22 Gently pound each pin into place.

23 Keep the bottom shelf assembly upside down.

24 Slowly lower the skirt assembly onto the bottom shelf assembly. Ensure that all pins and posts line up correctly.

25 Lock the two assemblies together.

26 Set this assembly aside.

27 Locate the middle shelf pieces.

28 Gently hammer the wooden pins into the middle shelf end pieces.

29 Lower the middle shelf into place on the two end pieces.

30 Insert screws into the bottom of the middle shelf. Fasten these screws with the provided tool.

31 Bring the skirt and bottom shelf assembly back and attach mounting posts into the middle shelf.

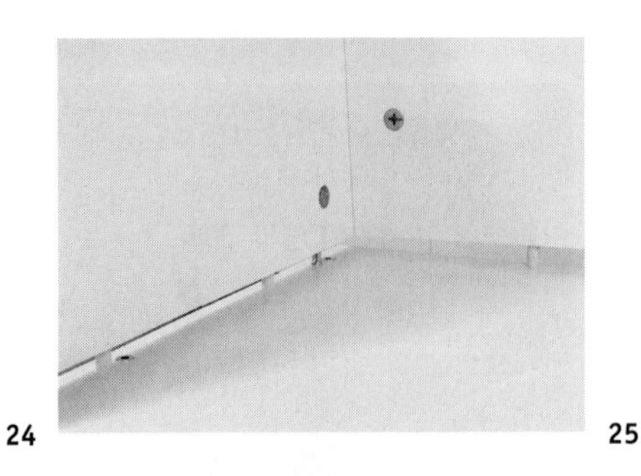
24

25

26

27

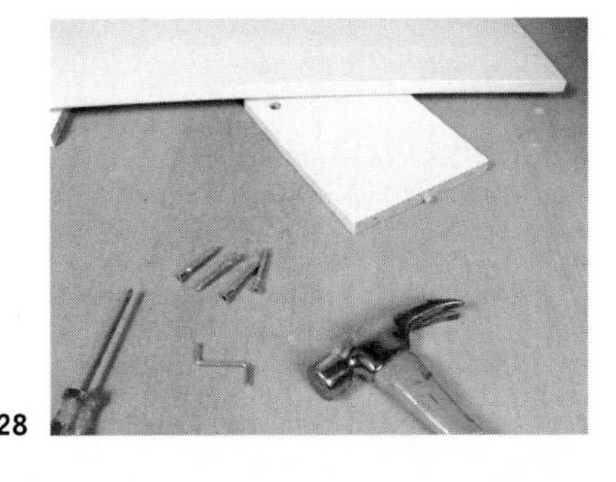
28

29

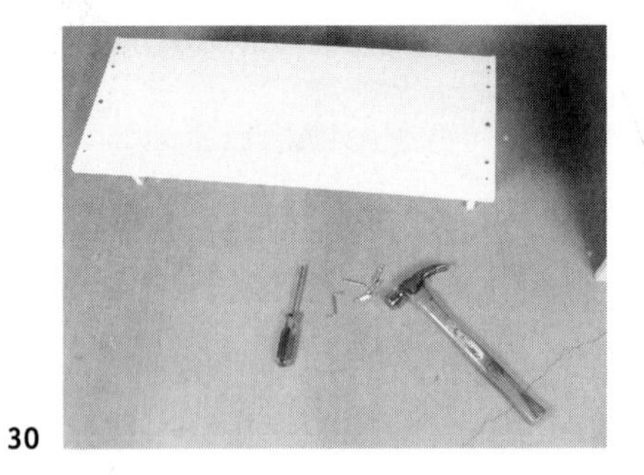
30

31

32

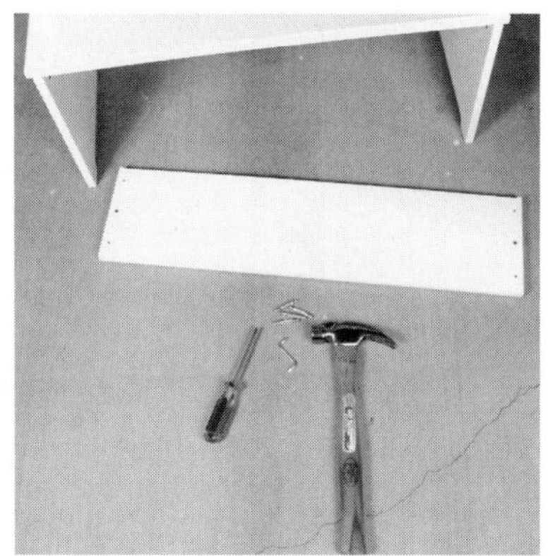

33

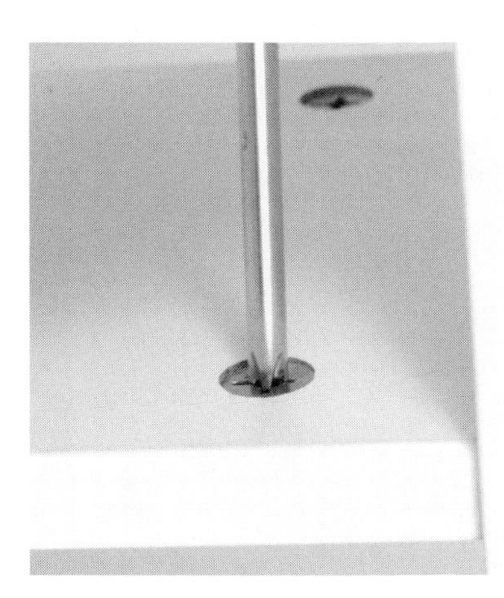

34

35

36

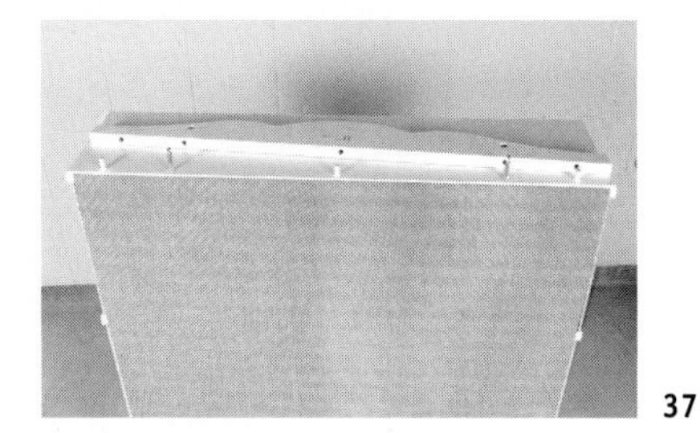

37

32 Lower the middle shelf down onto the skirt and bottom shelf assembly and lock it into place.

33 Attach mounting posts into the top shelf.

34 Place the top shelf on the middle shelf end pieces and lock it into place.

35 Hold the back panel firmly against the back of the bookcase and nail it into place. Make sure that the back panel fits smoothly inside the back edge rabbet.

36 Evenly space all of the nails around the perimeter of the back panel.

37 Attach the top shelf back plane trim.

38 A completed IKEA Hensvik.

39 Now Katherine's room has a better feeling.

38

39

CHAPTER 19

Courage! I'd Rather Not

Boy, if you enjoy watching the news readers lecture you on the news as they see fit, then March 9, 2005 was a dark day in televised news coverage. That's the day that Dan Rather of *CBS Evening News with Dan Rather* bade farewell to his viewing audience.

Granted, Dan Rather was no Edward R. Murrow nor was he equivalent to Walter Cronkite, rather Rather was, well, Rather. During his 24-year stint as the CBS News anchor, Rather would occasionally *become* the news.

Who can forget Rather's signoff for the news on September 2, 1985? On this memorable date, Rather ended the broadcast with a single word: "courage." This strange signoff persisted and became an interesting footnote marking Rather's career.

A more serious mark was left on Rather's body in 1986. During a walk along Park Avenue on October 5th, Rather was confronted by a man, later identified as William Tager. Tager purportedly asked Rather, "Kenneth, what is the frequency?" As Rather attempted

The Wrong Child

The musical group R.E.M. immortalized Dan Rather's mugging at the hands, and feet, of William Tager with the song, "What's the Frequency, Kenneth?"

to flee, Tager punched him from behind, knocking Rather to the ground. Tager continued his assault as Rather ran into a building, kicking him in the back. In 1994, Tager was arrested and convicted for killing an NBC stagehand outside the *Today* show studio.

Even that mugging incident couldn't top Rather's tempestuous walkoff on September 11, 1987 when a Steffi Graf v. Lori McNeil U.S. Open tennis semi-final match ran two minutes into the scheduled broadcast of *CBS Evening News with Dan Rather*. Rather did not return to his anchor desk until six minutes had elapsed. This six-minute absence prompted Johnny Carson to spoof that the six-minute lapse became CBS' highest-rated show of the year and that the empty screen had just been signed for 13 weeks.

Rather wasn't always the subject of the nightly news. From his joining of CBS News in 1962, Rather was referred to as "the hardest working man in broadcast journalism." Who can forget his round-the-clock coverage of the Presidential Election race in 2000? From 6 PM on November 7, 2000 until 10 AM on November 8th, Rather clung onto the story like a "hanging chad."

Educated with a journalism degree from Sam Housteon State Teachers College, this Class of '53 grad jumped into the national spotlight during his coverage of the assassination of John F. Kennedy on November 22, 1963. From that moment on, whether it was wars, presidents, civil rights, or hurricanes, Rather was there. Hurricanes? Yes, Rather had a strange self-admitted penchant for hurricanes and "big winds."

Probably the single, most defining moment in shaping Dan Rather's career was the date of his birth: October 31, 1931—Halloween; Trick or Treat.

And That's the Way it Was

Walter Cronkite became the anchor for the CBS EVENING NEWS on April 16, 1962. Just shy of his twentieth anniversary, Cronkite stepped down as the CBS News anchor and Dan Rather was named as his successor. During his tenure as anchor, Cronkite was referred to as "the most trusted man in America." On Cronkite's final broadcast his signoff was, "old anchormen don't go away, they keep coming back for more."

HOW TO TURN AN OLD MONITOR INTO A MODERN TV

1 ADS Tech TV tuner products.

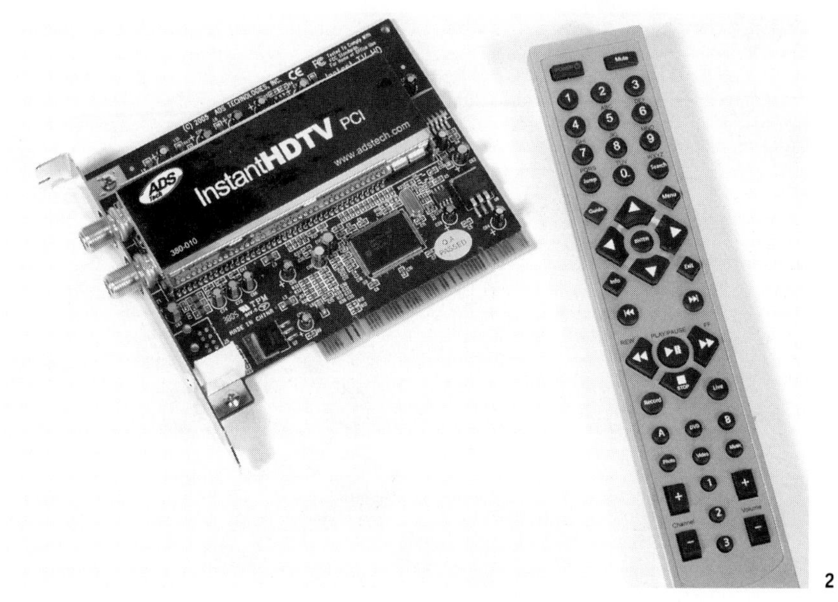

2 If you've got a spare PC, you can add HDTV to your monitor with ADS Tech PTV-380.

Courage! I'd Rather Not

3 Just slide this card into an open PC slot.

4 An IR sensor cable enables your PC to function with the included HDTV remote control.

5 An HDTV remote control.

6 If you don't have a spare PC, then ADS Tech PTV-360 PTV Tuner can turn an old monitor into a modern TV.

7 Status lights inform you about the TV's status. Likewise, you can turn on this system even if you lose the remote control.

8 Hide your cable clutter—all connections are made to the back of the PTV Tuner.

3

4

5

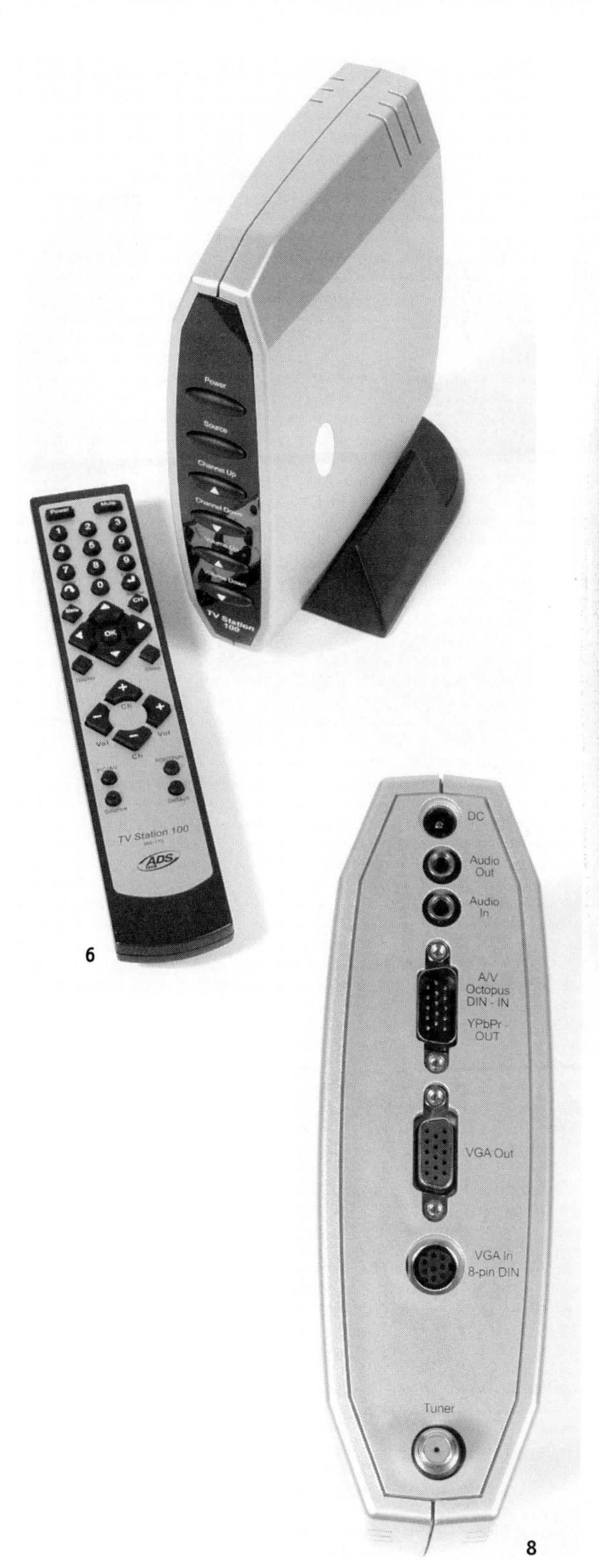
Power
Source
Channel Up
TV Station 100
DC
Audio Out
Audio In
A/V Octopus DIN - IN
YPbPr - OUT
VGA Out
VGA In 8-pin DIN
Tuner

6

8

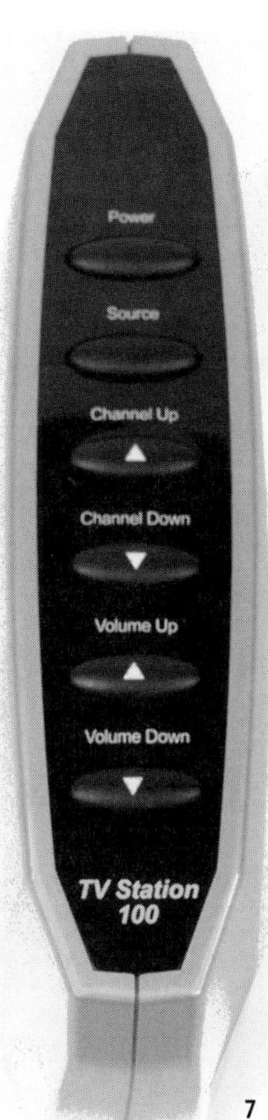
Power
Source
Channel Up
Channel Down
Volume Up
Volume Down
TV Station 100

7

9 All of the necessary cables are included.

10 The PTV Tuner comes with an IR remote control for armchair viewing. This type of “recycled” TV is great for any area where a new TV isn’t cost effective—cabin, mobile home, and spare bedroom.

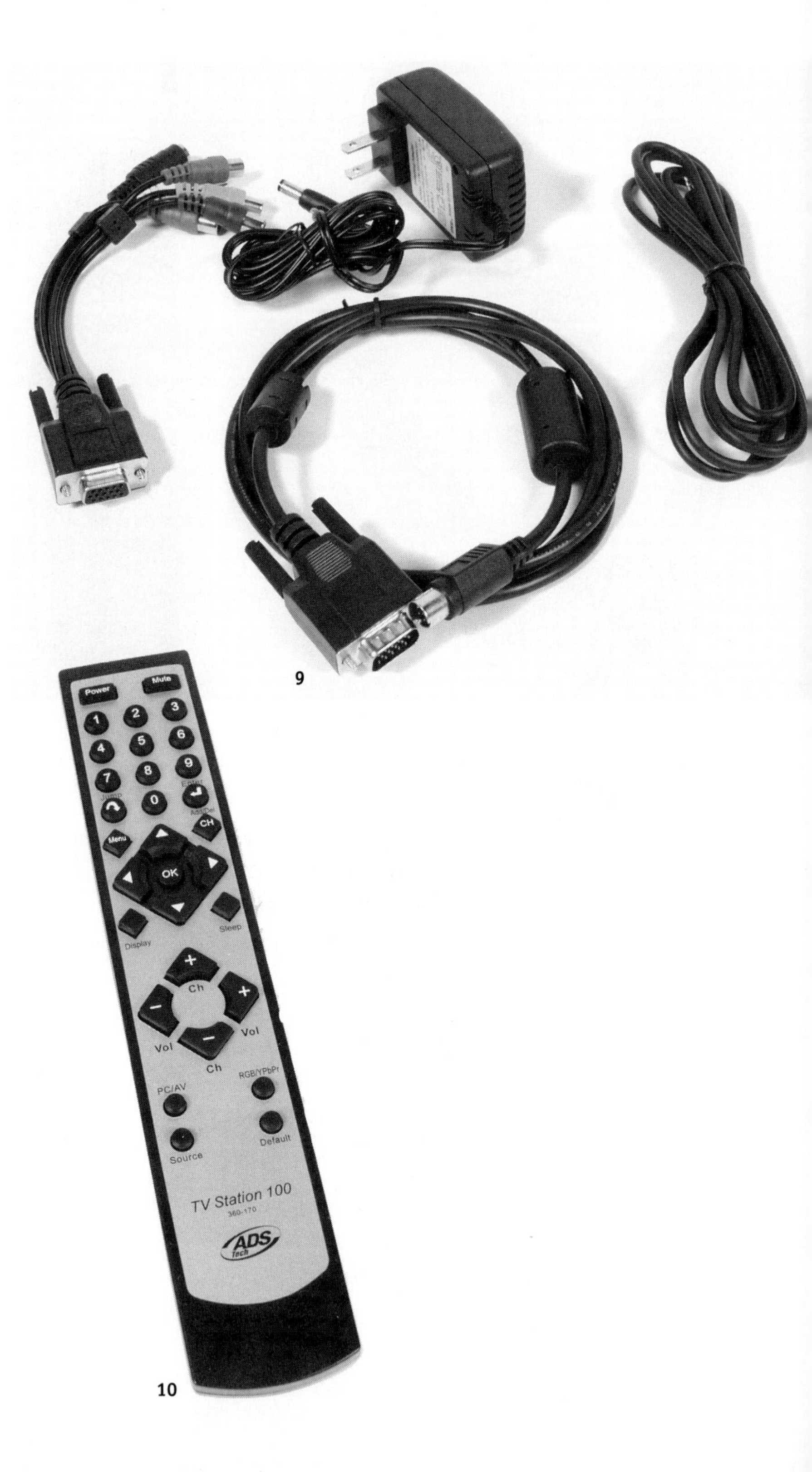

CHAPTER 20

’Tis the Season

According to the International Council of Shopping Centers (ICSC), 2002 Holiday Watch, the biggest sales day between 1996 and 2001 was the Saturday before Christmas. In a similar 2002 ICSC survey, based on the responses from approximately 230 enclosed malls, the survey said:

- Most malls began decorating for the holidays on November 1.
- The average number of days it took to decorate a center for the holiday season was 8.
- The average amount of money spent decorating for the holiday season was $22,998.
- The denomination of gift certificate that was sold most frequently was for $25.
- Santa Claus arrived at most malls on November 16.
- The holiday song played most frequently in 2001 was Jingle Bells.
- The percentage of malls that advertised for the 2002 holiday season on the Internet was 56 percent.

As for inside the home, the U.S. Census Bureau reported that for the 2000 holiday season there were $473 million in Christmas tree sales. This amount was an increase of 14 percent from 1995. Over 25 percent of the total tree sales were from tree farms in Oregon.

Now once you have your tree, you have to decorate it, right? During this same season, the U.S. Census Bureau reported $592 million in imports of ornaments from China (see Fig. 20-1). Ironically, China is also the leading source for artificial Christmas trees ($62 million) and lighting sets ($143 million). Truly, Christmas is an international season in the United States.

20-1 The Dunker lamp/room divider by IKEA is a great holiday lighting project. Just replace the lamps with the appropriate colored bulbs, adjust the built-in dimmer, and you have suitable lighting matched to any holiday. (*Photograph courtesy of IKEA; www.ikea.com; 877-345-4532.*)

Holiday Lights

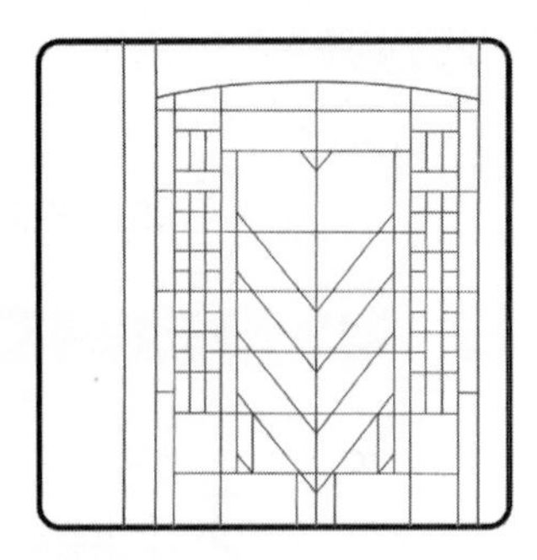

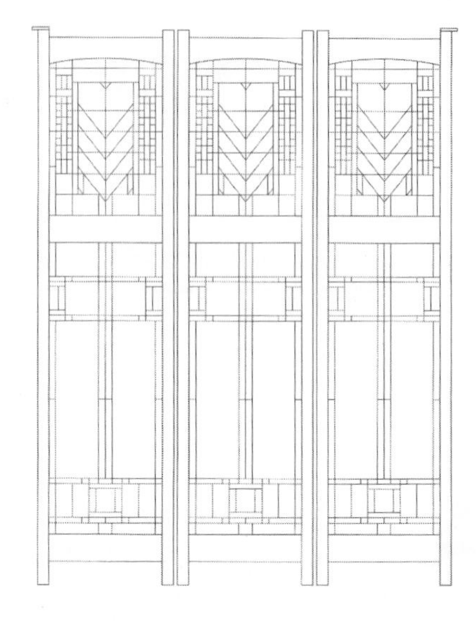

Frank Lloyd Wright inspired screen

The Dunker lamp/room divider by IKEA

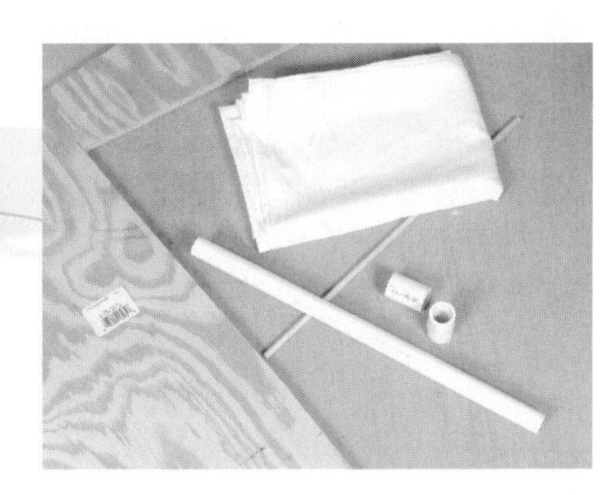

Make your own illuminated screen

Feeling a bit Scrooge-like? How about Valentine's Day? Valentine's Day is the number 1 holiday for florists. According to an IPSOS FloralTrends, 2005 survey,

- 65 percent of Valentine's Day floral purchases are made by men.
- 35 percent of purchases are made by women.

Hold on to your hat, dude, whereas 84 percent of those purchases by men are for their "wife or significant other," 32 percent of the sales to women go to their mother, followed by 24 percent for their husband or significant other. Oh, and in case you're wondering, 22 percent of floral sales to women on Valentine's Day go to themselves.

Returning to the U.S. Census Bureau, 180 million cards are exchanged on Valentine's Day. And you'd better include some candy with that card. In 2004, Americans ate 24.7 pounds worth of candy, per person.

Finally, let's not forget Halloween—a holiday that is marked by pumpkins, door-to-door solicitations, and fanciful costumes. Those good folks at the U.S. Census Bureau have some interesting statistics about this holiday that could make your hair stand on end:

- 998 million pounds of pumpkins were processed for 2004.
- Illinois pumped its pumpkin production to 457 million pounds in 2004.
- Approximately 106 million domiciles were visited on Halloween 2004.
- 36.4 million kids between ages 5 and 13 participated in Halloween 2004.

HOW TO SPREAD SOME HOLIDAY CHEER

1 Some greeting cards have a built-in sound circuit that plays seasonal music, greetings, or noises.

2 Remove this circuit from the card, add a new battery, and rig it up for remote playback.

3 Any thin piece of insulation (e.g., a fabric cord or plastic card) can be placed underneath the trigger switch. Then remove this insulation and the recording will be played.

1

2

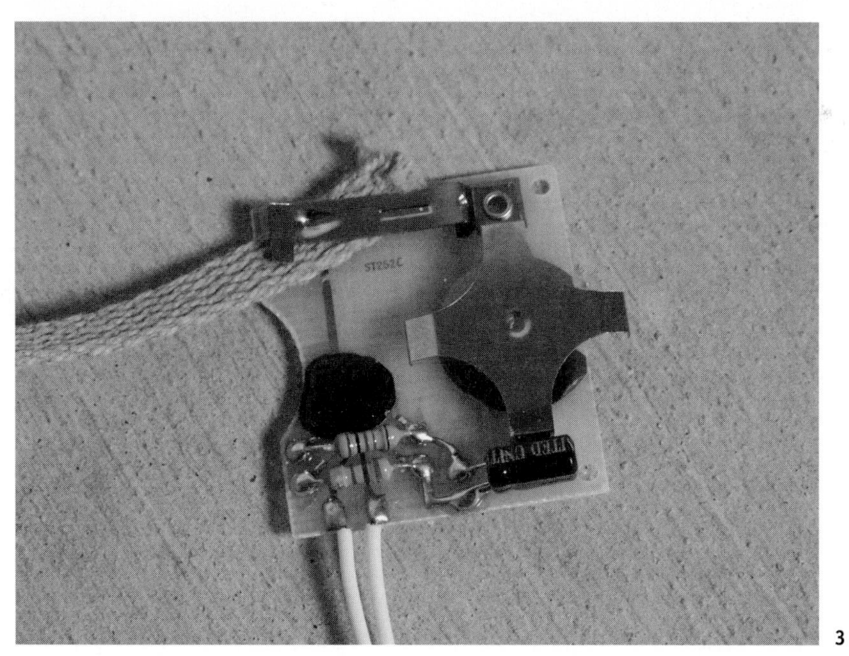

3

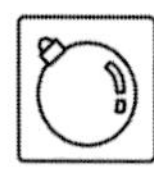

CHAPTER 21

Tail-e-Essen

How can you encourage children to consider the pursuit of a career that is "old world," stuffy, and predominately occupied by men? Such is the dilemma facing today's architecture schools. If there isn't any grassroots interest in architectural design, then these unfortunate tendencies become status quo and continue to dominate our mainstream educational curriculum. And this is a sad legacy for our future generations.

In order to counteract this dogma and dawn a new age in design appreciation, young adults must be enlightened about the possibilities for significant societal contributions that are inherent with a career in architecture. Who can argue that the Transamerica Pyramid in San Francisco isn't inspiring? Or, challenge the historical merits of New York City's Chrysler Building? Or, dislike the serene beauty of Minneapolis' community neighborhoods? These are all (as well as, many more) examples of architecture's subtle, yet powerful, effect on our daily lives.

So the challenge remains: How can we instill respect for the architect's profession, when we can't seem to make a strong enough case to students for selecting architecture as a career choice? Well, first of all, we must show that architecture is a valid occupation open to all people, not just the select membership of the "all rich men's club." But to debunk such a strong myth requires an indirect approach where education is cloaked inside a nonthreatening, exciting, and, most of all, fun package.

Secondly, architecture must be elevated to a noble professional status that is typically reserved for careers in medicine, politics, law, and teaching. Whereas this occupational nobility has become a silent "birthright" for these time-honored professions, it is a completely new approach for promoting a career in architecture.

21-1 Meet Feathers Lloyd Wing.

21-2 Meet Frankie Lloyd Wonder.

21-3 He's an architect, you know. Or, at least he wants to be one, when he grows up.

21-4 One day Frankie met Feathers. Feathers had a problem. He needed a new house.

21-5 "Aha, a client," exclaimed Frankie.

21-6 Frankie set to work to design the best birdhouse ever.

21-7 Frankie called his design, Tail-e-Essen.

Both of these challenges can be satisfied by *The Adventures of Feathers Lloyd Wing: A Bird's Eye View of Architecture*; in which a wayward bird enlists the design skills of some architects for rebuilding its nest (see Fig. 21-1). But these aren't just any architects—these are some of the most important pioneers in the annals of American design (see Fig. 21-2).

21-1

Frank Lloyd Wright, Louis Kahn, Phillip Johnson, Denise Scott Brown, and Rem Koolhaas band together to build the "perfect birdhouse." (See Fig. 21-3.) Along the way, this group interacts with Mies Van Der Rohe, Le Corbusier, and Louis Henri Sullivan, while striving to maintain their client's trust and deliver a dwelling with a difference. *The Adventures of Feathers Lloyd Wing: A Bird's Eye View of Architecture* isn't some staid historical fiction—how would that stimulate a student to explore a career in architecture? (See Fig. 21-4.)

21-2

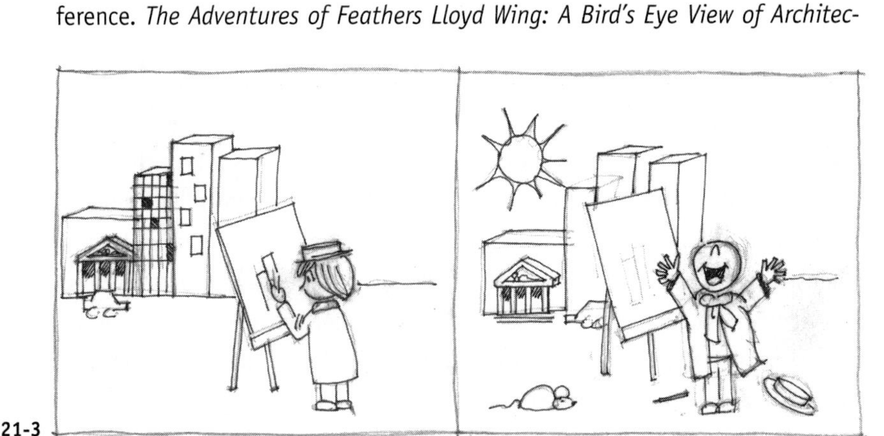

21-3

In the first of several plot twists, the characters in *The Adventures of Feathers Lloyd Wing: A Bird's Eye View of Architecture* are actually a group of "typical" neighborhood kids who have an interest in both helping their new bird friend and in building a new birdhouse (see Fig. 21-5). As such, the cast of characters from the book, Frankie, Louie, Phil, Denise, and, the foreign exchange student, Remie, help Feathers Lloyd Wing build a new birdhouse. Furthermore, Mieses the Mouse provides thoughtful guidance

21-4

21-5

to our small band of birdhouse builders, while Le Crow and Luis-s-s the Serpent try to foil their design (see Fig. 21-6).

21-6

The Adventures of Feathers Lloyd Wing: A Bird's Eye View of Architecture will supply several key role models, inspire team work, express compassion in helping others, and, most of all, develop strong design skills (see Figs. 21-7 through 21-9).

THE ADVENTURES OF FEATHERS LLOYD WING: A BIRD'S EYE VIEW OF ARCHITECTURE

21-7

Feathers's adventures begin at home. While Mama is away trying to find a tasty worm for dinner, Feathers is busy arranging and rearranging their home.

Unknown to Feathers, Luis-s-s the Serpent has slithered up the tree and lays coiled below his nest.

"Feathers-s-s-s," hisses the serpent, "did you know that good design form follows-s-s a basic foundation in function?"

"Who said that?" queries the curious Feathers glancing quickly over and around the neatly arranged nest.

"Why come closer, here, under your nest where I can dissolve your architectural illusions-s-s and awaken your inner pupil with a series of chats-s-s," lures Luis-s-s.

Continuing his tempting talk, Luis-s-s promises Feathers, "... freedom in exploring his-s-s creativity..." and "... s-s-spiritual awareness that can only be achieved through harmony in good design."

Driven by a desire for achieving a purer aesthetic appreciation of life, Feathers carefully creeps over to the edge of his nest. "Is the solution to my mother's nest's arrangement really an issue of good design?" asks young Feathers.

"It's-s-s more a reflection of nature's own growth, decadence, and infinite...," a hesitant lunge by Luis-s-s gives Feathers enough time to jump clear of the serpent's deadly attack. Alas, poor Feathers slowly falls and flaps helplessly to the ground below.

21-8 **A plan for a bird feeder.**

3"
Roof above
3 1/2"
4 1/2"
3 1/2"
3/4"
4"
3"
3 1/4"
3 3/4"
3 1/2"
10"

Plan View

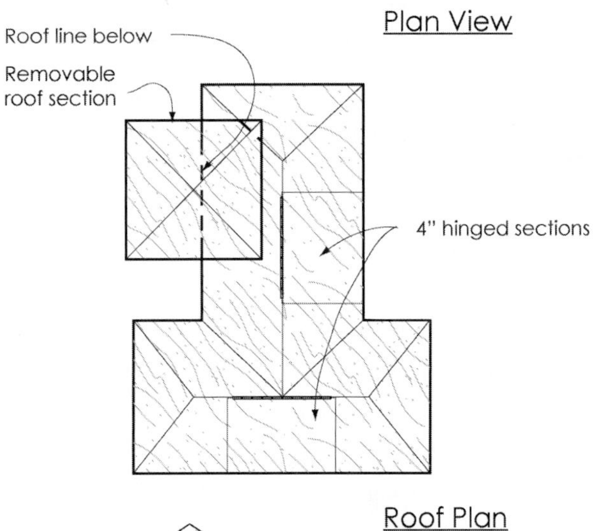

Roof Plan

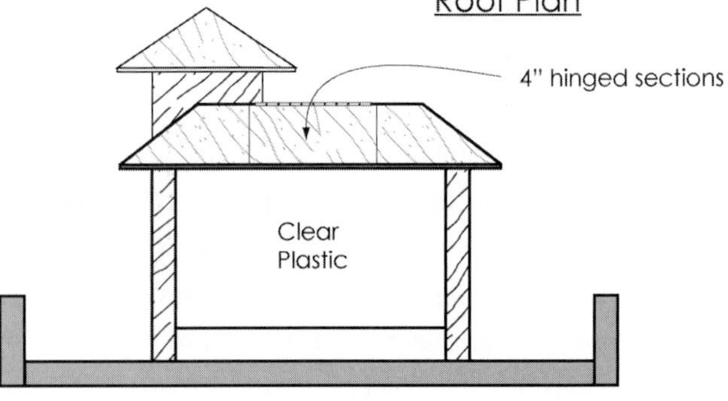

South View

Tail-e-Essen

21-8

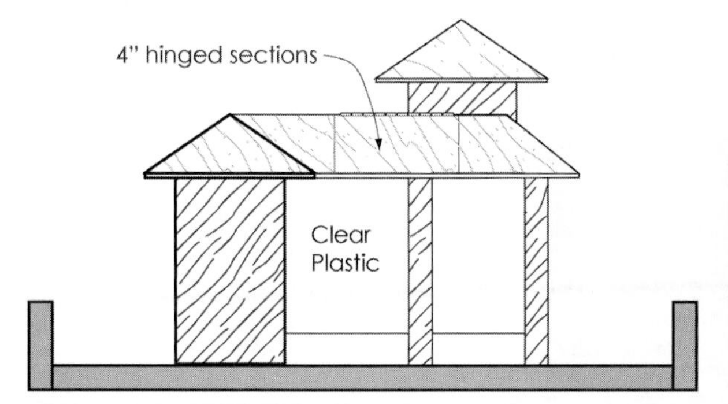

East View

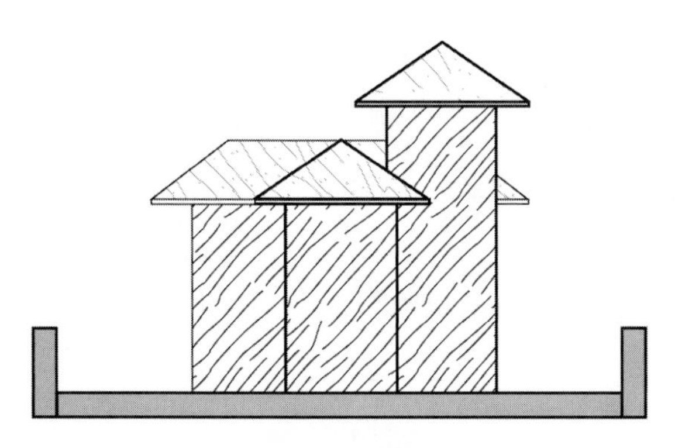

North View

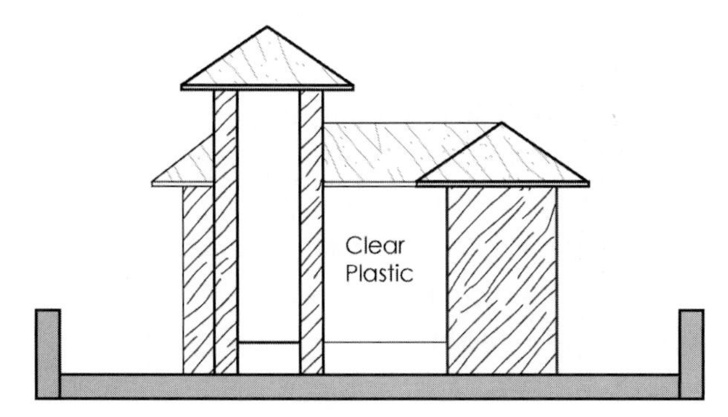

West View

Notes:

1. All wall boards are 3/4" stock.
   - (2) 2 3/4" x 6"
   - (1) 4 1/2" x 6"
   - (1) 1 1/2" x 6"
   - (1) 2 1/4" x 6"
   - (1) 1" x 6"
   - (2) 3 1/2" x 9"
   - (1) 2" x 9"
   - (1) 3 3/4" x 6"

2. All plastic sheets are 3/16" material. Router wood boards to accommodate plastic.
   - (2) 3 1/2" x 5"
   - (1) 8" x 5"
   - (1) 2 3/4" x 5"
   - (1) 2 1/2" x8"

3. Fix plastic to the top of all walls to allow for seed dispensing.

4. Roof boards are 1/2" stock.

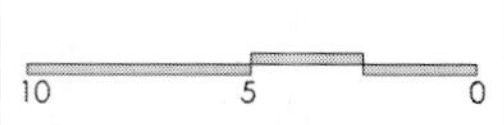

21-9 A plan for a birdhouse.

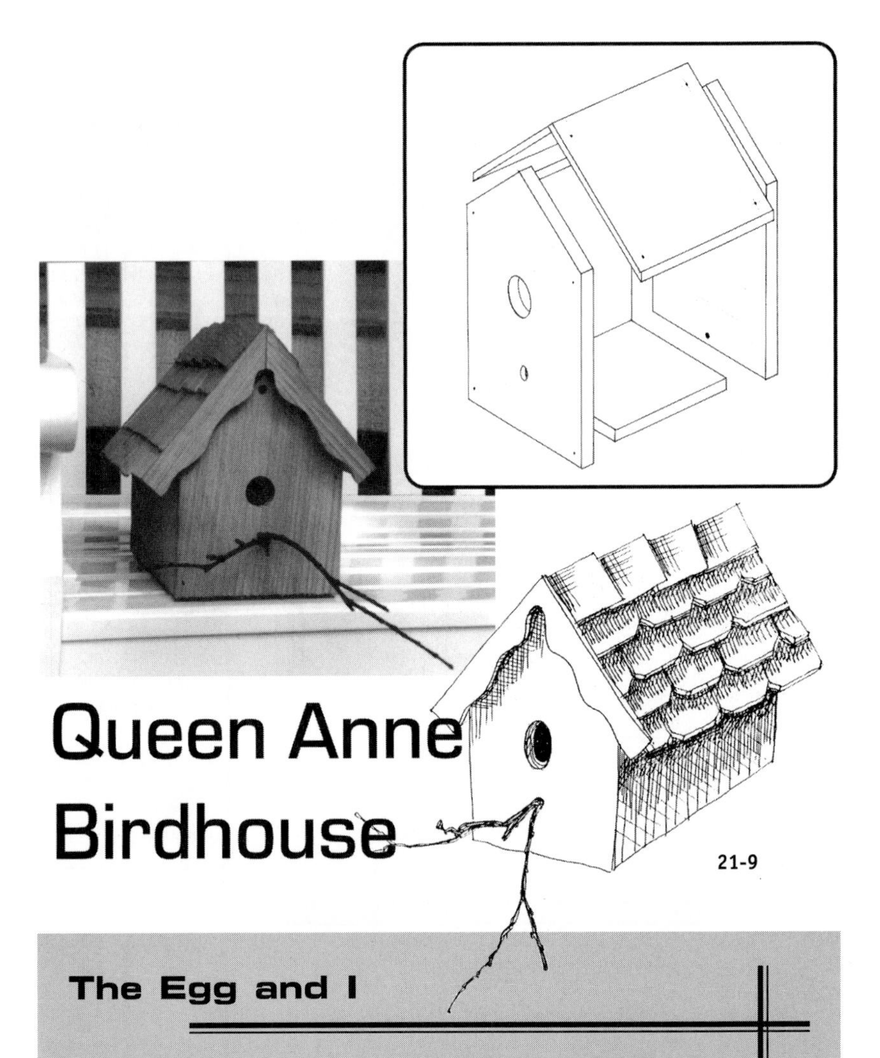

Queen Anne Birdhouse

The Egg and I

In 2000, Jim Schatz and Dan Llaurado created the first Egg Lamp in their ceramic studio located in New York City. Later in 2005, J. Schatz, as the studio was now known, relocated to Greene, New York. While their designs of sensual shapes were soothing to the soul, each design also exhibited an underlying function. Such was the case of the Egg Bird Feeder.

Handcrafted from ceramic earthenware these Egg Bird Feeders attracted more than just birds. They also caught the eye of FORTUNE MAGAZINE which named them one of the "25 Best Products of 2004."

HOW TO BUILD A REMOTE WARNING SYSTEM

1 A cheap RC car can be easily modified into becoming a remote warning system.

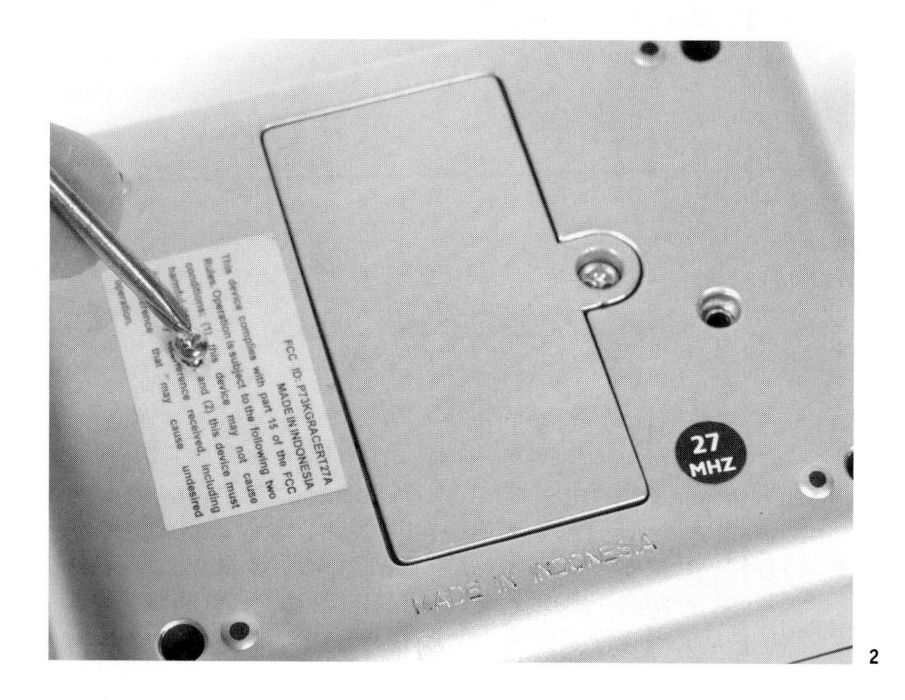

2 Remove the batteries.

3 Open up the transmitter.

4 This touch circuit from the transmitter will be your remote sensor.

5 Open up the RC car.

6 Remove the motor and replace it with a buzzer or LED/resistor combination. Now you can attach your remote warning system, for example, to your Tail-e-Essen bird feeder and have it buzz or blink you when it runs out of bird food.

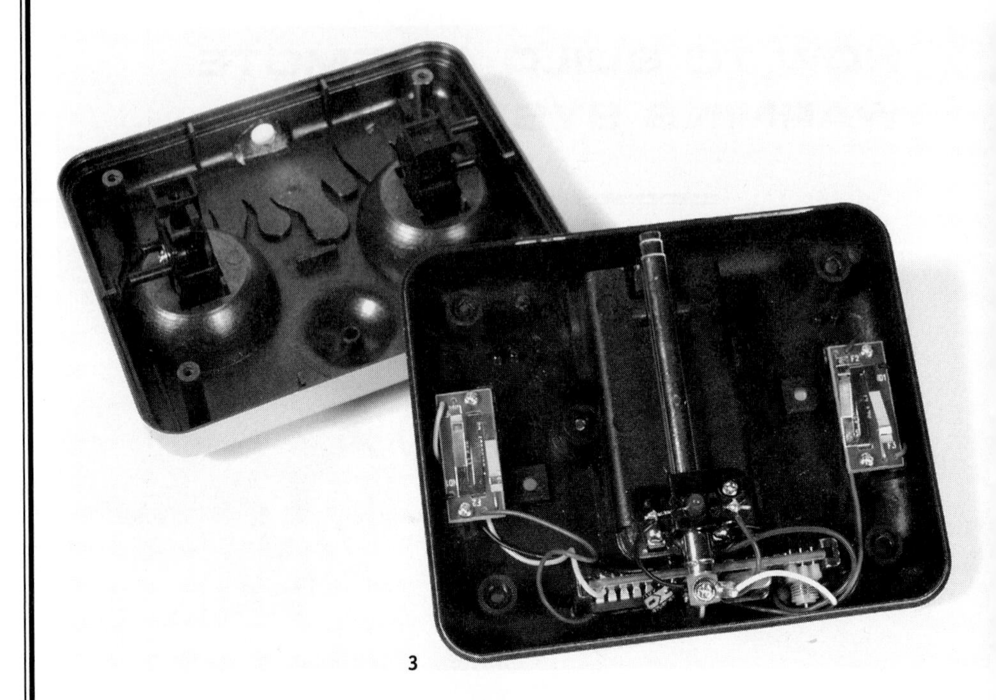

3

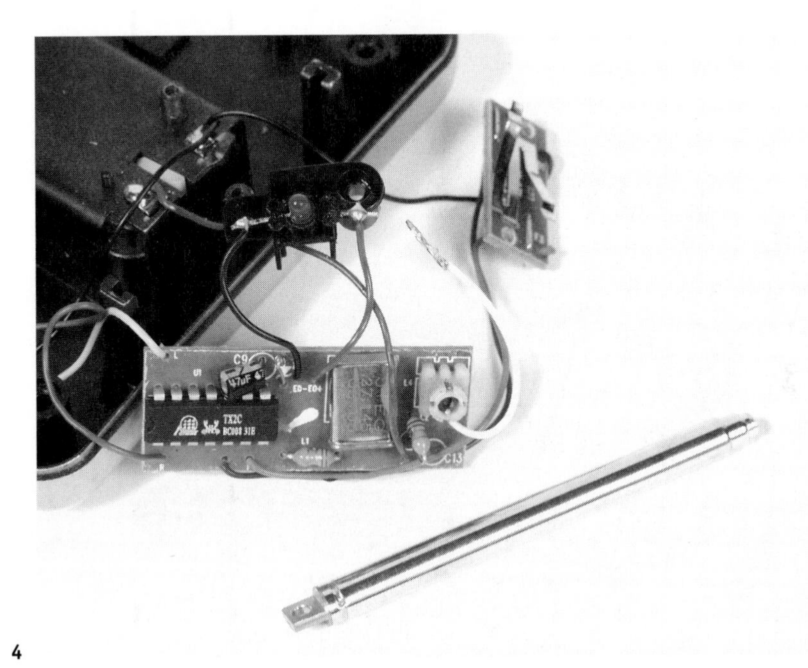

4

5

6

Tail-e-Essen

CHAPTER 22

Perk Me Up

Every morning the ritual begins anew. Attempting to brew that elusive perfect cup of coffee. Such a simple-sounding task, yet one that few of us ever master. Like sighting the Loch Ness Monster, enjoying the perfect cup of coffee is believed to exist, but have you ever tasted one?

If you ask the National Federation of Coffee Growers of Colombia how to brew the perfect cup of coffee, you will receive the vague guidance to boil six ounces of water and mix in two level tablespoons of 100 percent Colombian coffee. OK, if that's all it takes, how come the coffee at my local greasy spoon doesn't taste "perfect"?

Then along came pods and everything changed. No, not iPods (remarkably enough, this is an event that wasn't shaped by these diminutive music players), pod coffee makers or brewers.

The pod coffee brewer is the ultimate way to attempt to brew the perfect single-serve cup of coffee. Why are pod brewers so, well, hot? They minimize waste by using self-contained, single-serve packets of ground coffee that are simply dropped into a holder which is inserted into the brewer. Some of these packets are robust, paper, pancake-like discs filled with the perfect amount of coffee, while others are small cups which hold the exact amount of coffee needed for brewing that perfect cup of coffee.

Bean There Done That

Before there were pods, however, there were just beans...coffee beans. Remember Juan Valdez who taught you that the best coffee beans come from Colombia? Well, they do, and the two most common beans are:

- **COFFEA ARABICA**—also known as Arabica. These are a flavorful and aromatic bean that is low in caffeine and acidity. Along with these outstanding qualities these beans are also more expensive.
- **COFFEA ROBUSTA**—also known as Robusta. These are less flavorful and less aromatic than Arabica beans, but they are also less expensive. Robusta beans have twice the caffeine as Arabica beans and are also high in acidity. Typically, *Coffea Robusta* is not grown in Colombia.

You also might see a coffee bean that is not from Colombia but it is called Colombian. This variety is known as Washed Arabica. Yes, it is an Arabica bean, but, in spite of its name, it probably isn't from Colombia. Why? All Colombian Arabica beans are washed. This process helps to lessen the acidity, as well as remove impurities. Coffee beans from outside of Colombia use this name to imply that their beans are from Colombia, when, in fact, they are not.

Once you've picked the beans, you must roast them to remove the water and convert the natural sugars in oil through caramelization. This roasting process also will increase the size of the bean and change its color to brown. You can control the roasting time of the bean which will, in turn, affect the color of the bean—the longer the roasting process, the darker the bean will

Poor Man's Pod Brewer

Before pod brewers became the rage, the coolest coffee pot in town was the French press or coffee press (see Fig. 22-1). Vogue in the 1920s, these glass cylinders with metal plungers steep loose ground coffee in hot water for four minutes. The metal plunger is pushed down, trapping the grounds against the bottom of the glass cylinder. The coffee is then decanted into your cup. A French press by Bodum® was one of the more popular contemporary models. In fact, I used a Bodum French press for years before I discovered the simplehuman pod brewer.

22-1

be. Coffee roasting takes about 10 to 20 minutes with temperatures ranging between 400 and 425°F.

Yes, dark roasted beans will have less acid and less caffeine, but they will not result in a stronger, richer cup of coffee. Your brewing technique will determine that result. Furthermore, a darker roasted bean can mask an inferior bean. Conversely, a lightly roasted bean will retain much of the bean's original flavor and, therefore, expose the bean's quality. Because it is so easy to judge a bean's quality based on its roasting process, most light roasts are limited to only the higher quality beans.

As a guide to determining the roast level that is right for you, use the following summary from the National Federation of Coffee Growers of Colombia:

NAME	ROAST LEVEL	FLAVOR
Cinnamon Roast	light roast	nut-like flavor, high acidity
American Roast	medium roast	caramel-like flavor
City Roast	medium roast	full coffee flavor, some loss of acidity
Full City Roast	medium roast	full coffee flavor, good balance
Vienna	dark roast	dark roast flavor
French Roast	dark roast	bitter, smoky taste, pungent aroma
Italian	dark roast	burnt flavor
Espresso	dark roast	burnt flavor that is strong and sweet

It's Simple, Human

Founded in 2000, simplehuman® has a mission statement that sounds a lot like similar statements made by other companies: "to create tools for efficient living." Unlike these other companies, however, simplehuman has found a way to touch people's lives without costing an arm and a leg.

How? By subtly combining rock-solid functionality with innovative design, simplehuman is able to transform those common, everyday items like disc racks, towel holders, and even spoon rests into works of art. Earning a fistful of international design awards has helped make believers of their simple design credo.

In November 2005, simplehuman entered the competitive pod brewer market with its take on the perfect cup of coffee. While better known for their line of award-winning stainless steel "butterfly" step trash cans, simplehuman created the simplehuman pod brewer as a single-serve brewer with a flavor extraction system that meets connoisseur standards for brewing and flavor.

The designers at simplehuman knew that this task wasn't going to be easy. Incorporating an affordable price point (MSRP $129.99), brewing flexibility (either 8- or 5-ounce cups), and a common pod disc format (thin and thick European-style pods), the simplehuman pod brewer is able to deliver a superior cup of coffee within five minutes of unpacking the unit. And these results can be reliably repeated day after day.

A key feature of the simplehuman pod brewer is the flavor extraction system during which water is infused over the pod evenly at the ideal brewing temperature (200°F). The brew chamber optimizes pressure between the coffee and the water to maintain a consistent flow. Customized channels in the pod holder prevent water from flowing through the pod too quickly and weakening the flavor.

In addition to improving the coffee extraction process, the brewer also includes features that make it easier to use, easier to clean, and more durable: a space-efficient stainless steel body; an easy to reach and refill water tank; a bright LCD display that shows brewer status, settings, and time; and a smooth, easy to operate handle for loading pods. The simplehuman pod brewer uses a pod format that is widely accepted as the single-serving coffee standard in Europe. Therefore, you can use any brand of coffee that is sold in these standard size pods.

Coffee can be brewed for either an 8-ounce cup or a smaller, more robust 5-ounce cup. The brewer comes with two pod holders, one for thicker 9-gram Molto pods, and another for thinner coffee or tea pods.

Wet Your Whistle

Hack that pod; no tools required. The key to this hack is to moisten the pod prior to extraction. Depending on your pod brewer, the best way to achieve this moistening step is by running hot water through the pod holder. Then drop in your pod and begin making your cup of coffee.

Another nifty trick for achieving a better cup of coffee is to run hot water through the brewer before you start the coffee brewing process.

HOW TO MAKE YOUR OWN COFFEE POD

1 The simplehuman® pod brewer.

2 A key feature of the simplehuman pod brewer is the flavor extraction system.

3 The brew chamber balances the pressure between the coffee and the water.

4 Molto Coffee brand pods are recommended by simplehuman for this brewer. Although other brands *might* work, the Aloha Island Coffee Kona-Pod™ pods that I tested did *not* work (i.e., extreme pressure buildup with only about 2 oz. of coffee produced).

5 Channels in the pod holder prevent water from flowing too quickly through the coffee.

6 Pods sit in the pod holder for proper coffee brewing.

7

8

9

7 Unfortunately, not every flavor or brand of coffee is available in pods. French Market® chicory coffee, for example, isn't made for pods. So, if you want chicory coffee, you're going to have to make your own pods.

8 A good candidate for making your own pods is a sealed tea bag.

9 Open the tea bag and empty the contents. Refill the empty bag with your selected coffee. Place your newly created pod into the brewer's pod holder.

10 Seat the pod holder inside the brew chamber.

11 Do a test brew.

12 Adjust your coffee concentration until you get the best flavor.

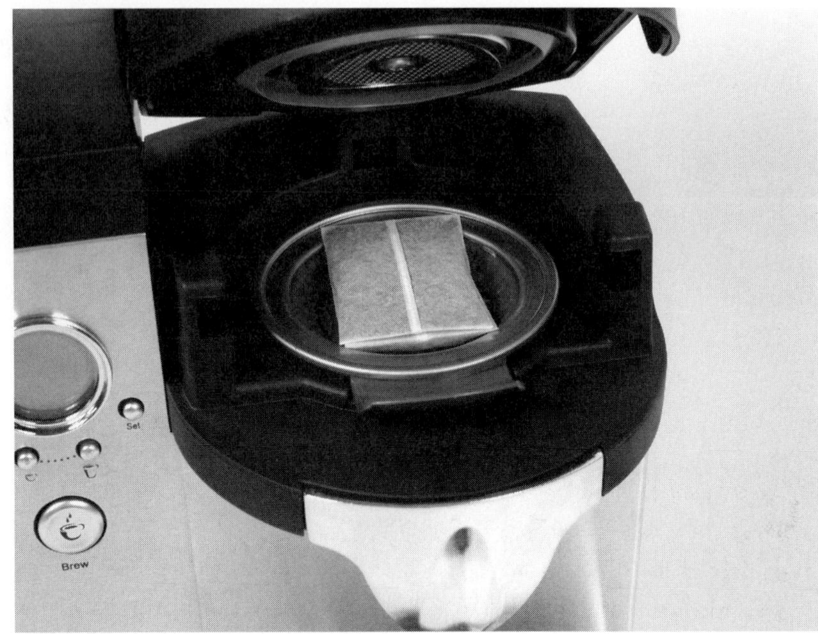

10

11

12

CHAPTER 23

Call Me

No piece of technology is so deeply entrenched in our daily lives as the cell phone. Just look at the driver sitting next to you in traffic; yup, he's on a cell phone. How about waiting in line anywhere for anything? Chances are every seventh person is using a cell phone—probably telling the party on the other end that they are, in fact, standing in a line. Or, my all-time favorite example of cell phone pervasiveness, the November 18, 2003 story reported by *Expatica Belgium News* about a cell phone ringing in a dead man's coffin while his family grieved during a private ceremony prior to the actual funeral. Talk about "reach out and touch some *body*."

Along with this newfound technology comes another, more ugly, aspect of having the entire world in your coat pocket. That's the splashing, cramming, slamming, "bluesnarfing," spoofing, and "bluebugging" attacks on your cell phone that can happen anytime and anywhere.

Splish, Splashing

When you place a long-distance call from a remote public phone, your call can get routed to a distant call center before being "handed off" to your chosen long-distance carrier. Your preferred long-distance carrier might then bill you as if your call originated from the distant call center, rather than from the actual location of the public phone. As a result, you may be charged higher long-distance rates for the call than you expected. This is called "call splashing," and it can be in violation of Federal Communications Commission (FCC) rules.

No Need to Cram for this Exam

"Cramming" is when unauthorized, misleading, or deceptive charges are added to your monthly telephone bill. Cramming companies rely on "hiding" these charges inside your telephone bill (e.g., a long list of charges won't look too suspicious when appended to your typical long list of charges) in order to mislead consumers into paying for services they did not authorize or receive.

Local telephone companies often bill their customers for long distance and other services that other companies provide. When the local company, the long-distance telephone company, or another type of service provider sends inaccurate billing data to be included on the consumer's local telephone bill, cramming can occur.

Cramming also occurs when a local or long-distance company, or another type of service provider does not clearly or accurately describe all of the relevant charges to you when marketing the service. Although you did authorize the service, the charge is still considered "cramming" because you were misled.

Slam You Up Side the Head

"Slamming" is the illegal practice of changing a consumer's land-based (e.g., terrestrial or wired) telephone service without permission. New consumer protection rules created by the FCC provide a remedy if you've been slammed.

Snarf Up Some Bluetooth

In a *WIRED* article from December 2004 by Annalee Newitz ("They've Got Your Number ..."), the CSO from a London security firm, Adam Laurie, demonstrated his "bluesnarfing" program. During operation this program looks for and latches

R U a Phreaking Freak?

Hacking telecommunication systems is called phreaking. The odd spelling of this hacking variant is derived from the "ph" of phone.

onto Bluetooth-enabled cell phones operating on the 2.4-GHz frequency range that haven't been secured from outside snoopers. Within seconds this program can locate a Bluetooth device and copy the entire contents of its address book, calendar, and account information—that is, if you fail to secure your Bluetooth connection. In other words, turn off Bluetooth when you aren't near your own computer.

Gone in a Spoof

Organizations like Telephreak offer an unbelievable array of "phreak" PBX services that are free to use. Voicemail, conference calls, and direct calls are all possible from this service. The Telephreak system consists of Direct Inward Dial (DID) phone numbers, freeworlddialup.com (FWD) gateways, and Voice over Internet Protocol (VoIP). Using a PBX phreak server you can easily spoof or mask your caller ID. Likewise, these same PBX phreaks can unmask "hidden" caller ID phone numbers.

All Bugged Up and Blue

In that same *WIRED* article mentioned above, Newitz describes how Laurie is able to completely commandeer a Bluetooth-enabled cell phone and force it to dialup another phone without the cell phone's owner being either aware or involved with the call.

Leave Me Alone

It's enough to make you scream. During dinner, your favorite television program, or just when you start to drift off to sleep, the phone rings, the voice on the other end mangles the pronunciation of your surname and then begins a lengthy spiel about something that you couldn't care less about. Well, the FCC has had an ear full of these types of complaints, too.

The Telephone Consumer Protection Act (TCPA) of 1991 was created in response to consumer complaints about the growing number of unsolicited telephone marketing calls to their homes and the increasing use of automated and prerecorded messages. You spoke and the FCC listened, sort of. On June 26, 2003, the FCC revised its rules implementing the TCPA and established, together with the Federal Trade Commission (FTC), a national Do-Not-Call (DNC)

Registry. The FCC also adopted restrictions on the number of abandoned calls that are permissible.

Simply put, a "telephone solicitation" is an advertisement. Even if you have an unlisted or telephone number that isn't published in your local directory, you may still receive unsolicited telephone calls.

In its attempt at "helping" you the consumer, FCC rules prohibit telephone solicitation calls to your home before 8 AM or after 9 PM. Additionally, anyone making a telephone solicitation call to your home must provide his or her name, the name of the person or entity on whose behalf the call is being made, and a telephone number or address at which that person or entity may be contacted.

Another piece of weak legislation aimed at stemming the tide of telephone solicitations, the FCC and the FTC have established a National DNC Registry. The registry applies to some telemarketers (with the exception of certain nonprofit organizations), and covers both interstate (from one state to another) and intrastate (within state) telemarketing calls. Commercial telemarketers are not allowed to call you if your number is on the registry subject to certain exceptions. And that's the rub. Oh, if that wasn't enough, your fax machine also can be deluged with spam facsimiles.

On July 9, 2005, Congress enacted the Junk Fax Prevention Act. The Act amends the TCPA by permitting businesses or entities to send unsolicited advertisements to consumers and businesses with whom the sender has an established business relationship. It also requires senders of fax advertisements to include a notice and contact information on the first page of the fax informing the recipient how to "opt-out" of any future fax advertisements from the sender. Thanks, I guess.

On January 19, 2006, the FTC made the unusual move of releasing a comment about an e-mail message that was sent around the country. Basically, this FTC press release stated that consumers should not be "concerned that their cell phone numbers will be released to telemarketers in the near future, and that it is not necessary to register cell phone numbers on the National DNC Registry to be protected from most telemarketing calls to cell phones." What does "most" mean? I don't know and they didn't explain it, either.

This press release goes on to say that FTC regulations prohibit telemarketers from using automated dialers to call cell phone numbers. Automated dialers are standard in the industry, so most telemarketers are barred from calling consumers on their cell phones without their consent. Then there was this

little crumb of false hope stated by the government: "The national associations representing telemarketers have stated that their clients do not intend to start calling consumers' cell phones." Yeah, like I believe and trust telemarketers to police their own actions.

According to the FTC press release, a nationwide wireless 411 directory (e.g., this registry is being created by Qsent) will have to be "opt-in" (i.e., you will have to consciously sign up for this service) and this wireless 411 directory would not be available in a printed, electronic, or Internet list for telemarketers. This government release ends with a Web link reference for more information about a planned "wireless 411" directory (www.qsent.com/wireless 411/index.shtml).

Sure enough, Qsent echoes much of the same information included in the FTC press release, specifically that this directory will offer wireless users a database that features "choice, privacy and security." Ironically, Qsent hints at the absence of these same features for land-based phones. Furthermore, unlike the FCC/FTC DNC Registry users of the wireless 411 directory will be able to remove their names and cell phone numbers at any time claiming that "no residual personally identifiable information will remain anywhere."

You Can't Take It with You

Under the FCC's wireless local number portability (LNP) rules, you can switch wireless carriers within the same geographic area and keep your existing phone number. If you are moving from one geographic area to another, you may not be able to take your number with you. In other words, if you change area codes, forget it. You'll need a new phone number.

In addition to switching from one wireless carrier to another, in most cases, you will be able to switch from a land-based carrier to a wireless carrier, or from a wireless carrier to a land-based carrier, and still keep your phone number.

Ether Phone

Many savvy computer users are embracing Internet Voice, also known as Voice over Internet Protocol (VoIP). Gone are the early days when this technology was equivalent to a bad citizen band (CB) radio call. Today VoIP is a technology that allows you to make telephone calls using an Internet connection instead of a regular (or analog) phone line. Some services using VoIP may only allow you to call other people using the same service, but others may allow

you to call anyone who has a telephone number—including local, long-distance, mobile, and international numbers. Also, while some services only work over your computer or a special VoIP phone, other services allow you to use a traditional phone through an adapter.

Get the Scoop on Skype

Skype is a free program that uses Peer-to-Peer (P2P) technology to help get the word out—literally, all around the world. While P2P sounds awfully similar to that "file-swiping" technology fostered by Napster, with Skype, P2P is a system "where all nodes in a network join together dynamically to participate in traffic routing-, processing-, and bandwidth-intensive tasks that would otherwise be handled by central servers." Skype claims that this technology is third-generation P2P, or 3G P2P.

Specifically, Skype is a P2P telephony network that is able to achieve a voice quality that is roughly equivalent to a Plain Old Telephony System (POTS). Additionally, Skype is able to surmount typical Internet obstacles with an encrypted message that can easily be routed through gateways and firewalls.

HOW TO RECYCLE THAT OLD CELL PHONE

1 Don't throw that old cell phone away. Change its battery and yank its Subscriber Identity Module (SIM; if needed)—voilà a great toy, address book, PDA, and clock/calendar.

2 Most cell phones have easy access to both the battery and the SIM card.

3 In some models, you can extract the liner for the cell phone's antenna and use it to push the SIM card out.

4 Keep this card. Some cell phones won't work without one.

5 An old pager can also be repurposed as a clock/calendar, phone number lookup, or toy for the kids.

6 Old alphanumeric pagers are the best for recycling.

7 Usually no SIM card is required, just a fresh battery.

3

4

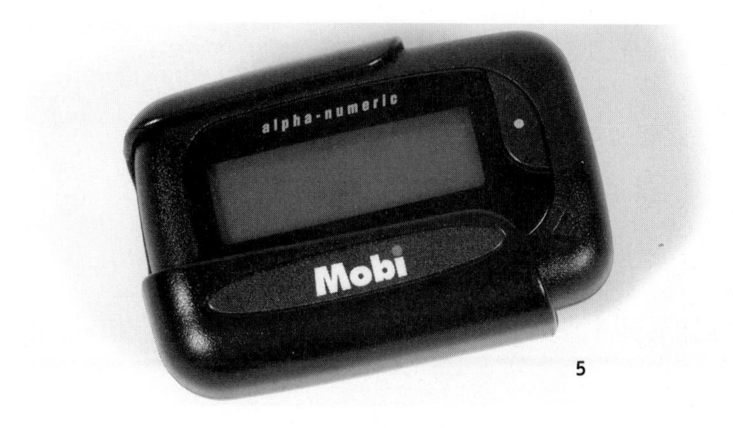
alpha-numeric
Mobi

5

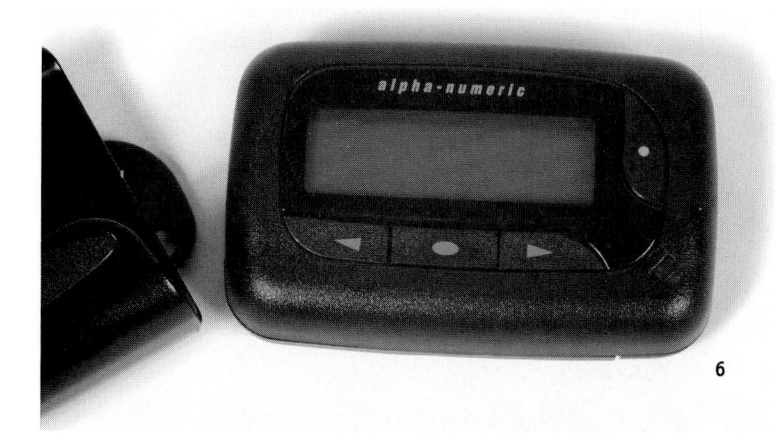
alpha-numeric

6

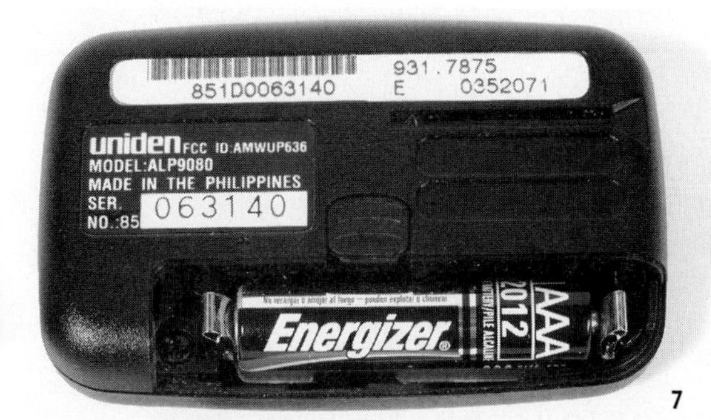
931.7875
851D0063140
E 0352071
uniden FCC ID:AMWUP636
MODEL:ALP9080
MADE IN THE PHILIPPINES
SER.
NO.:85
063140
Energizer
AAA
2012

7

CHAPTER 24

Get da Case on Deez Goods

Furniture is as varied in its construction technique as it is in its style; for every authentic mahogany and satinwood Hepplewhite sideboard replica there are hundreds of pine imitations. Usually these inexpensive counterfeits are stained with period colors in the hopes of convincing the discriminating buyer that beauty is only finish deep.

The construction techniques, historical perspectives, and product resources described in this book are an attempt at debunking the hype surrounding the confidence game pitch spun by disreputable antique dealers and larcenous furniture showrooms who espouse the virtue of purchasing "reproduction" furniture versus costly "originals." Furthermore, the confusing, inconsistent usage of these terms strengthens a case questioning the veracity of these sales agents.

Quality is Job 1 Through 5

There are four descriptive qualities that are generally used in defining furniture construction: original, reproduction, copy, and replica. Some connoisseurs include the term "fake" into this listing as a catchall for labeling all nonoriginal furniture. Each of these four qualities, however, has a specific definition that is rigidly adhered to by dealers and collectors alike, in judging the quality of construction used in building a piece of furniture.

1. **ORIGINAL.** This is a source furniture piece. All other pieces derived from the original's shape, style, and construction technique are either reproductions, replicas, or copies. Erroneously, many buyers believe that all period furniture

pieces are originals; this simply isn't true. For example, original eighteenth-century lowboys were built almost exclusively from walnut (with the occasional oddball constructed from mahogany). Conversely, nineteenth-century lowboys built from cherry, incorrectly labeled originals, are more accurately either reproductions or replicas. Granted, both lowboys are period furniture pieces (with the eighteenth-century original being either Queen Anne or Chippendale period, while the nineteenth-century reproduction is a Victorian period piece), but the eighteenth-century walnut lowboy is considered to be the historical ancestor from which all other lowboys sprang.

2. **REPRODUCTION.** This is an exact or close imitation of an original. Next to original, this is the second most commonly applied, and often misused, description for furniture. In the previous example, the nineteenth-century cherry lowboy is a reproduction of the original eighteenth-century walnut lowboy. This determination is based on the employment of cherry wood in the construction of the nineteenth-century version.

3. **COPY.** A multiple number of reproductions are usually machine manufactured. Although copy is an accurate description of their product line, many furniture manufacturers shun the usage of copy during marketing. Instead, they opt for labeling their pieces as "historically inspired," "traditional," "character," "authentic," "classic," and, representing the most gross misuse of furniture nomenclature ever "collector's heirloom" or "classic antique." This slightly tongue-in-cheek accounting of the furniture trade should not be construed as a condemnation of their craft; rather, this is an illustration of how furniture copies can be camouflaged through creative advertising.

4. **REPLICA.** An exact reproduction derived from the style, wood type, and construction techniques found in the original. The heart and soul in achieving this goal is through the application of extensive historical research, laborious production drawing design, and exacting construction techniques.

5. **OTHERS.** There are several other descriptive terms that are infrequently used in the assessment of furniture pieces. Although these terms are generally restricted to usage in the antique trade, they can be equally

applied to the construction of fine furniture. These other terms include: fake, interpretation, marriage, conversion, and alteration.

A. FAKE. This is furniture that is intentionally produced for the purpose of misleading the buyer. Even fake furniture can be categorized as either "poor" or "good." A poor fake is usually a replica that suffers from an attempt to pass it off as an original piece. The easiest means for correctly identifying this type of fake is through a careful examination of the wood. Poor fakes are usually constructed from contemporary woods that lack the proper degree of stress and age. Therefore, a pristine furniture piece that is being marketed as an original with a highly polished surface devoid of discoloration and lacking hardware position outlines might be a prime candidate for a fake replica.

Good fakes, on the other hand, can be harder to detect. Expert fakes are usually replicas that have been constructed from aged woods along with dated hardware and vintage escutcheons. These fakes can only be exposed through a careful examination of the furniture piece. Several key features to look for in exposing a fake are:

- Old veneer applied over new wood. This clever dodge can usually be discovered by looking for evidence of discoloration, stains, or holes that indicate the salvage of the veneer from an older, dissimilar piece.
- New decorative features or inlays added to a pedestrian carcass. The difference in aging is an easy means of fingering this form of fake.
- Inconsistent damage. In this case, specific areas or landmark portions of the furniture are irregularly damaged from cracks, stains, or splits in the wood finish. This localized damage represents a partial repair with modern veneers, stains, or fillers.
- Incorrect hardware mounting holes. Drawer handle and escutcheon mounting hardware occasionally leave telltale holes on both the outside, as well as the inside of drawers. Whereas the outside of the drawer face can be easily covered with veneer, the inside mounting holes are hidden with plugs or left exposed. In either case, this fake can be identified by examin-

ing the inside of each drawer for the incorrect placement of mounting hardware holes.

B. **INTERPRETATION.** Furniture that becomes an adaptation of a design is an interpretation of the original. Whether the piece is made from another material or built with a different construction technique, the result is an interpretation. The Grand Confort Petit was originally designed by the Swiss architect Le Corbusier in 1929 and built from tubular steel with down-filled cushions. An interpretation of The Grand Confort Petit might be constructed from PVC tubing with web straps on a foam polyurethane cushion foundation. Therefore, the interpreted design follows the same style and form as the original Le Corbusier chair, but uses a different building material.

C. **MARRIAGE.** This is large furniture, like cabinets, libraries, and chests that are the result of assembling or "marrying" several smaller units together. As with human marriage, a final furniture marriage is only as good as the individual pieces. Therefore, joining pieces with incompatible styles, wood types, or construction techniques can result in an inferior marriage. One of the most common furniture marriages is the highboy or chest-on-chest. This combination should be carefully scrutinized by the furniture buyer due to the frequency of incompatible chest-on-chest marriages. Usually, the style and wood type are similar in this highboy marriage. It is the application of unmatched moldings, escutcheons, carvings, or unbalanced proportions that spell doom for this piece.

D. **CONVERSION.** Whenever a change in function is requisite in furniture, a conversion is created. This change in function can range from converting screens or trays into tables to modifying a linen press or armoire into an entertainment center. Since most conversions require the physical modification of the piece, a precipitous decrease in the value of the furniture is likely.

E. **ALTERATION.** This is the enlargement or reduction in the size or dimensions of a furniture piece. While an alteration can be performed for practical reasons (fitting a tall bookcase into a room with a low ceiling), most alterations are executed on larger, less valuable pieces. These cosmetic alterations in height or width are attempts at increasing the value of an inexpensive behemoth by reducing its proportions and forming a diminutive counterfeit.

Historical Style

In addition to learning the descriptive terminology of period furniture, one also must gain a working knowledge of the historical furniture periods themselves. Therein lies the first problem—overcoming the misconception that a piece's period is its style.

Generally, a period is a fixed timeframe defined in relationship to our recorded history. In other words, periods are specific years that span a division or age in time. Periods in furniture are further divided into countries. This added element of confusion, for example, renders the British Late Georgian Period (circa 1760–1811) coincident with the French Louis XVI (ca. 1774–1793), Directoire (1794–1799), and Empire (ca. 1800–1815) periods. In defining these periods, only the major manufacturers of furniture are commonly represented. Therefore, in addition to England and France, Italy, Germany, and United States also share their own unique period-naming conventions.

Within the United States, there are nine generally accepted periods: Colonial (ca. 1607–1780), Federal and Empire (ca. 1780–1850), Victorian (ca. 1840–1910), Arts and Crafts (ca. 1860–1925), Art Nouveau (ca. 1888–1905), American Beaux Arts (ca. 1870–1920), Art Deco (ca. 1920–1939), Modernism (ca. 1920–1965), and Post-Modern (ca. 1950–1990). Both the naming and dating of these periods are hotly debated topics among historians.

In contrast to period, is style. Trying to properly define a style generates a contentious debate rivaling the establishment of an accurate period chronology. Similar to the license taken with the definition of the nine U.S. periods, 35 different major styles can be attributed to American furniture. Much of the scholarly work expended on the research of American furniture styles was conducted by Milo M. Naeve. Mr. Naeve worked as the Curator of American Arts at the Art Institute of Chicago where he published his treatise on identifying 35 major American furniture styles (*Identifying American Furniture: A Pictorial Guide to Styles and Terms, Colonial to Contemporary*).

There, now you know more about furniture design than you ever wanted to know. How about building or modifying the perfect Wi-Fi case good? In this case, a replica Louis XVI étagère.

Étagère

Although contrasts in shapes usually generate enough conflict to limit a style's longevity, by balancing rectangles with ovals and arcs, straight lines with tapers and flutes, and panels with moldings and beadings, the Louis XVI style was able to endure over 60 years of immense popularity. Founded by both the English and the French in 1850 with the moniker "Marie Antoinette style" applied in 1853, Louis XVI enjoyed a long life with numerous variations in workmanship, design variations, and hardwood selections.

Elaborate carvings, inlays, and marquetry designs were initially incorporated into Louis XVI furniture. Bolder designs replaced more of the craftsmanship during the waning years of this style. Likewise, while rosewood and ebony were the predominant woods in the mid-nineteenth century, walnut and black walnut replaced these exotic hardwoods in the early twentieth century. Other significant attributes of Louis XVI furniture include:

- ebonized hardwood
- egg-and-dart molding
- gilding decoration
- ormolu
- silk damask
- tapered and fluted leg

Other furniture styles evolved from the design advances sponsored by the flourishing reign of Louis XVI. Renaissance Revival and Neo-Greek sported adaptations of tapered and fluted legs, as well as the numerous wreath, urn, and lyre motifs. The result is an influence that spawned extensions of the original contrasts in shapes.

An accurate depiction of the delicate balance between complexity in design and simplicity in shape is presented in the Louis XVI étagère interpretation. Crafted from rosewood, rosewood veneer plywood, and western red cedar, this étagère stands as a fitting example of mid- to late-nineteenth century workmanship. Hardwood deviations from vintage Louis XVI, like the veneer plywood and the western red cedar serve as modern substitutes for the shelving and the carcass back, respectively. Highlights of these design tributes include rosewood columns, Doric column capitals, silk damask curtains, gold leaf edging, and bun feet.

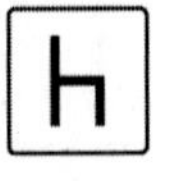

Keeping in mind this goal of interpreting the Louis XVI style, special construction techniques were employed in the preparation and building of the étagère. The rosewood columns, for example, were turned on a lathe with rings added to both ends. A router with a beading bit is then used for adding the repetitive beads that run the length of each column. This beading procedure requires the use of a jig for securely holding the turned column and an adjustable track running parallel to the column. The track is used for holding and guiding the router as it is moved along the length of each column. One pass down the track and the column is rotated for accommodating another router pass. Each pass should make another bead directly adjacent to the previous one.

Another candidate for turning is the bun foot. Four bun feet are needed for the étagère. These characteristic feet are shaped by turning on a lathe similar to the procedure used for the rosewood columns. The main difference between the two turnings, however, is that all four bun feet are turned from one piece of stock. Conversely, the four columns are turned from individual stock. At the conclusion of the shaping, the four feet are cut from the stock.

Enclosing the lower half of the étagère are two curtained doors. These doors are shielded by two pleated silk damask curtains. The curtains serve as barriers for obscuring the contents behind the doors, yet provide a lightness that is unavailable from typical hardwood panels. Each door is framed with rosewood muntins that act as support for each curtain's span. Gold leaf edging is used as a highlight for each door's frame. No pattern other than linear strokes is used for this highlighting.

Gold leaf is also added to the rosewood Doric column capitals. Patent gold leaf is applied as an accent to the carved detail of each capital. The gilded accent for each column is a precise region that is defined by the small bead ringing the underside of each capital. One particularly nasty problem that rears its ugly head during the gilding of a curved ring is placing the gold leaf into the valley that divides the bead from the capital. This task is best accomplished through a piecemeal placement of gold leaf fragments that are manipulated gingerly on the bristles of a small squirrel-hair brush. Following the placement of these fragments, they must be collectively burnished with compressed, absorbent cotton balls. A diligent burnishing of these column capital beadings will ensure that the gold leaf will not flake off during the final varnishing.

THE PLANS FOR BUILDING AN ÉTAGÈRE

1 **Étagère front.**

2 **Étagère back.**

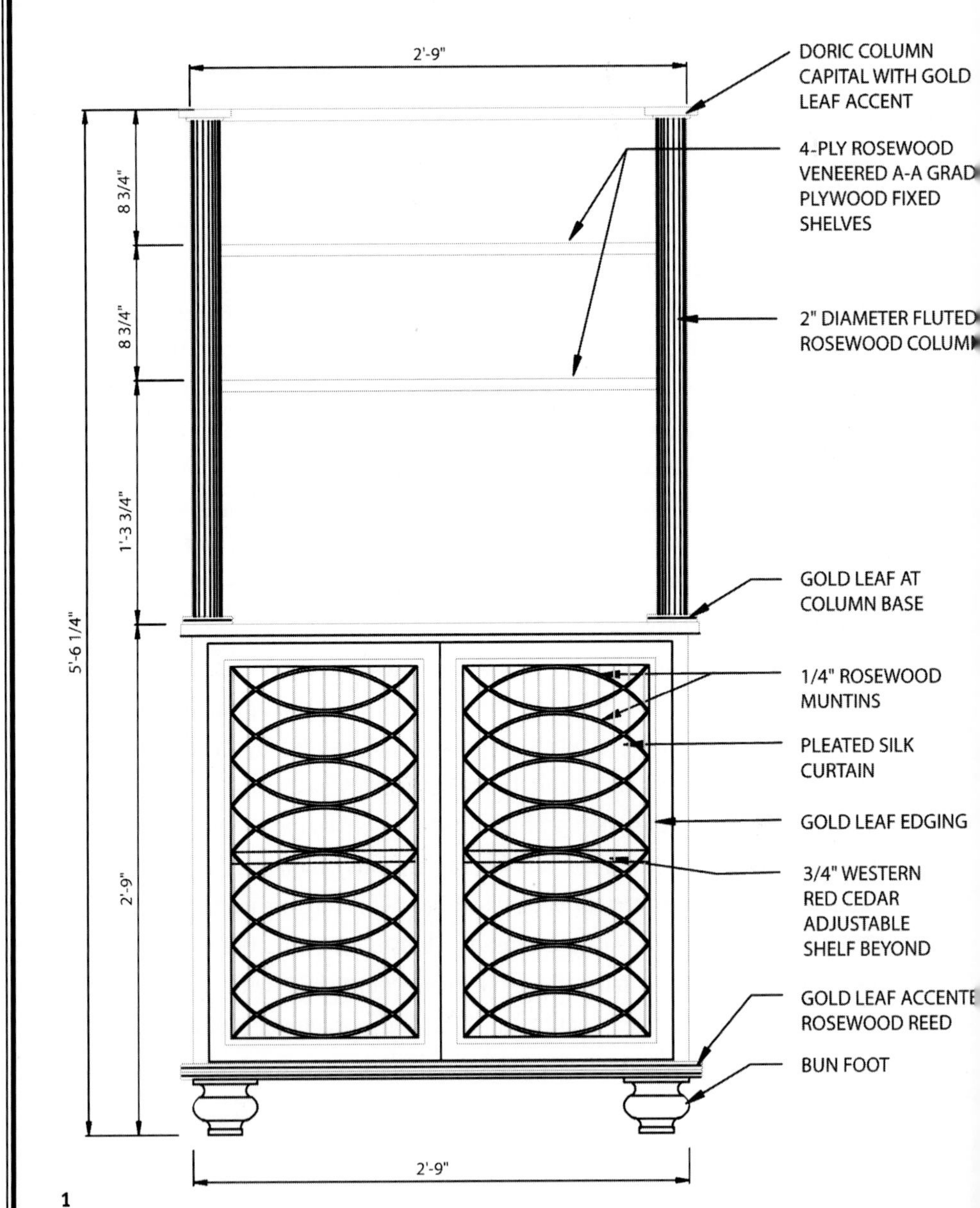

1

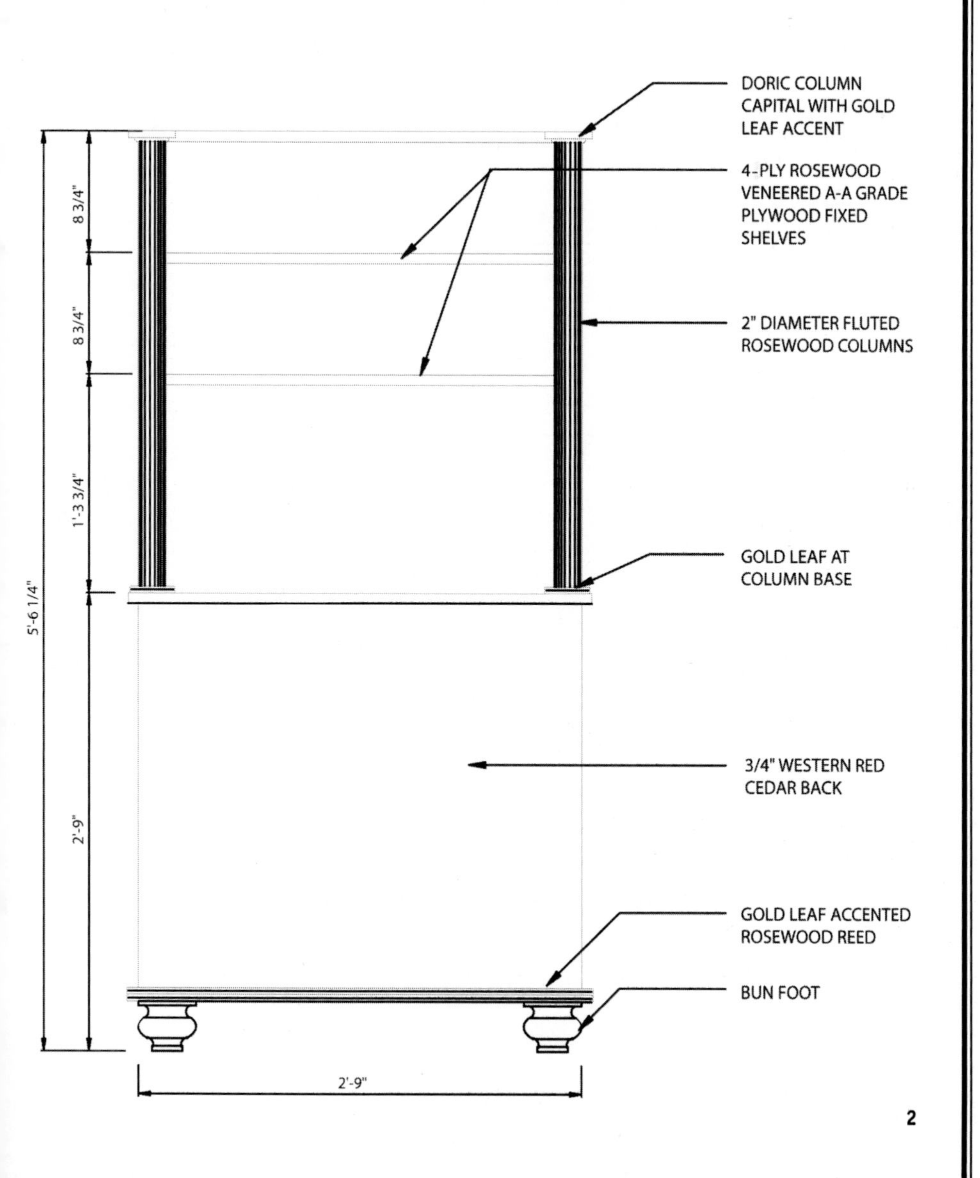
DORIC COLUMN CAPITAL WITH GOLD LEAF ACCENT
4-PLY ROSEWOOD VENEERED A-A GRADE PLYWOOD FIXED SHELVES
2" DIAMETER FLUTED ROSEWOOD COLUMNS
GOLD LEAF AT COLUMN BASE
3/4" WESTERN RED CEDAR BACK
GOLD LEAF ACCENTED ROSEWOOD REED
BUN FOOT
8 3/4"
8 3/4"
1'-3 3/4"
2'-9"
5'-6 1/4"
2'-9"

2

Get da Case on Deez Goods

3 Étagère side.

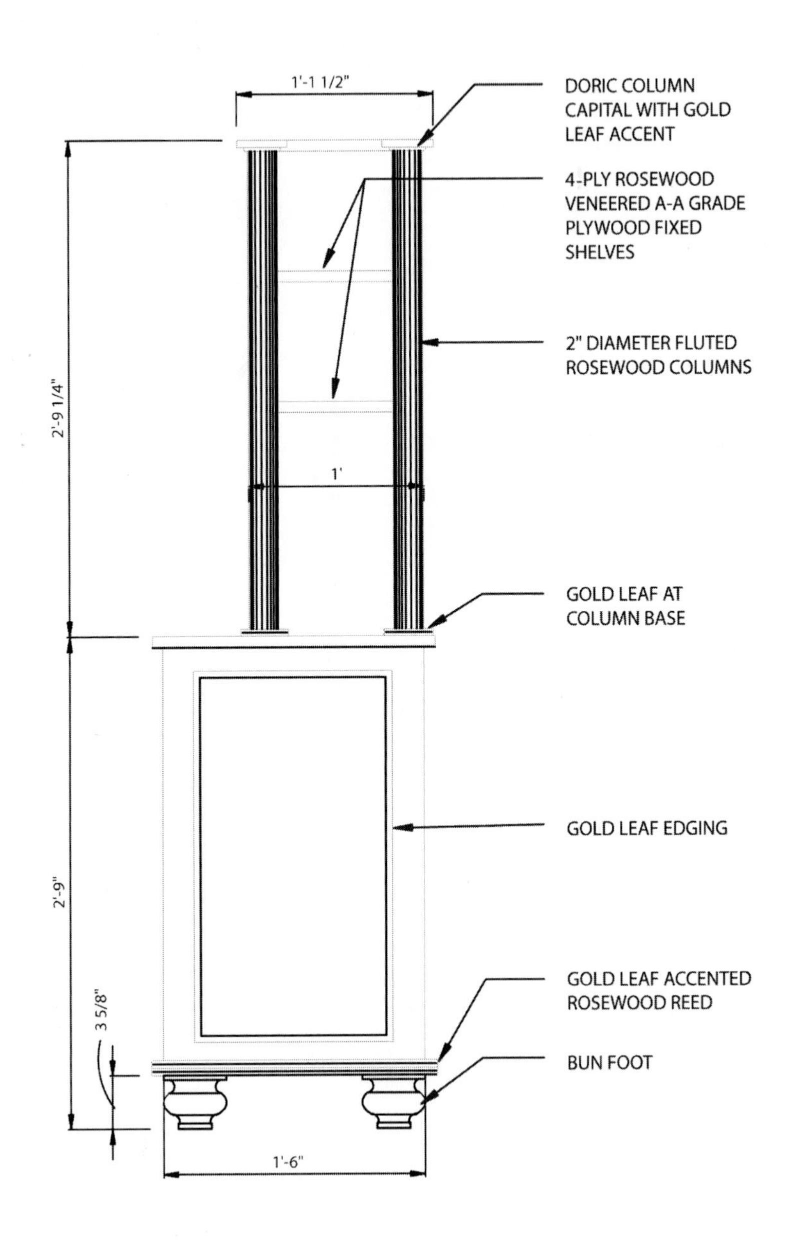
1'-1 1/2"
DORIC COLUMN CAPITAL WITH GOLD LEAF ACCENT
4-PLY ROSEWOOD VENEERED A-A GRADE PLYWOOD FIXED SHELVES
2" DIAMETER FLUTED ROSEWOOD COLUMNS
2'-9 1/4"
1'
GOLD LEAF AT COLUMN BASE
GOLD LEAF EDGING
2'-9"
GOLD LEAF ACCENTED ROSEWOOD REED
3 5/8"
BUN FOOT
1'-6"

3

h

HOW TO LIGHT UP YOUR LIFE

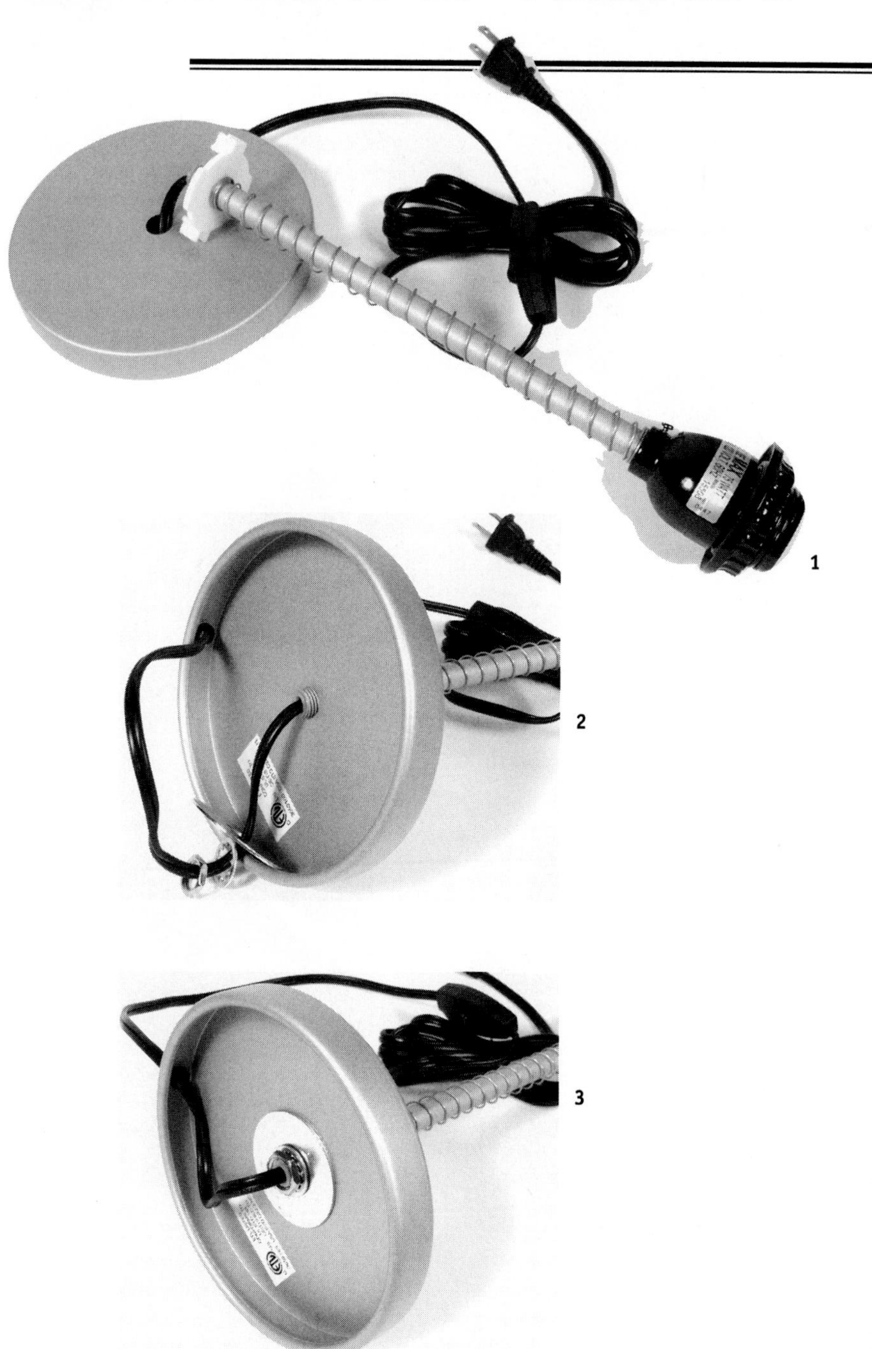

1 The *Storm* tabletop lamp from IKEA.

2 Turn the base over.

3 Reduce the slack in the electrical cord.

4 Fix the stem to the base with a crescent wrench.

5 Remove the lamp shade retaining ring.

6 Add the lamp shade frame and thread the retaining ring back into place.

7 Thread the lamp shade stem into the top of the frame.

8 Remove the stem cap.

9 Expand the lamp shade over the frame, fix it under the lamp shade hold-down on the base, and reattach the stem cap.

4

5

6

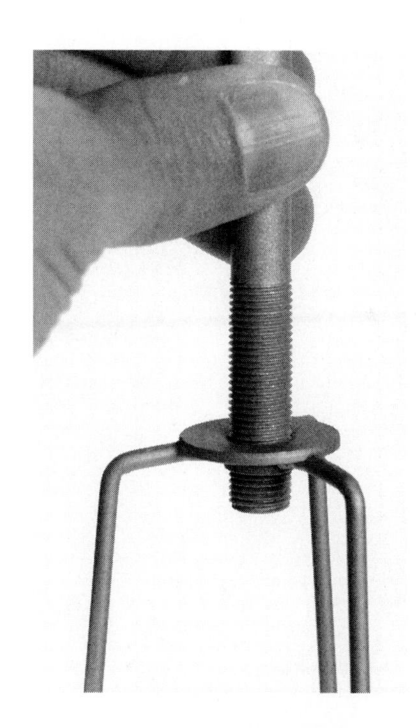

7

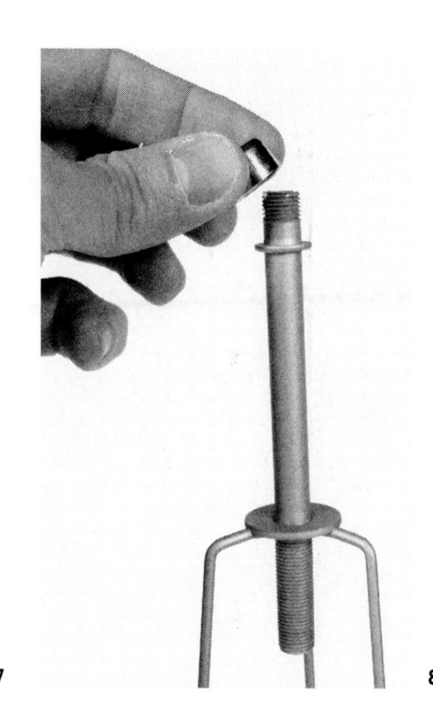

8

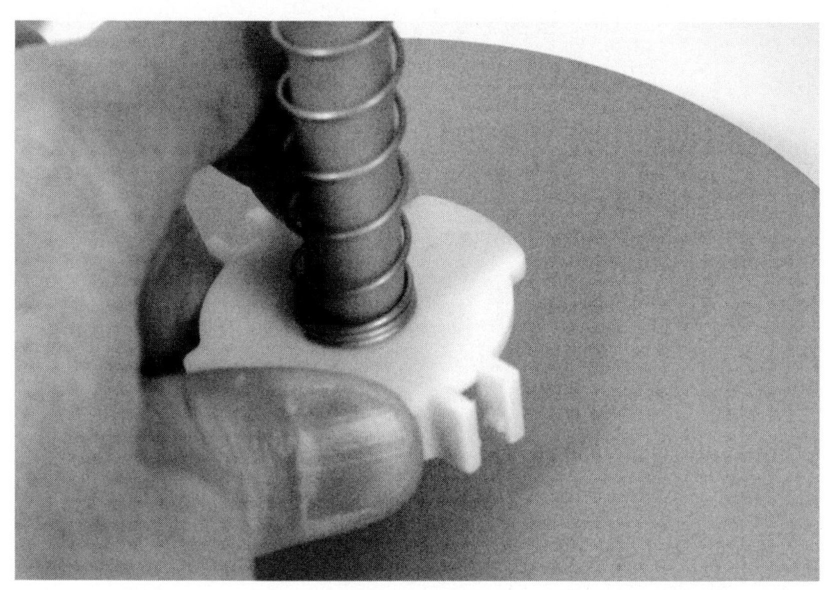

9

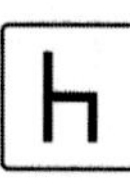

10 See the light.

10

CHAPTER 25

Red Flag, Top Gun

If there is one aircraft design that has become representative of all of the aerial warplanes of the First World War, that aeroplane has to be the Fokker Dr.I triplane. Forever linked with the legendary Manfred von Richthofen and his "Flying Circus," even though less than a scant 23 percent of his aerial kills were achieved at the controls of a Fokker triplane, this odd three-winged fighter actually began its short-lived dog-fighting career as a limited-production prototype. Three models of this initial triplane prototype were built by Fokker. They were designated F.I 101/17, F.I 102/17, and F.I 103/17.

Feeling that he had designed the ultimate fighter-pilot's fighter, Fokker personally flew and presented prototype F.I 102/17 to the "Red Baron" himself, Manfred von Richthofen on August 28, 1917. As for the other two prototypes—the first prototype (F.I 101/17) stayed at the Fokker factory for performance tests, while the third, and final, prototype (F.I 103/17) was presented to Richthofen's closest competitor, Werner Voss. Both of these presentations to Richthofen and Voss would prove to be premature.

In less than 30 days, both of the prototypes had been shot down and their pilots killed. Even though the second prototype (F.I 102/17) was an aircraft that had been presented to Richthofen, he wasn't flying the triplane on the fateful day in early September. He was on leave. Rather, his acting *staffel* commander, Kurt Wolff, was at the controls when he was shot down by a Sopwith F.1 Camel.

A similar fate awaited Werner Voss. Functioning as the *staffel* commander of Jasta 10, Voss met his match against a formation of SE.5as. Thus, in less than one month all three of the Fokker triplane prototypes had been destroyed. (Note: Undocumented reports suggest that the original prototype, F.I 101/17, was destroyed during stress tests at the Fokker factory.) This inauspicious beginning, however, didn't dissuade the

German Army Air Service from ordering 320 Fokker triplanes, newly designated Dr.I and featuring a reshaped horizontal stabilizer and wingtip skids. Even a series of deadly crashes with brand new Dr.Is which were attributed to a total upper wing failure didn't tarnish either the air service commander's or its pilot's favorable opinion of this nimble aerobatic fighter.

Is That A Peanut in Your Pocket?

If you've never built a Pistachio scale aircraft, now is the time to start with this model of the Fokker F.I triplane. Spanning 8 inches from wingtip to wingtip, this petite Pistachio can be built from 1/16- and 1/32-inch balsa sheet, bond paper, and tissue. These materials can be ordered from suppliers like Peck-Polymers. Likewise, the plan contains enough optional equipment (e.g., twin machine guns, wheel cover, cockpit seat) that you can build a super-detailed triplane that will fly as well as it looks.

Begin the construction of the Pistachio Fokker by cutting out all of the solid balsa sheet pieces and the 52 wing ribs. While most of these solid pieces, like the wing tips and fuselage formers, are fairly straightforward in their shape, the wing ribs can be a tricky exercise in delicate cutting. Therefore, I would recommend that you make a rib template from 1/32-inch plywood. Make this template so that it follows the upper line of the plan's rib shape.

Unlike typical model-building practices, this Fokker's fuselage is built from full cross-section formers. In other words, you don't need to build half of the fuselage and then flip it over for attaching its second half. By using these full formers, you can assemble the complete fuselage in a matter of minutes.

For powering your triplane, make two loose loops of 1/16-inch rubber and knot the stray ends together. Slip one end of these loops over the rear motor-holding shaft and the other ends onto the main propeller's shaft. Remember, this step must be done prior to covering the Fokker.

The coloring and markings are added to this Pistachio design's tissue *before* the tissue is added to the model. Any technique for coloring the tissue can be used—chalk, colored tissue, decals, ink, markers, and, of course, paint. Regardless of the technique, make sure that the tissue is smooth and flat prior to attaching to the model's framework. Use a fine-tipped ink marker for adding the control surface outlines. These are the dashed line markings that are on the plan. Finally, use either a glue stick or diluted white glue for attaching the tissue to the Fokker model.

There are three exciting paint schemes that can be applied to the Fokker F.I. The one illustrated (Fig. 3 of Fokker Spread) is the standard "factory finish" that was carried by F.I 101/17. In order to obtain the natural metal finish for the triplane's nose, silver leaf was glued to the sanded balsa cowling cover. Silver leaf is a microthin metal coating that can be obtained from most large hobby/craft stores.

The markings for the other two triplane prototypes feature either bright-red tissue coverings (e.g., F.I 102/17) or a stunning chrome yellow nose (e.g., F.I 103/17). Furthermore, don't hesitate to embellish your Pistachio Fokker with the armaments, motor cylinders, wheel covers, wire foot step, and handhold handle accoutrements that will transform the ordinary into the extraordinary.

Add the balsa propeller to the cowling's thrust bearing. Make sure that the spinning propeller will clear the ground. Now attach the elevator and rudder to the fuselage. Make sure that both control surfaces are level and square to the fuselage.

The wings are attached to the fuselage beginning with the bottom wing. Spend some extra time to align these wings perfectly with the fuselage and embed the main spar into former F6. Now run each 1/32-inch interplane strut through the bottom wings. The remaining wings will slide into place on these struts. Similarly, the middle wings are glued to the interplane struts, as well as to the top forward decking of the fuselage. Lastly, the top wing is delicately slid onto the struts and glued into place. Four cabane struts are carefully slid into position between the forward top fuselage decking and the upper wing. All three of these ribs should mate tightly and hold the 1 1/2-inch dihedral that was built into each wing's root rib.

Your triplane is best trimmed for a gentle climbing left spiral. This effect can be achieved by adding a small amount of right rudder and installing a trim tab along the trailing edge of the left wing. Bend this tab *down*. Likewise, you can add another trim tab to the right wing; this tab should be bent *up*. Also, make sure that the model balances *exactly* over the lower wing's main spar.

Launch the model into the wind or attempt to raise it from the ground. If everything is trimmed and balanced correctly, the little triplane should quickly climb upward with a slight turn to the left and then, when the motor stops, drop like a rock back to the ground. Make as many test flights as needed to get all adjustments perfect. Add approximately 200 to 300 turns for maximum flight performance. A flight duration of at least 30 seconds should be possible under calm wind conditions.

Bird of a Different Color

As the Allied air forces fought the Nazis' Luftwaffe for aerial supremacy in the skies over Europe, a small group of U.S. Army Air Corps airmen fought an additional enemy—racial segregation. These pilots, ground support, and maintenance personnel were African-Americans who, while expressing their passion for patriotism and democracy, were forced by the color of their skin to train in a segregated air base near Tuskegee, Alabama. These were the Tuskegee Airmen.

Enduring "second-rate" flight instruction and racially motivated verbal abuse, 926 Tuskegee Airmen completed their training and were organized into the 99th Fighter Squadron. Ironically, the 99th FS was ordered into combat on April 1, 1943—April Fool's Day. Operating under the auspices of the 15th Air Force, the 99th FS arrived in Tunisia, North Africa and began ground attack operations near Sicily. The first "kill" for the Tuskegee Airmen was scored by Lt. Charles Hall in a Curtiss P-40 against a Focke-Wulf FW-190 on July 2, 1943.

A switch in operations for the 99th FS occurred in January 1944, when 15th Air Force Commander General Ira Eaker reorganized the 99th FS into the 332nd Fighter Group and ordered the group's commander, Colonel Benjamin O. Davis, to change his pilots' mission to bomber escort duties. There was a caveat with this order, however. Davis's pilots were required to protect the bombers at "all costs." Thus was born the mystic of the 332nd FG. During 15,553 sorties and 1,578 missions, the 332nd FG never lost a single Allied bomber to enemy fighters. In fact, based on their easily recognizable red-painted rudders and stabilizers, the "Red Tails" of the 332nd became known as the "Red-Tailed Angels" to bomber crews of the 15th AF.

Despite their impressive combat record, the "Red Tails," as well as the other 15th AF fighter groups, were typically last in line for aircraft replacements. For example, when 8th AF fighter groups were supplied with North American P-51D Mustangs, the "cast-off" P-51B and P-51C models were transferred to the pilots of the 332nd FG. Therefore, the impressive P-51D model didn't make strong inroads into the 15th AF fighter ranks until early 1945. Not surprisingly, it was after the appearance of the D model that on one of the longest escort missions of the war, 59 "Red Tail" pilots engaged Messerschmitt Me-262 jet fighters attacking their B-17 bomber formations and shot down three of them.

These extraordinary exploits of the Tuskegee Airmen are captured in this free flight model of a P-51B/C. Designed for either electric or rubber power, this model displays the distinctive red tail of the Tuskegee Airmen proudly and

A FIELD GUIDE TO THE IDENTIFICATION OF STARFIGHTERS

Use this guide to help identify a Starfighter:

1. No ventral fin 2
   Ventral fin present 3
2. No sonic shock cone XF-104
   Sonic shock cone YF-104A
3. Two-seat cockpit 4
   Single-seat cockpit 5
4. No clear connecting panel between canopies............... F-104B
   Clear connecting panel between canopies F-104D
5. Narrow chord fin and rudder (see Fig. 25-1) 6
   Wide chord fin and rudder F-104G
6. No arrestor hook.. F-104C
   Arrestor hook present.. F-104A

OTHER VARIANTS:

CF-104—deleted M61 cannon and added internal avionics with various radar system differences

F-104J—export variant with license agreement through Mitsubishi

F-104S—joint Lockheed/Italian variant with two ventral sub-fins added to each side of fuselage

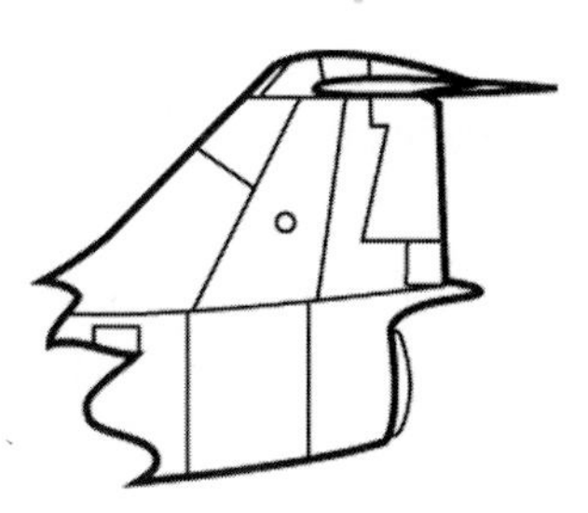

25-1 Narrow chord fin and rudder of F-104A and F-104C.

flies as safely and confidently as the 15th AF bomber crews who were escorted by these remarkable pilots.

This P-51B/C flies great, whether you're using either electric or rubber power. As a general guideline, the rubber-powered Mustang will require four strands of 3/16-inch rubber approximately 12 to 13 inches long. While a 6-inch propeller will work, a 9 1/2-inch propeller is much better for ensuring a slower, stable flight characteristic. In fact, coupling a right rudder deflection of about 3/16 inch with a slight amount of right thrust will produce a smooth right-hand turn that is perfect for the initial climb of the free flight profile.

On the other hand, if you are going to install electric power for this P-51B/C, a KP01 will provide sufficient power for a 3- to 4-minute flight profile. Combining a good charge with a solid hand-tossed launch will result in impressive altitudes of 30 to 50 feet off the ground. Be forewarned, however, this P-51B/C doesn't have a strong glide angle and altitudes above 40 feet can result in both a hard and destructive landing. One way of minimizing this damaging possibility is through the use of a lightweight micro R/C system for controlling the rudder and electric motor. With this control system, you can gradually turn back toward your launch point, throttle the motor back, and save enough current for making a powered landing.

Granted, the P-51 Mustang is a rather "overworked" subject, but the P-51B/C model is only occasionally seen and it is rarely built as a free-flight design. Furthermore, this Mustang design is extremely easy to build. There are only two major construction components: fuselage and wing. Throw in the rudder and elevator and you can build this Mustang in less than two weeks' time.

Prior to beginning construction of the P-51B/C, you should transfer all keel, former, and rib patterns onto high-quality lightweight 1/16-inch balsa sheet wood. Now cut out, label, and set these pieces aside until needed. Now, it becomes a simple process of building the two main components that comprise this red-tailed Mustang.

Unlike typical model-building practices, this P-51B/C 's fuselage is built from full cross-section formers. In other words, you don't need to build half of the fuselage and then flip it over for attaching its second half. By using these full formers, you can assemble the complete fuselage in a matter of minutes.

Simply take keels K1, K2, K3, and K4, lay them over the plan, and mark the locations for formers F1 through F9. Repeat this layout and marking procedure with keels K5 and K6. You now have a set of templates for attaching and aligning each of the fuselage formers.

Rather than attempting to glue the formers directly into place on these keel templates, each former should be held in place with tape. Just square and align each former with its respective keels and then set them into place.

After the glue on the fuselage has dried, you begin adding 1/16-inch square stringers between each former. There are no notches provided on the plan for the placement of these stringers. Therefore, you can "notch as you go" and add the number of stringers to your P-51B/C per your building tastes. Remember, build light.

There are various points where scale refinements can be made. Adding gun ports that are made from rolled paper tubes, exhaust stacks formed from discarded drinking straws, and forming a carburetor air intake scoop from balsa will make this P-51B/C look realistic in its upcoming flight testing.

Painting the Tuskegee Airmen

The flamboyant red rudder and stabilizer of the Tuskegee Airmen's 332nd FG Mustangs makes for an attractive color scheme. Even better, the bulk of these aircrafts' camouflage can be duplicated with colored tissue and, therefore, reduce weight in the model.

The P-51B/C subject chosen for this model was the mount of Lt. Charles Bailey of the 99th FS. Nicknamed "My Buddy," Bailey's P-51B/C featured a checked nose, along with yellow theater wing bands. Like many of the 332nd FG's aircraft, the trim tab on Bailey's Mustang was painted with the squadron's color (i.e., insignia blue for the 99th FS).

By using light-gray tissue for the bulk of Bailey's Mustang, you will just have to add yellow tissue stripes for the theater bands, red tissue for the rudder and elevator, and black tissue for the anti-glare panel between the propeller and the canopy. Then with just a modest application of paint and decals you can finish this war bird into an authentic, scale 332nd FG trophy.

Trimming and flying this P-51B/C is actually easier with the rubber-powered version versus the electric one. Right off the building board, this Mustang was perfectly balanced for rubber flight. As a precaution, however, the rudder was built as an articulated variety. In other words, the rudder was divided into two pieces on the building board with spar 1/16-inch stock used for adding additional framing and strength. After removal from the building board, two small holes were drilled and mated into both the fixed vertical stabilizer and the now-movable rudder. Two small lengths of copper electric wire were used for linking the rudder to the rigid vertical stabilizer. Now the P-51B/C had a working rudder for helping with flight trimming.

For the initial flight tests, crank about 100 to 150 turns into the Mustang's rubber motor. Coupled with a gentle forward toss, the P-51B/C should glide smoothly in a flat, slightly right-turn glide. While the wing area of this P-51B/C is adequate for powered flight, nonpowered glides can be short and direct. Therefore, always try to fly over a grassy field rather than from harder surfaces. Once the initial test flight and glide have been successfully completed, start to add more turns to the rubber motor. As a suitable break-in strategy (no pun intended) consider this plan for adding increased thrust to each test: Begin with 200 to 300 turns, next try 400 to 500 turns, and finally increase the

FAST STARFIGHTER FACTS

- Famed test pilot Charles "Chuck" Yeager was almost killed in the crash of an NF-104A equipped with a special rocket motor.
- The original F-104A suffered from a "super stall" condition where the aircraft remains pitched upward in a stalled attitude.
- On September 20, 1965 USAF Major Philip Smith was shot down in his F-104C over Hainan Island, China by Chinese F-6 fighters.
- Starfighters flew 2,269 combat sorties in Southeast Asia during the Vietnam War.
- Both F-104As and F-104Cs had downward-ejection escape systems for the pilot.
- Starfighters were deployed during the Quemoy/Matsu Crisis of 1958, the Berlin Crisis of 1961, and the Cuban Missile Crisis of 1962.
- Pakistani F-104As were involved in dogfights with various Indian Air Force (IAF) aircraft between 1965 and 1971. The final count: Starfighters 6 and IAF MiGs 3 (probable 4). Ironically, none of the F-104 kills was a MiG.

motor's windings to 500 to 800. At this last stage of testing, you can easily achieve 1- to 2-minute flights with altitudes of 70 feet in an active thermal.

When flying with electric power, however, you will need to make some critical adjustments during the construction phase of building this P-51B/C. First, depending on the weight of the selected motor, the model will be nose heavy. Even a KP01 will require a substantial amount of counterbalance. One of the best methods for achieving this compensation for the motor's excessive amount of nose weight is by moving the 3 × 50-mAh battery pack (or, 3 × 110 mAh for greater power and motor run duration), switch, and wiring as far back as possible. As a general rule of thumb, any compartment behind former F6 would be a great location for installing these counterbalancing elements. Finally, a small pellet of lead shot *might* have to be glued to the inside of former F9 for achieving perfect balance.

Unlike the initial flight testing phase of the rubber-powered P-51B/C, the electric version can be simpler to trim due to the predictable results of rechargeable battery power. Similar to the rubber-powered testing, however, begin the electric flight testing with less than a full battery charge—say a 30-second charge. After a stable climb, cruise, and glide have been achieved, increase the battery charge up to 3 minutes. This should result in a 1-minute motor run with good flight performance.

Either power system will yield a great visual treat for everyone at your local flying field. And the red tail feathers of the P-51B/C serve as a tribute to the courageous pilots of the 332nd FG who proved that patriotism, democracy, and freedom should be color blind.

If you're looking for an airframe that's a bit more versatile, then you might want to consider the Air Hogs®™ Hydro Freak™ (see Fig. 25-2).

25-2 Air Hogs®™ Hydro Freak.™ Why take three remote control models to the park when the Air Hogs Hydro Freak is three-in-one? This tremendous toy will have your friends open-mouthed as it takes off from the pond to soar into the sky. When you've finished your aerobatics you can get back down to earth for some high-speed land action. A whole lot of fun for everyone.
Ages: 14+
MSRP: $99.99
Launch: Fall 2006
(Illustration courtesy of Spin Master Ltd.)

Jet Age Jet Set

Any youngster (or oldster) growing up near a NATO (North Atlantic Treaty Organization) airbase during the 1960s had the opportunity to watch flights of the first supersonic "fighter pilot's aircraft." Born from a desire to compete with the MiG-15 in a strictly defined dogfight role, the Lockheed F-104 Starfighter quickly became a test of both a pilot's flying skill and NATO's fiscal budget.

Designed by famed Lockheed engineer Clarence L. "Kelly" Johnson, the Starfighter looked more like a research missile than an air superiority fighter. In fact, some West German Luftwaffe pilots probably thought they were "flying" a missile when over 30 F-104G crashes occurred in 1965 alone. Unfortunately, this type of excessive attrition forced many NATO air forces to begin replacing the temperamental Starfighter with RF-4Es, F-16As, and Tornados. Oddly enough, this Starfighter replacement process took some countries approximately 20 years to complete. While this longevity in NATO service might seem contradictory to the urgency of relieving the crash-prone F-104 from service, the Starfighter was able to provide European air forces with a high-performance, quick reaction point-to-point "fast mover" interceptor that could rapidly counter any Warsaw Pact incursion.

Based on the NATO service record of the F-104G, it's tough to imagine adapting the F-104G to a flying scale model. Even more difficult to imagine is harnessing a realistic power plant inside the narrow-diameter Starfighter fuselage. Ironically, it's the ingenuity of a former Warsaw Pact nation that makes it possible for once again flying the Starfighter through the skies of Europe.

Rocket Power

A small Czechoslovakian disposable rocket engine marketed as Rapier is the ideal power plant for making this scale model of the F-104G fly with the same flight characteristics as its full-size namesake. Manufactured in four levels of thrust (L-1 through L-4), the Rapier is a solid-fuel rocket motor that is started with a fuse. Once lit, the Rapier quickly builds up thrust for a burn time duration of 20 to 25 seconds.

Following the conventional design scheme for Rapier-powered scale models, the Starfighter is equipped with a motor trough. In other words, the rocket motor hangs from the bottom of the aircraft model inside a hollowed out cavity that is obscured by the fuselage's side profile. This design element facilitates the rapid insertion and subsequent exchange of Rapier motors before and after each flight.

Although somewhat of a departure from true scale, the trough is an elegant option to the alternative of using intake/thrust augmentation tubing and mounting the rocket motor inside the F-104's fuselage. In this case, access to the motor would necessitate the building of a removable hatch. Likewise, adequate cooling for the motor during flight would be paramount for preventing damage to the aircraft.

While the design of this Starfighter is predicated on the installation of a Rapier L-2 rocket motor, reusable Jetex motors can be substituted for the Czech-made motor. One benefit from electing to use Jetex motors is that you won't need to construct a motor mount. This paper and wire container is essential for holding Rapier motors during flight. Conversely, the Jetex motor is equipped with a metal clip that is capable to holding the motor in flight. Regardless of whether you use Rapier or Jetex power, you still will need to incorporate the motor trough system into your Starfighter. Likewise, make sure that you strive to balance your Jetex-powered F-104 at the same center of gravity point as indicated on the plans for the Rapier L-2 motor.

In addition to the motor trough, there is one other "slight" deviation from true scale: wing dihedral. You see, the F-104 was the first USAF supersonic jet fighter that was designed with an extreme negative wing dihedral angle. And, while I was able to design and successfully fly a Starfighter prototype model with negative dihedral, I deemed it too unstable for publication. So I opted for a slight positive wing dihedral angle (3/8-inch per wing) along with an increase in the wing's overall airfoil shape. This modification greatly enhanced the Starfighter's stability and reliability in flight without sacrificing too much scale detail.

This F-104G Starfighter is extremely easy to build. There are only three major construction components: fuselage, wing, and wingtip tanks. Prior to beginning construction of the Starfighter, you should transfer all keel, former, and rib patterns, and the rudder and elevator outlines onto high-quality 1/16-inch balsa sheet wood. Now cut out, label, and set these pieces aside until needed. With that drudgery out of the way, it is now a simple process of building the three components that comprise this Starfighter. Let's begin with the fuselage.

Unlike typical model-building practices, this F-104G's fuselage is built from full cross-section formers. In other words, you don't need to build half of the fuselage and then flip it over for attaching its second half. By using these full formers, you can assemble the complete fuselage in a matter of minutes. Here's how:

The fuselage itself is divided into two subassemblies: the nose subassembly and the main fuselage subassembly. Beginning with the main fuselage subassembly, take keel K3 and lay it over the plan. Now mark the locations for formers F4 through F8. Repeat this layout and marking procedure with keels K5, K6, and K7. You now have a set of templates for attaching and aligning each of the main fuselage subassembly's formers.

Rather than attempting to glue the formers directly into place within this set of keel templates, each former should be held in place with tape. Either cellophane or masking tape will work for temporarily holding the formers in place. Now square and align each former with the keels. When you're satisfied with the integrity of this subassembly, glue each former into place.

Prior to lining the trough, you should build and install the Rapier motor mount. This paper tube is glued to the bottom of former F6 and the back of former F5. Use 1/32-inch balsa sheet to "frame" a motor mount housing, if desired. Additionally, make sure that the wire-retaining clip on the motor mount is facing toward the bottom of the trough before you glue the mount into place. You can now finish the building of the fuselage by adding the solid balsa nose and rudder supports. Now on to the wing construction....

Life as a Tuskegee Airman

In today's climate of emerging racial equality, it's tough to imagine a government who would treat their veterans with the disdain that the United States displayed toward the returning Tuskegee Airmen. First, members of the 332nd FG weren't rotated stateside after 50 missions. Rather they languished throughout the Mediterranean Theater of Operations (MTO) until the conclusion of the war. Many of the pilots, like Lt. Bailey, flew over 130 missions.

Secondly, following the war, officers from the 332nd FG were denied entrance into officers' clubs, whereas German POW officers were granted access. Furthermore, white members from other 15th AF fighter groups regressed to discriminatory practices after they returned to the States. Finally, President Harry Truman attempted to bridge this divide in July 1948 by issuing Executive Order No. 9981 mandating equality of treatment and opportunity for all members of the U.S. armed forces. Sadly, it took the rest of the country over 10 more years to adopt this progressive attitude.

For such a large, high-performance aircraft, the F-104G had a remarkably short wing span—less than 22 feet. Similarly, the plan for the F-104G Starfighter results in a model with a wing span over 6 inches! Therefore, you must ensure that the final wing is square to the fuselage and its airfoil shape is true to the plan.

Be forewarned, the construction of the Starfighter 's two wingtip fuel drop tanks is a labor of love...and determination. The 1/32-inch formers and 1/32-inch square stringers are difficult to cut, manipulate, and glue into place. If you're unable to successfully master the construction of these tanks, you can opt for converting the F-104G into an F-104A. If you elect to go this route, make sure that you narrow the rudder's chord accordingly.

Neat-O NATO

Now that all of the major components of the Starfighter are completed, it's time to sand all surfaces and cover them with tissue. Even the solid balsa rudder/elevator combination should be covered with tissue. This uniformity in coverage will guarantee that the finish on the final F-104G is both smooth and consistent—resulting in a better aerodynamic aircraft.

There are four areas that demand some extra attention prior to tissue covering. First, the motor mount area must be lined with bond paper and then covered with a heavy-duty heat-resistant aluminum foil. Both the paper and the final foil layer must be applied as single, continuous sheets.

Next, a small paper ring should be added to the engine exhaust opening located in former F9. This ring simulates the thrust augmentation system that was employed on the F-104G. Make sure to paint the inside of the ring black and mark the outside with a series of parallel lines.

Likewise, the third area that should be dealt with before applying the tissue is the definition of the two engine air intake openings. From some scrap balsa, fashion two shock cones. These cones will be glued "inside" each air intake *after* the tissue has been applied to the fuselage. Furthermore, a thin ring of bond paper is glued to the front lip (left and right) of former F5. This paper ring adds a shadow line to the outline of each air intake and makes them look more realistic.

Finally, the last finishing touch that needs to be addressed is the creation of the Starfighter's canopy. The best way for creating this canopy is by carving a balsa replica and then pushing it into a sheet of warm, soft acetate. In order to hold the heated acetate, you should build a plywood frame and staple the sheet of acetate to the frame. Then, after heating the acetate/frame combination, push the balsa canopy form into the soft acetate. The result should be a nice, tight, clear canopy for painting and mounting over the Starfighter's cockpit.

Now that all of the components have been covered with tissue, it's time for the fun part—assembling the F-104G. First of all, glue the elevator to the rudder; be sure that the elevator is absolutely and precisely square with the rudder. Attention to this detail at this stage in assembly will save you countless headaches during the initial flight trials...and could even prevent your precious Starfighter from crashing. Next slide the rudder/elevator combination into the slot at the rear of the fuselage that is created by the rudder supports. Once again, square everything up and glue it home.

All that remains for completing the F-104G is the attachment of the wings to the fuselage. This is one of the more difficult steps in assembling the Starfighter. Delicately slice the tissue on the fuselage for receiving the wing mounts W4 and W5. Each wing panel should be glued to the underside of keel K7 and then connected to each other via the wing dihedral mounts W4 and W5. Be sure that a 3/8-inch dihedral is still maintained at each wing tip. Finally, add the wingtip fuel tanks to rib W3 on each wing. The tanks should hang a little low. Also, make sure that the stabilizing fins of the drop tanks are aligned with the plane of the wing and *not* the rudder/elevator. You are now ready to paint your Starfighter.

The color scheme used for the F-104G depicted in Fig. 1 of F-104 Spread represents a West German Navy Starfighter with the Marinefliegergeschwader 2 (MFG-2) used in an anti-shipping maritime strike role. In this scheme, the upper surfaces were a dark-gray, while the lower surfaces were a light-gray. Additional markings worth noting are the Dayglo Orange bands wrapped around each wingtip tank. In theory, these markings were to have been deleted during wartime, but served as safety indicators during normal day-to-day operations.

Taming the Starfighter

Flying the F-104G Starfighter is anything but simple. This is an advanced rocket-powered model that is best flown by only experienced modelers. Case in point was the first two flights of the Starfighter prototype model.

The prototype actually attempted to fly with the negative wing dihedral angle that was found on the full-size F-104G. In the initial flight, the Starfighter leapt out of my hand and streaked skyward, reminiscent of an actual F-104. The second flight, however, proved to be the model's undoing, as it dove full-power into the ground.

Adding the 3/8-inch wing dihedral and trimming the F-104G with movable stabilizer fins on the wingtip drop tanks helped tame this Starfighter. It still requires a firm and solid launch toss, as well as a Rapier L-2 motor. Basically, just slide a new L-2 motor into place, put the fuse loosely inside the motor's exhaust vent, light the fuse, wait for thrust to build to maximum power (usually 2 to 3 seconds), and fling the Starfighter with a strong, steady, straight-ahead toss. Then sit back and watch as your F-104G attempts to set an altitude record for Rapier-powered aircraft. Due to the heavy wing loading of this Starfighter, after the rocket motor exhausts its fuel, the return glide back to Earth can be extremely fast. Therefore, make sure that this needle-nosed "fighter pilot's aircraft" doesn't skewer either you or your neighbor. Otherwise, you might end your Starfighter's career like 30 of its real-life West German counterparts did in 1965—as a smokin' hole in the ground.

HOW TO BUILD A FOKKER F.I RUBBER-POWERED MODEL

1 The Fokker F.I triplane prototype makes a beautiful Pistachio scale model.

2 A large number of scale features (e.g., guns, motor cylinders, wheel covers, wire foot step, and handhold handle) can be added without a large cost in overall model weight.

3 There are three great color choices for marking the Fokker F.I triplane prototypes.

1

2

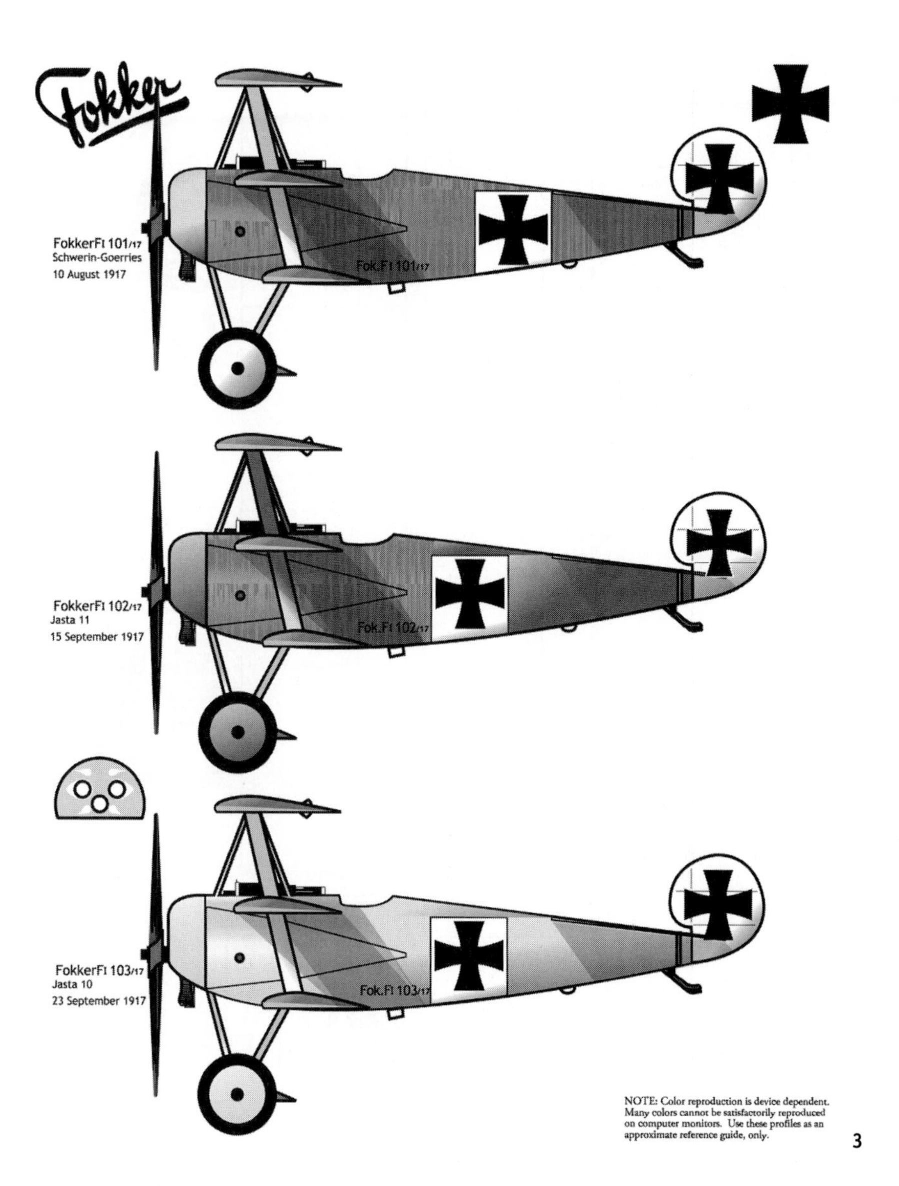

NOTE: Color reproduction is device dependent. Many colors cannot be satisfactorily reproduced on computer monitors. Use these profiles as an approximate reference guide, only.

3

4 **Complete plans for building your own Pistachio scale Fokker F.I triplane.**

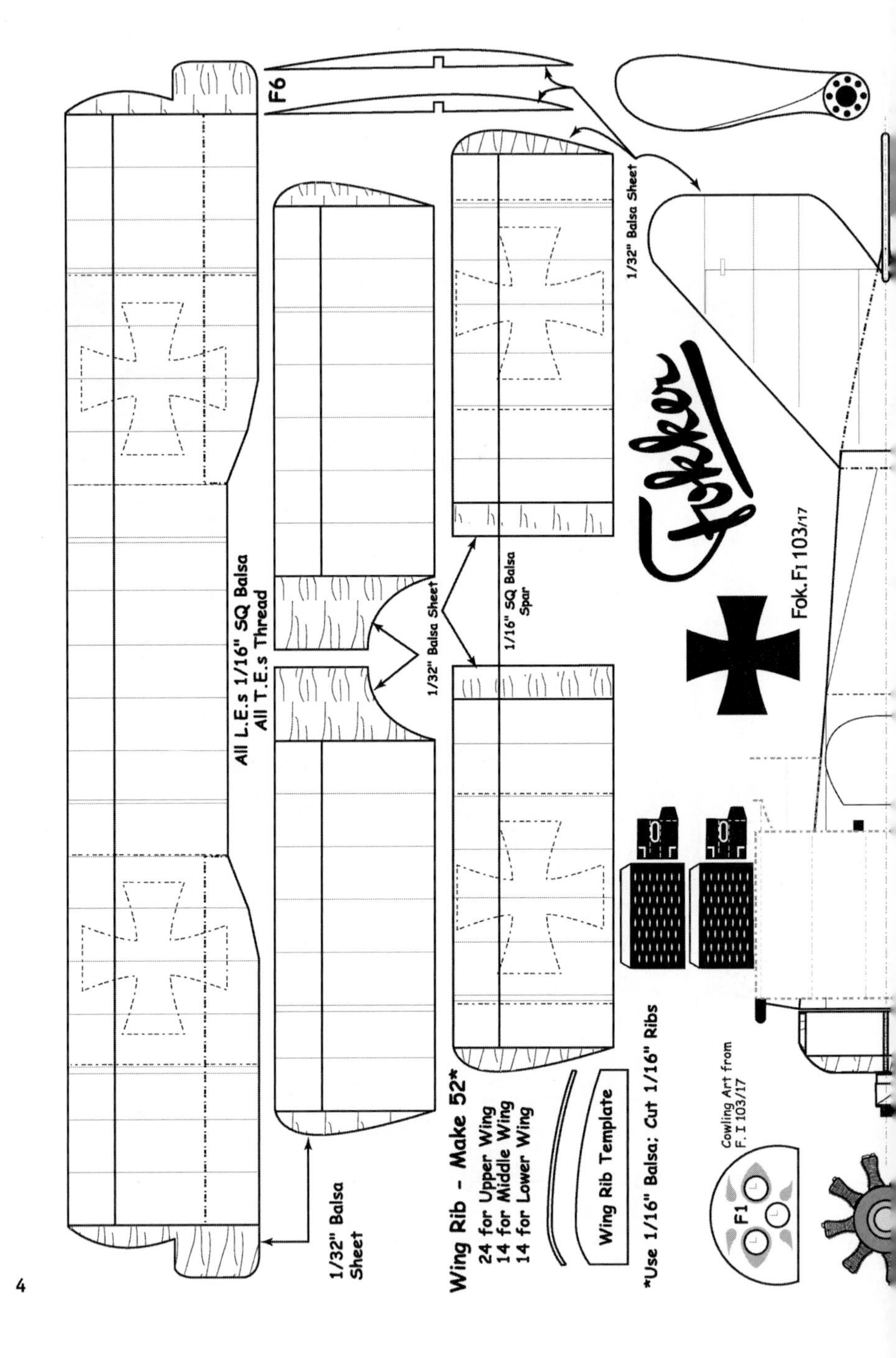

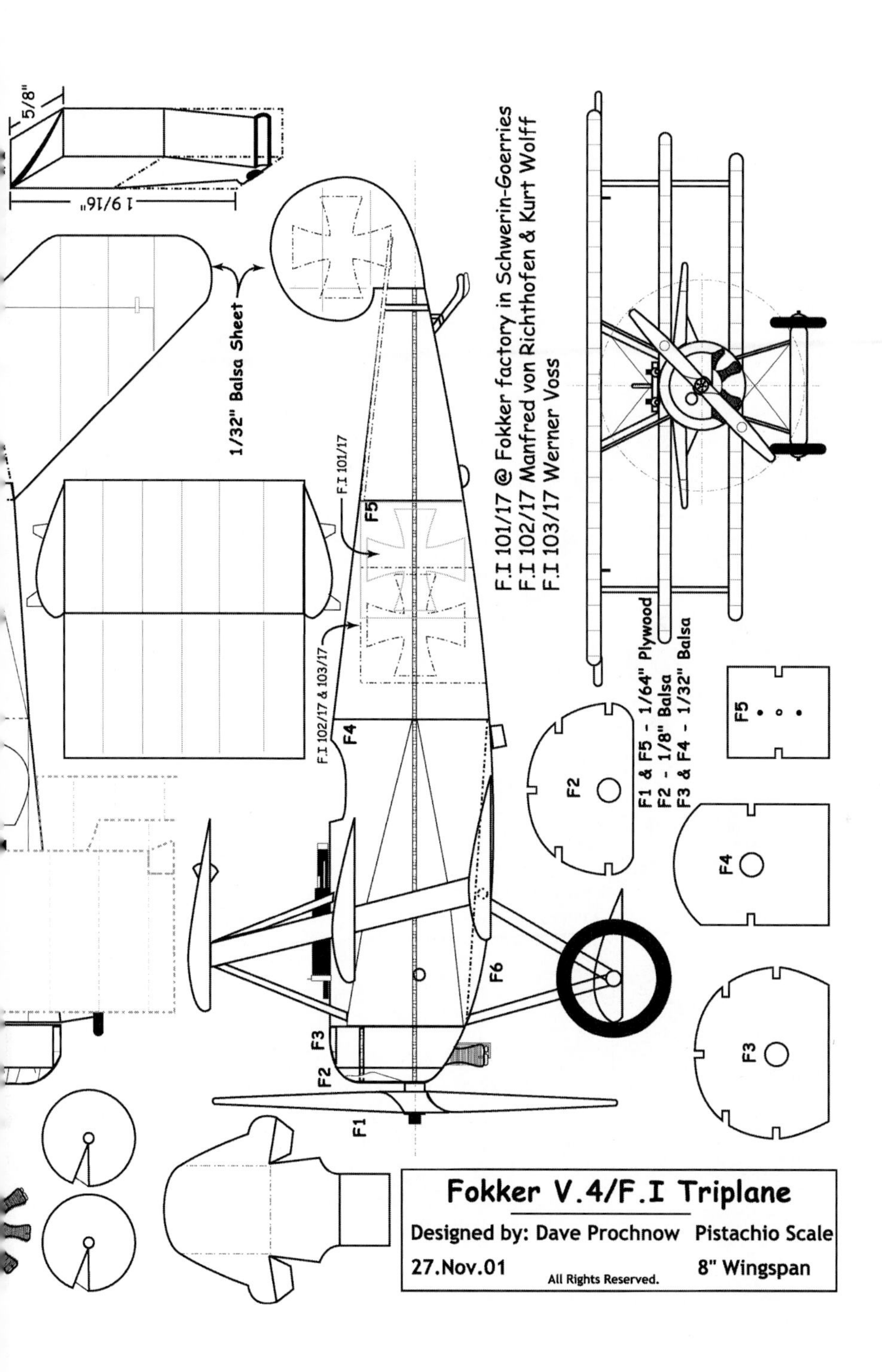

Fokker V.4/F.I Triplane
Designed by: Dave Prochnow
Pistachio Scale
27.Nov.01
All Rights Reserved.
8" Wingspan
F.I 101/17 @ Fokker factory in Schwerin-Goerries
F.I 102/17 Manfred von Richthofen & Kurt Wolff
F.I 103/17 Werner Voss
F1 & F5 - 1/64" Plywood
F2 - 1/8" Balsa
F3 & F4 - 1/32" Balsa
1/32" Balsa Sheet
5/8"
1 9/16"
F.I 101/17
F.I 102/17 & 103/17
F1
F2
F3
F4
F5
F6

HOW TO BUILD A P-51B/C ELECTRIC-POWERED MODEL

1 This junior "Airman" is ready for launching her P-51B/C.

2 A rubber-powered P-51B/C decked out in the markings of Lt. Charles Bailey with the 99th FS.

3 Make sure that you install your electric motor *before* you add the stringers.

4 Features like gun ports can be easily added to the P-51B/C prior to covering with tissue.

1

Chapter 25

2

3

4

5 **Complete plan for building a P-51B/C fuselage.**

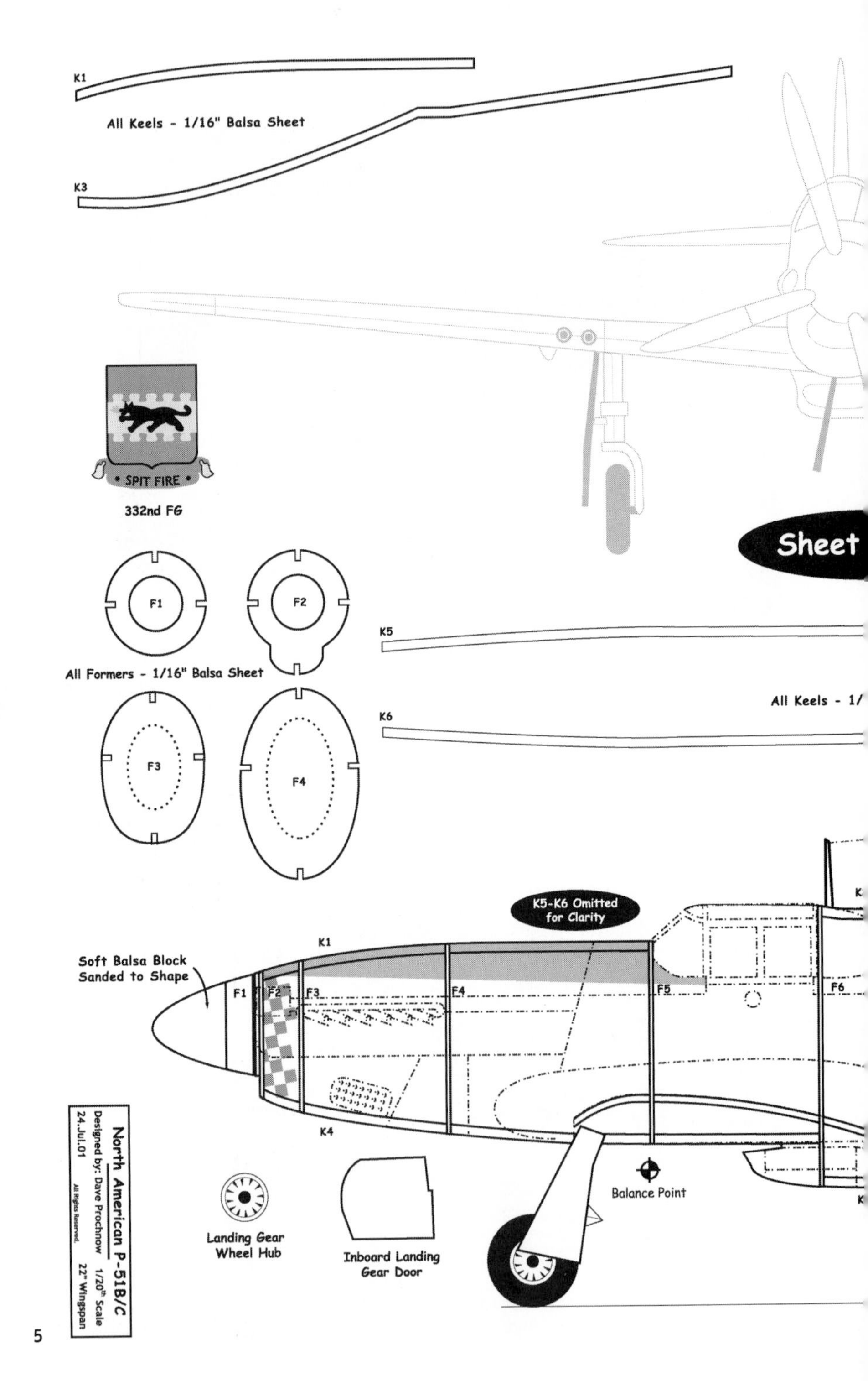

5

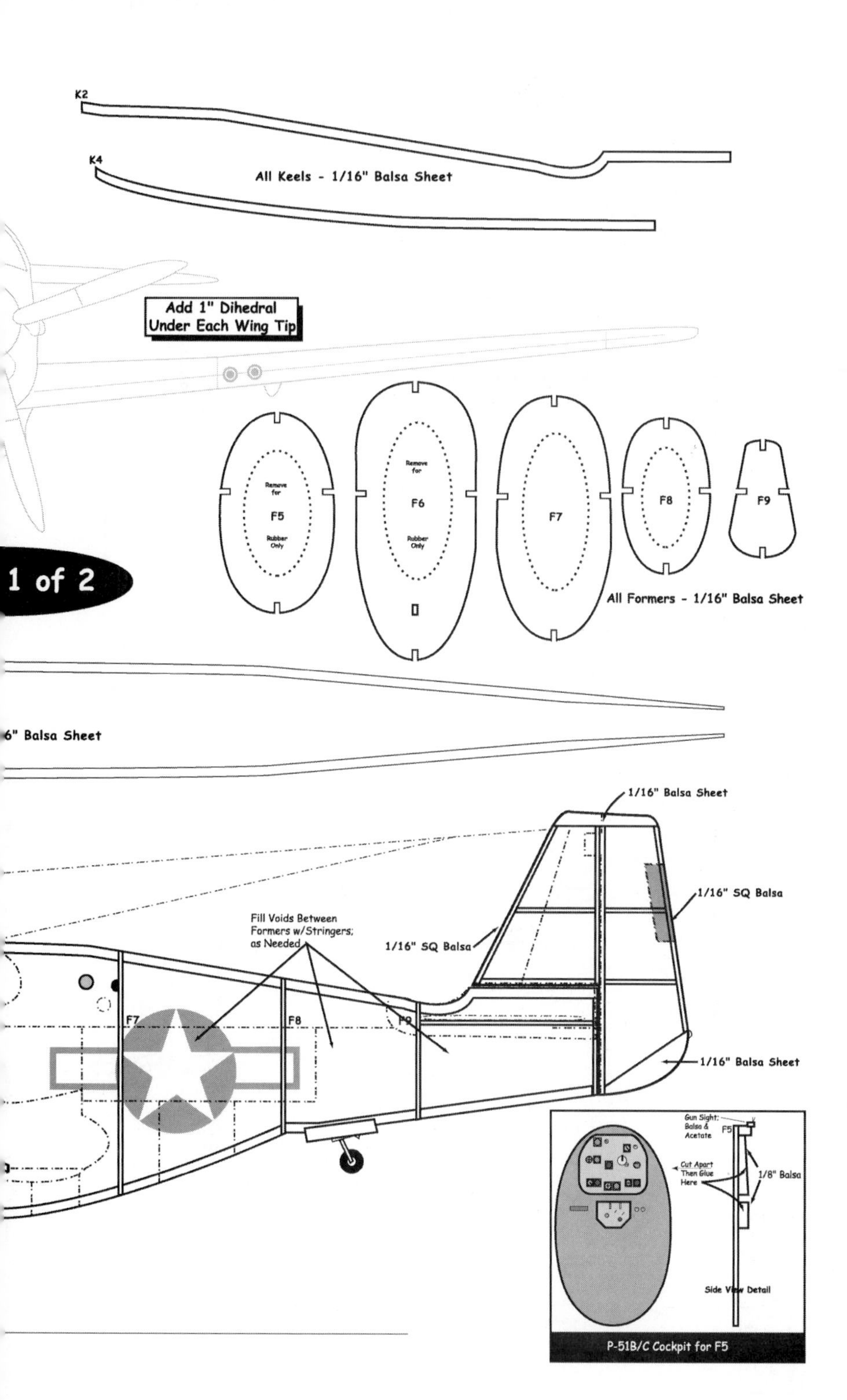
K2
K4
All Keels - 1/16" Balsa Sheet
Add 1" Dihedral Under Each Wing Tip
F5
F6
F7
F8
F9
All Formers - 1/16" Balsa Sheet
1 of 2
6" Balsa Sheet
1/16" Balsa Sheet
1/16" SQ Balsa
1/16" SQ Balsa
Fill Voids Between Formers w/Stringers; as Needed
F7
F8
F9
1/16" Balsa Sheet
Gun Sight; Balsa & Acetate
F5
Cut Apart Then Glue Here
1/8" Balsa
Side View Detail
P-51B/C Cockpit for F5

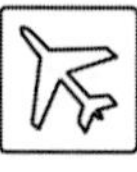

6 **Assemble your wing directly over this plan.**

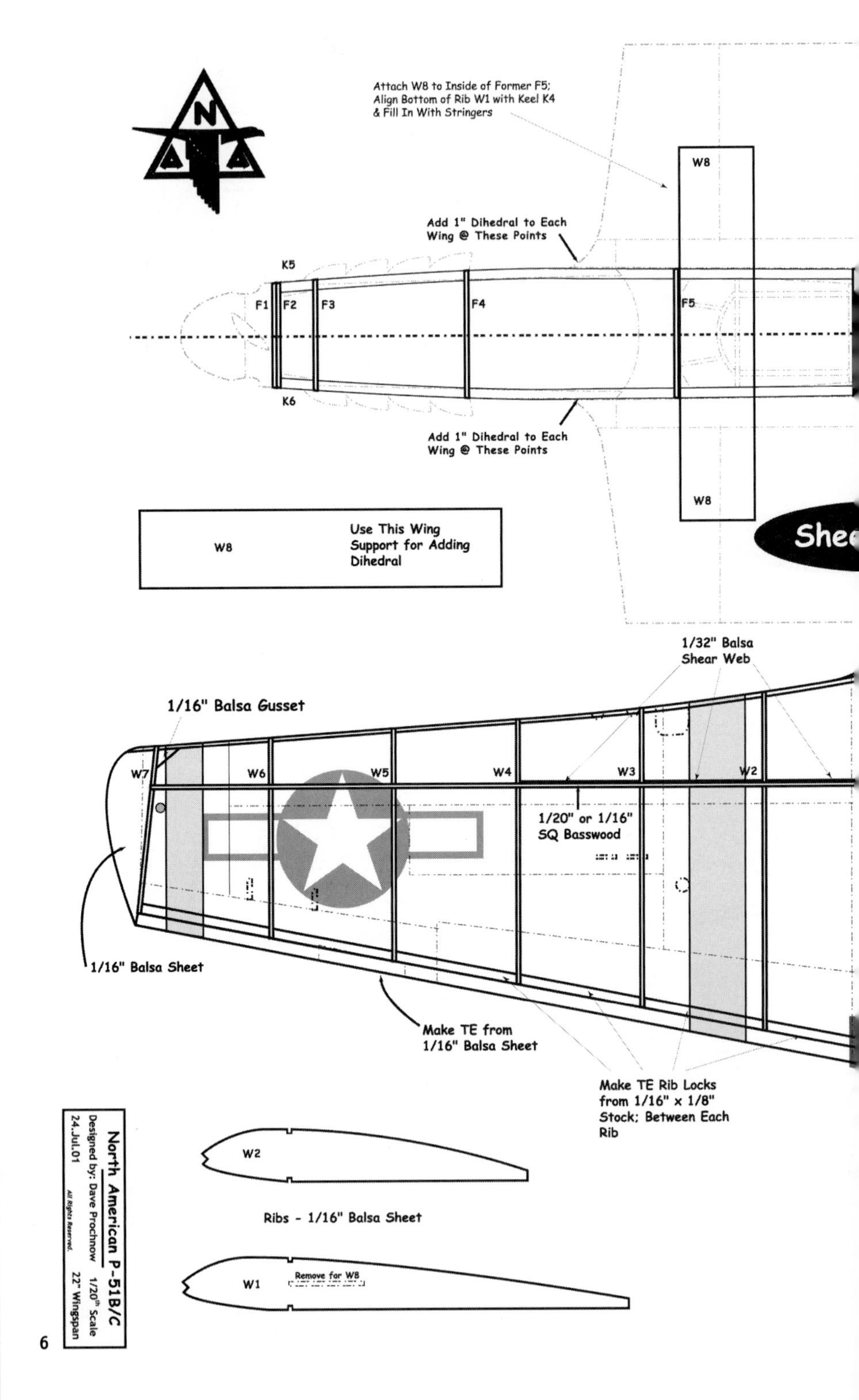

Attach W8 to Inside of Former F5; Align Bottom of Rib W1 with Keel K4 & Fill In With Stringers
W8
Add 1" Dihedral to Each Wing @ These Points
K5
F1
F2
F3
F4
F5
K6
Add 1" Dihedral to Each Wing @ These Points
W8
W8
Use This Wing Support for Adding Dihedral
1/32" Balsa Shear Web
1/16" Balsa Gusset
W7
W6
W5
W4
W3
W2
1/20" or 1/16" SQ Basswood
1/16" Balsa Sheet
Make TE from 1/16" Balsa Sheet
Make TE Rib Locks from 1/16" x 1/8" Stock; Between Each Rib
W2
Ribs - 1/16" Balsa Sheet
W1
Remove for W8
North American P-51B/C
Designed by: Dave Prochnow
1/20th Scale
24.Jul.01
22" Wingspan
All Rights Reserved.
6

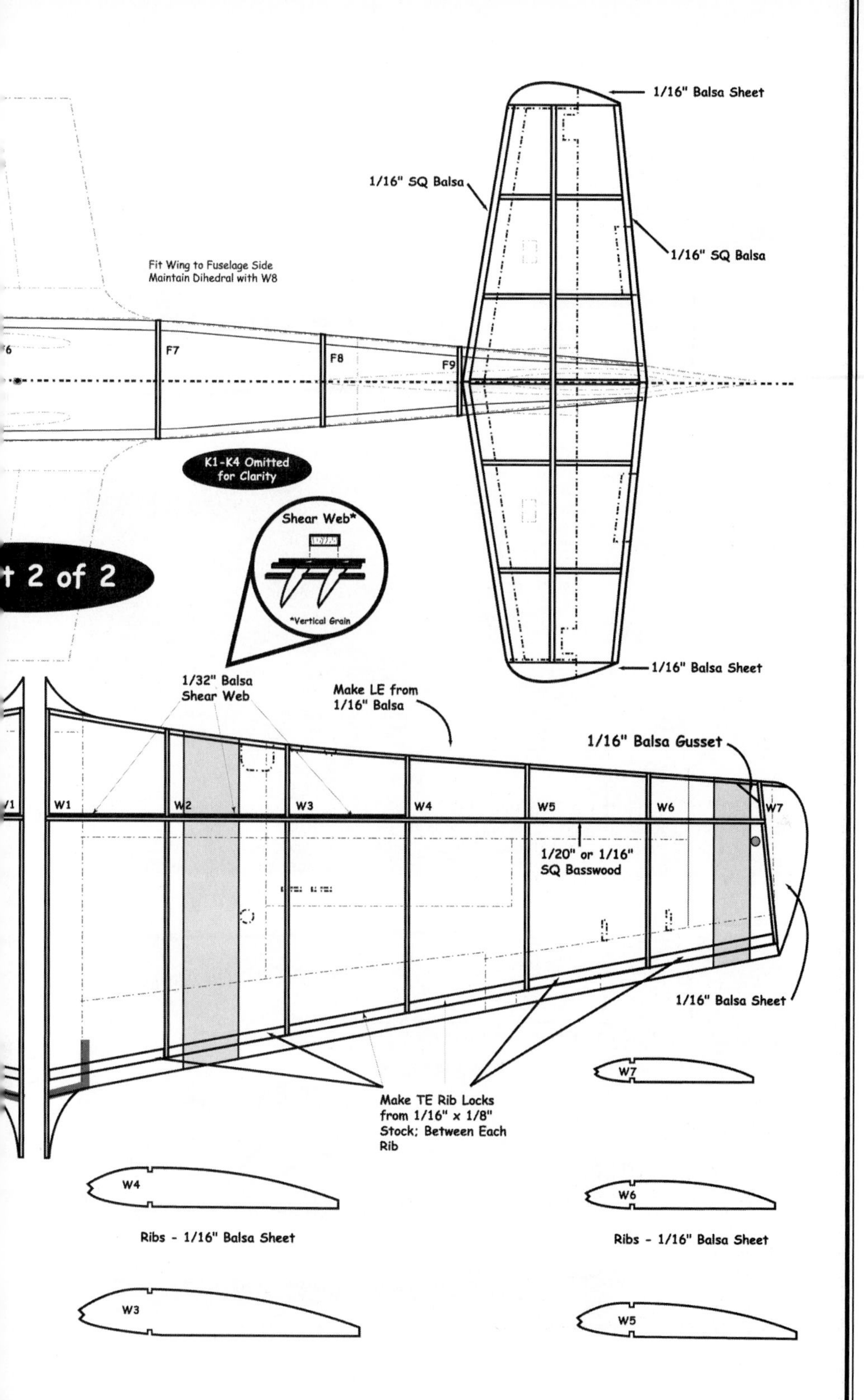
1/16" Balsa Sheet
1/16" SQ Balsa
1/16" SQ Balsa
Fit Wing to Fuselage Side
Maintain Dihedral with W8
F7
F8
F9
K1-K4 Omitted for Clarity
Shear Web*
*Vertical Grain
t 2 of 2
1/16" Balsa Sheet
1/32" Balsa Shear Web
Make LE from 1/16" Balsa
1/16" Balsa Gusset
W1
W2
W3
W4
W5
W6
W7
1/20" or 1/16" SQ Basswood
1/16" Balsa Sheet
Make TE Rib Locks from 1/16" x 1/8" Stock; Between Each Rib
W7
W4
W6
Ribs - 1/16" Balsa Sheet
Ribs - 1/16" Balsa Sheet
W3
W5

7 The P-51B/C of Lt. Charles Bailey as flown with the 99th FS during the summer of 1944.

8 During the summer 2001 air show season, the American Airpower Heritage Foundation, Inc. unveiled their completely restored, flying P-51C. Painted as a 302nd FS "Tuskegee Airmen" aircraft.

9 The distinctive red tail of this restored P-51C has become the hallmark of the "Red Tail Project." This nonprofit group was formed to provide the funding needed for completing this restoration.

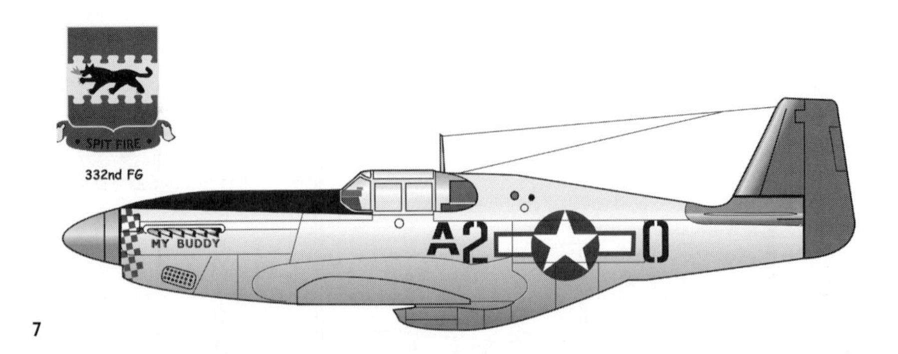

7

8

9

HOW TO BUILD AN F-104 ROCKET-POWERED MODEL

1

This 16" span model is designed for the Jet-X 50 Rocket motor. It can also be flown as a high flying catapult glider
Contents include:
Printed sheet panels and stripwood
Full size plan with detailed building and flying instructions
Covering tissue
Vacuum formed canopy, pilot and wing fillets
Authentic decals
designed for JET-X
MESSERSCHMITT 163
German WWII Rocket powered fighter

2

1 This Rapier-powered Lockheed F-104G Starfighter is a joy to build but a terror to fly.

2 There are many kits available for all types of model aircraft. This rocket-powered model could be an alternative to the following F-104 plan.

3 Complete plan for rocket-powered F-104G. Both Rapier and JETEX rocket propulsion units are illustrated.

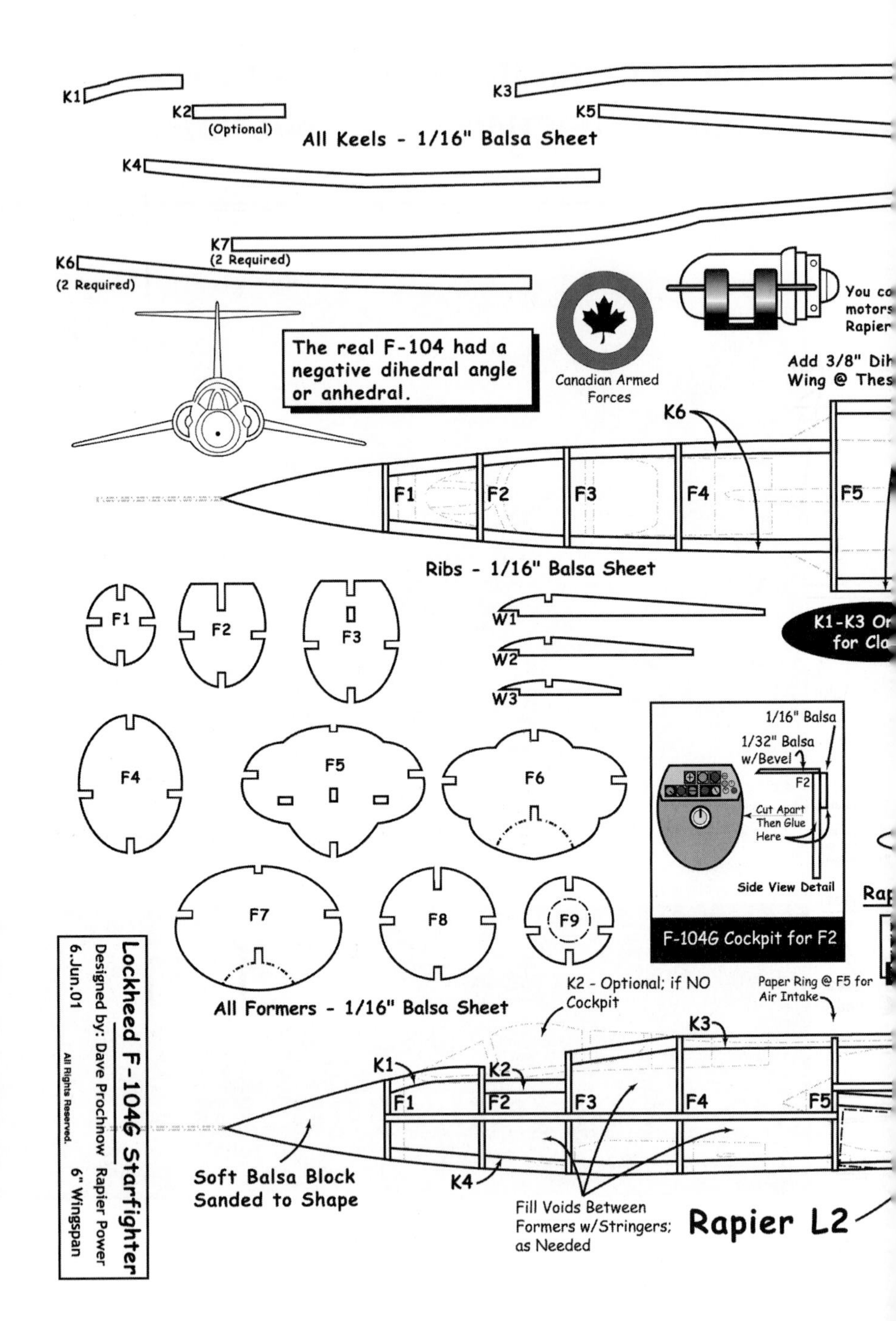

3

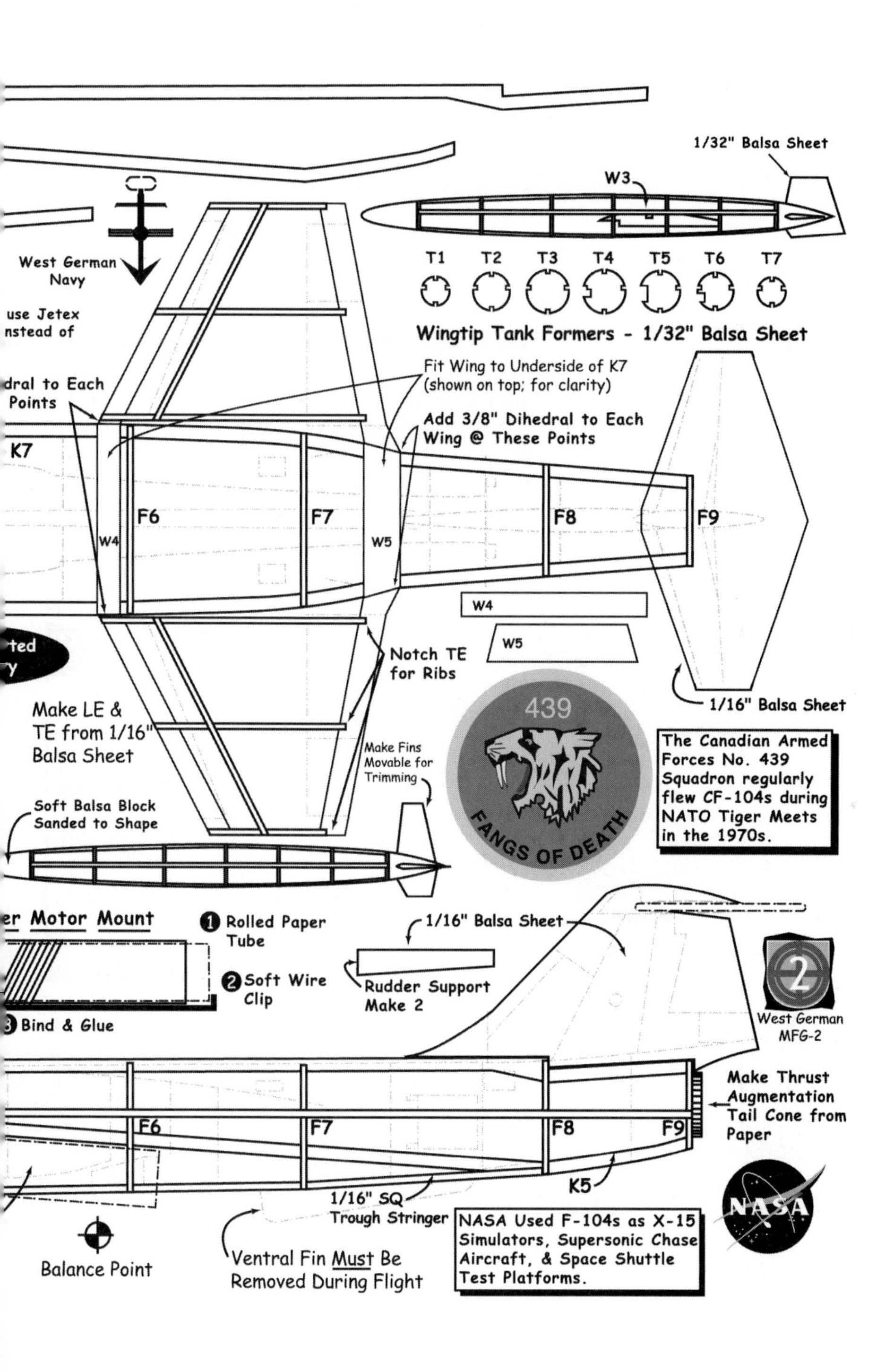
1/32" Balsa Sheet
W3
West German Navy
use Jetex
nstead of
T1
T2
T3
T4
T5
T6
T7
Wingtip Tank Formers - 1/32" Balsa Sheet
dral to Each
Points
Fit Wing to Underside of K7
(shown on top; for clarity)
Add 3/8" Dihedral to Each
Wing @ These Points
K7
F6
F7
F8
F9
W4
W5
W4
W5
Notch TE
for Ribs
Make LE &
TE from 1/16"
Balsa Sheet
1/16" Balsa Sheet
439
FANGS OF DEATH
The Canadian Armed Forces No. 439 Squadron regularly flew CF-104s during NATO Tiger Meets in the 1970s.
Make Fins
Movable for
Trimming
Soft Balsa Block
Sanded to Shape
er Motor Mount
1 Rolled Paper Tube
2 Soft Wire Clip
3 Bind & Glue
1/16" Balsa Sheet
Rudder Support
Make 2
West German
MFG-2
Make Thrust Augmentation Tail Cone from Paper
F6
F7
F8
F9
K5
1/16" SQ
Trough Stringer
NASA Used F-104s as X-15 Simulators, Supersonic Chase Aircraft, & Space Shuttle Test Platforms.
NASA
Balance Point
Ventral Fin Must Be
Removed During Flight

4 An F-104G from the Jagd Bomber Geschwader (JaBoG) 34 "Allgäu." The last flight of a JaBoG 34 F-104 occurred on August 21, 1987.

5 The Canadian Armed Forces entry into the NATO Tiger Meets of the late 1970s featured the Sabre Toothed Tigers of No. 439 Squadron. In 1977 the CAF entry was 104838. This aircraft crashed on March 7, 1978.

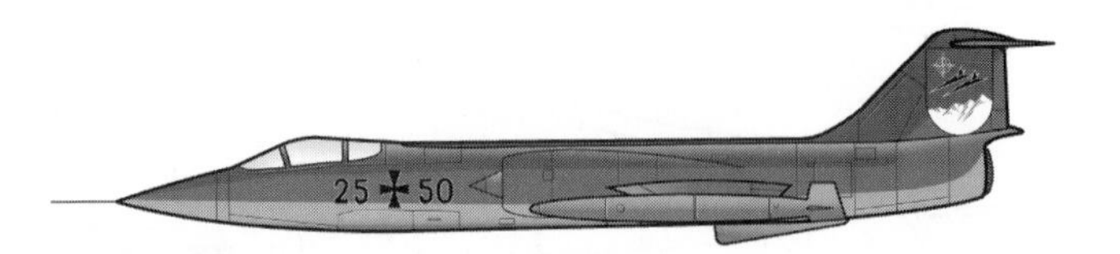

4

5

CHAPTER 26

Tattoo You; and Me, Too

There is a tremendous amount of confusion and misinformation about one of today's most exciting technologies. You've probably already seen this technology in action—by waving an identification badge in front of a reader panel at work or a MasterCard® PayPass® Tap N Go™ payment transaction at your local gas station. In each of these cases, small embedded chips with an antenna are able to transmit a unique identifier (ID) that has been registered to you or your account. The reader relays this ID to a central computer database which looks up your ID and, in our two examples, you are granted admission or your account is debited.

The magic that makes all of this happen is called Radio Frequency Identification or RFID. RFID is a wireless data collection technology that uses electronic tags for storing a reference ID. Based on this ID a database lookup can provide a wealth of information about the tag's holder.

Many sources cite the origin of RFID technology as a form of Identification Friend or Foe (IFF) transponder that was developed by the British in 1939. While some sources feel that this linkage is somewhat inappropriate, active RFID tags could be considered as derivatives of IFF (i.e., where aircraft actively transmit a signal called squawking to a receiver). Nonetheless, RFID is being used more and more frequently in toll booths, personal identification cards, and anti-theft tags on expensive clothing and electronics in retail stores.

A typical RFID system consists of hardware and software. The hardware includes RFID tags, which can be "active" (i.e., they have a battery-powered transmitter embedded

in the tag) or "passive" (the tag only transmits its data when energized by antenna radiation from a reader).

An RFID hardware system contains basically two components:

- tag or transponder
- reader/interrogator

RFID tags come in many shapes and sizes and are broadcast in three different frequency ranges:

- high frequency (850 to 950 MHz and 2.4 to 5 GHz)
- intermediate frequency (10 to 15 MHz)
- low frequency (100 to 500 kHz)

The transponder, or tag, is placed on or inside the host object (e.g., book, CD, dog, human). As this object moves into the reader's area of interrogation, the reader is activated and begins signaling via electromagnetic waves (radio frequency waves).

The transponder subsequently transmits its unique ID information to the reader, which in turn converts it, through the software technology, into useful information. This information is not restricted to the location of the object, and can include specific detailed information concerning the object itself.

RFID readers are matched to these tag types and translate the information on the tag into a digital form that is converted for use in software. The software sends the information to the business system, which stores the information in a database or displays it on a screen for human decision making.

Yeah, but what can you do with an ID number, isn't that just a fancy barcode? The tags sold by Parallax, for example, contain a unique ID number (that Parallax claims can have 1^240, or 1,099,511,627,776 possible different values) that is read by the RFID Reader Module and transmitted to the host via a simple serial interface.

Also, barcodes must be seen to be read. RFID transponders can transmit their information without actually being visible. This attribute is exciting for embedding tags into things like clothing, pets, and human beings.

Yes, as reported in an Associated Press article published by *CNN.com*, in 2004, "some employees" of the Mexico attorney general's office had RFID transponders embedded in their forearms. These embedded tags were going to be used for limiting access to sensitive areas.

Now if clothing is more of a suitable site for your tag installation, then according to the *RFID Journal* the "laundry and uniform rental business" use a plastic coated transponder that transmits a 13.56-MHz frequency to a distance of under 3 feet.

One other major benefit that RFID has over barcodes can be seen at your local grocery store checkout lane. Have you ever seen the clerk repeatedly run your gallon of milk or carton of ice cream through the barcode reader? The reason for these false "reads" could be attributed to moisture or dirt that obscured part or all of the barcode. Not so with RFID. RFID transponders are resistant to dirt, water, vibrations, and other atmospheric conditions. Furthermore, you have to see the RFID to read it.

The one stumbling block that is slowing the widespread introduction of RFID is its cost. In a *WIRED News* article by Kendra Mayfield, the author quotes Kevin Ashton, executive director of MIT's Auto-ID Center as saying "I have little doubt that a five-cent tag is achievable in the next few years, if a market of the right size emerges." Unfortunately, Ashton then went on to prognosticate that, "we're probably looking at five cents (tags) in 2005." Oops, while prices have been dropping, as recent as early 2006 pricing (in bulk quantity) was still quite high:

T555715-PAE	Transponder R/W 330-BIT 330PF	by Atmel:	89¢
RI-I03-112A-03	Transponder in-lay 13.56-MHz	by TI RFID Systems:	54¢

RFID from a Parallax Galaxy Far, Far Away

What if you wanted to experiment with RFID technology? Where would you start? Parallax would be your best bet for building a complete, semi-professional RFID tag-reading system. Designed in cooperation with Grand Idea Studio, the reasonably priced RFID Reader Module is an integrated solution for reading passive RFID transponder tags from approximately 4 inches away. The RFID Reader Module can be applied to many different control activities: access control, automatic identification, robotics, navigation, inventory tracking, payment systems, and car immobilization.

This RFID Reader Module works exclusively with the EM4100-family of 125-kHz, passive, read-only transponder tags that are manufactured by EM Microelectronics-Marin SA. The most popular compatible tags also are available from

READ ALL ABOUT IT, LCDBUG

What a Boe-Bot beginner really needs is a low-cost, alphanumeric LCD unit that can be readily "plugged" into the Boe-Bot's breadboard. No debugging, no soldering, no programming—just plug and display.

There is one other important factor that must characterize this LCD unit. It has to be inexpensive. While the Parallax LCD Terminal Application Module (#29121) can be installed on the Boe-Bot, the $39 price tag is a little too hefty for those users who are looking for a simple display without terminal capability. For around half the price of this AppMod, Boe-Bot owners can find the ideal answer from BG Micro (www.bgmicro.com).

Known as the LCDBug, this inexpensive 2-line-by-8-character display is a terrific visual interface companion to the Boe-Bot. Housed on a standard 20-pin IC socket header, the LCDBug consumes +5V of power, uses an Atmel ATtiny26 microcontroller for providing the LCD firmware, and requires only one serial output pin from the Boe-Bot. Therefore, with just a simple three-pin connection and a couple of lines of PBASIC, virtually anyone can have a great robot display for less than $20.

Parallax. Each transponder contains a unique ID that is read by the RFID Reader Module and transmitted to the host via a simple serial interface. What the host computer or microcontroller does with this information is where the real fun begins.

Features of the Parallax RFID Reader Module:

- Fully integrated, low-cost method of reading passive RFID transponder tags
- 2400-baud TTL serial interface (output only) to PC, BASIC Stamp module and other processors
- Requires single +5VDC supply
- Bi-color LED for visual indication of activity
- 0.100. pin spacing for easy prototyping and integration

I've Got to Hand It to Ya

Some Common Knitting Abbreviations

BO = bind off
CO = cast on
dec = decrease
dpn = double pointed needles
inc = increase
K2tog = knit 2 together
K = knit
P = purl
pm = place marker
Rnd = round
sts = stitches
tog = together

These Fingerless Gloves will fit a medium woman's or small man's hands—so if you like being manhandled, beware. Finished hand circumference is 7.5 inches.

NOTE: Always make a test knitting or gauge before starting your project. This gauge will tell you how many stitches per inch you are capable of making with your yarn.

This project is for intermediate knitters.

MATERIALS:

- (2) skeins of sport weight yarn, 50 grams, in two colors, approximately 110 yards each. I chose navy blue color 308 and green color 46 from Blue Sky Alpacas. The gauge is 5 to 6 stitches per inch.
- (4) U.S.: 3–3.25-millimeter double pointed needles (dpn) for the cuff
- (4) U.S.: 6–4-millimeter double pointed needles for the hand

1. **CUFF.** Cast on (CO) 32 stitches (sts) onto 3 dpn U.S.: 3. Work knit (K) 1 purl (P) 1 ribbing until one color equals 1.5 inches, switch colors and continue ribbing with the other color for another 1.5 inches.
2. **HAND.** Switch needles to U.S.: 6 needles and switch color. Use a straight stitch and increase 1 stitch at the end of the round.
3. **THUMB** . Knit 16 sts across, mark your place with a paper clip or small safety pin, increase one stitch to the left, knit 1, increase one stitch to the right, mark, knit 1 round. Increase 2 sts inside markers in this manner. Complete 3 rounds. Increase 2 sts inside markers every 4 rounds for 3 times. After this step there should be 11 sts inside markers. Place sts on a spare piece of yarn, remove markers, use a backward CO loop to cast on 1 stitch over gap, rejoin, and knit to end of round. There should be a total of 33 sts. Work sts for 3.5 inches above ribbing sts. Change to other yarn color. Mark the opposite side of the thumb. This is the start of the little finger.
4. **LITTLE FINGER.** Work across 4 sts. Place 26 sts on a spare needle for Step 5. There should be 7 sts remaining on a needle to be worked. CO 1 backward loop to cast on 1 stitch over gap, rejoin, and knit to end of round. There are 8 total little-finger sts. Arrange sts on 2 dpn and work in the round until the sts are up to your first little-finger knuckle. Bind off (BO) all little-finger sts.
5. **UPPER HAND.** Slip 26 held sts from Step 4 on needles, join yarn, and pick up and knit 2 sts along CO edge at base of little finger. The total sts should be 28. Rejoin and knit in the round for .5 inch.
6. **RING FINGER.** Place first 5 sts on one dpn and last 5 sts on another dpn. Place remaining sts on a spare needle to be worked in Step 7. CO 1 backward loop to cast on 1 stitch over gap. There should be 11 total ring-finger sts. Work in round until the sts are up to your first ring-finger knuckle. BO all ring-finger sts.

7. **MIDDLE FINGER.** Place first 4 sts on one dpn and last 4 sts on another dpn. Repeat rounds similar to Step 6. There should be 11 total middle-finger sts. Work in round until the sts are up to your first middle-finger knuckle. BO all middle-finger sts.
8. **INDEX FINGER.** Place remaining 10 sts on dpn. Repeat rounds similar to Step 6. There should be 12 total sts. Arrange sts on 2 dpn and knit until the sts are up to your first index-finger knuckle. BO all index-finger sts.
9. **THUMB.** Place held sts on 3 dpn and pick up and knit 1 stitch along CO sts. Knit until the sts are up to your thumb knuckle. BO all thumb sts.

This is NOT London Calling, Radio-Free IDs

Before we can use an RFID tag, we have to know the tag's ID. The best way to gain access to an ID is with a simple PBASIC RFID tag reader program. Once this program displays the tag's ID, then you can code a different program that will perform a defined action based on the reading of a specific tag ID. First, however, here is a simple program for reading tag IDs:

Worth a Grain of Dale Wheat

After a close inspection of the LCDBug's underside, two names figure very prominently in the design and assembly of this display unit. The most visible of these two names is Hantronix. Anyone who has experimented with LCDs knows that Hantronix is a major manufacturer of display modules (e.g., HDM08216H-3 is the module used with the LCDBug; you can find ample technical documentation on the Hantronix Web site at www.hantronix.com/2_2.html). The other name, Dale Wheat, is a little more difficult to figure out. Thankfully, Mr. Wheat has his URL printed on the 'Bug's underside. It turns out that Dale Wheat is the inventive mind behind the LCDBug. Please visit his Web site (www.dalewheat.com) for more information about the LCDBug, as well as some of his other projects.

```
' Reads and displays RFID tag ID.

' ——-[ Definitions ]——————————————————————————-

Enable  PIN 0                  ' low = reader on
RX      PIN 1                  ' serial from reader
buffer  VAR Byte(10)           ' RFID buffer

' ——-[ Program Code ]———————————————————————————

Main:
   LOW Enable
   SERIN RX, T2400, [WAIT($0A), STR buffer\10]
   HIGH Enable

Display_Tag:
   DEBUG "Tag ID: ", buffer
   DEBUG CR
```

It's All in This Family

According to the EM Microelectronic-Marin SA datasheet for the EM4100 tag, this is a complementary metal-oxide semiconductor (CMOS) integrated circuit that is a passive read-only device with a factory-programmed 64-bit memory array. This programming step is performed with laser fusing of poly-silicon links which hold the unique data code on each tag.

Operating with a frequency of 100 to 150 kHz, the EM4100 will cycle through its ID code over and over again, until the reader is removed from its area of interrogation.

There are three forms of encoding for the memory array: Manchester, Bi-Phase, and PSK. In the more common Manchester and Bi-Phase encryptions, the EM4100 holds 64 bits of data that is divided into five groups: 9 bits for the header, 10 row parity bits (P0-P9), 4 column parity bits (PC0-PC3), 40 data bits (D00-D93), and one stop bit (logic 0).

TOMMY CAN YOU HEAR ME?

Regardless of whether you believe that IFF is the "father" of RFID, the actual history of IFF is very difficult to track down. In fact, during my extensive research for this book (and anyone who knows me really well, knows that my research efforts are both thorough and meticulous) I couldn't locate any direct correlation between IFF and RFID. Also, the actual historical development of IFF is overshadowed by the British development of "radar."

A search through the archives of the Transmitter Block, Bawdsey, Suffolk, England, turned up some exciting information about the Scot physicist Robert Watson-Watt. Watson-Watt, who was the supervisor of the national radio research laboratory, was related to the inventor of the steam engine, James Watt.

Although the notion that radio beams could be emitted, reflected off of an aircraft, and then received was conceived by Watson-Watt, the actual "proof of concept" test was developed along with Arnold Wilkins.

On February 26, 1935, Watson-Watt and Wilkins conducted a test of their project using a BBC transmitter and a "test" bomber target. This test was a success. Three months later in May of that same year, Watson-Watt and Wilkins performed a number of radio beam "interception" tests and the first working "radar" system was born.

```
  PAUSE 1000
  GOTO Main
```

OK, now that you've read (and stored) all of your tag IDs, it's time to do something meaningful with those data. In this example, we are going to control the navigation of a Parallax Boe-Bot™ via RFID tags:

```
' Drive your Boe-Bot with RFID tags

' ——-[ Definitions ]——————————————————————————-

Enable   PIN 0                       ' low = reader on
RX            PIN 1                       ' serial from reader
Buffer        VAR Byte(10)           ' RFID buffer
Tag1          DATA "0F01DD404F"
pulse_count   VAR Byte

' ——-[ Program Code ]——————————————————————————

Main:
  LOW Enable                                    ' activate the reader
  SERIN RX, T2400, [WAIT($0A), STR buffer\10]
  HIGH Enable

Check_ID:
  FOR idx = 0 TO 9                              ' scan bytes in tag
   READ (tagNum - 1 * 10 + idx), char           ' get tag data
    IF (char <> buffer(idx)) THEN Bad_Char      ' compare tag
  NEXT
  GOTO Tag_Found                                ' all bytes match!

Bad_Char:                                       ' end program
  END

Tag_Found:
  GOSUB Right_Turn                              ' nav path
  GOTO Main

  END
```

```
' ---[ Subroutines ]----------------------------

' Drives away

Right_Turn:                          ' Right turn routine
  FOR pulse_count = 1 to 25          ' 25 left rotate pulses
    PULSOUT 12, 1000                 ' 2.0 ms pulse right
    PULSOUT 13, 1000                 ' 2.0 ms pulse left
    PAUSE 20                         ' Pause for 20 ms
  NEXT
RETURN
```

NOTE: Only one RFID tag is used in this sample program.

If you're like me, then you hate being tethered to a PC for reading these tag IDs. A better solution is to add an LCD to the Boe-Bot and read the tag IDs. Also, this LCD interface is great for displaying the navigational path that the Boe-Bot is following. Then, at least, your robot won't be taking you for a ride.

Tag, You've Been Zapped

Not everyone is enamored with the RFID concept. One group trying to raise a collective consciousness about RFID is Preemptive Media. Started in 2002 by Beatriz da Costa, Jamie Schulte, and Brooke Singer, this gaggle of artists, activists, and technophiles has formed a workshop for helping the everyday person become more aware of RFID. Called Zapped!, this workshop shows attendees how to build an RFID detector. Basically, this keychain-sized device beeps whenever it comes within range of an RFID reader. But an underlying tone for this workshop also deals with how to "zap" an RFID tag—rendering it useless.

One interesting design element sponsored by Zapped! is the Zapped! Madagascan Jam & Hissing Roach. Carrying a reprogrammed RFID tag on its back, hordes of roaches are sent out during the night as "data couriers."

Follow the RFID Reader

You will need three jumper wires for connecting the LCDBug to the Boe-Bot. First, make sure that the three-position power switch on the Boe-Bot is off (i.e., position 0). Now attach one jumper from the power socket header (Vdd) on the BOE to pin 5 (you also can use pin 15 or both pin 15 and pin 5) of the LCDBug. Next connect another jumper from the BOE ground header (Vss) to pin 6 (you also can use pin 16 or pin 16 and pin 6) on the LCDBug.

The final connection between the Boe-Bot and the LCDBug is for a serial data input line. This connection corresponds to one of the BASIC Stamp I/O pins. Since this Boe-Bot is already using several of these lines for navigation inputs, I connected the LCDBug to I/O pin P7. Any pin that you have open will work just fine, however. Just run a jumper wire from pin 9 of the LCDBug and connect it to I/O pin P7 (or, your alternately selected I/O pin) of the Boe-Bot.

Once you've completed all of the hardware connections for the LCDBug, only one PBASIC command is needed to drive output to the LCD. Use SEROUT for sending text to the LCDBug. For example,

```
SEROUT 7, 84, ["McGraw-Hill"]
```

where,

SEROUT = PBASIC command for serial output
7 = Tpin; the I/O pin we used in Step 6
84 = Baudmode; baud rate for LCDBug; 9600, 8-bit, no-parity, true
["McGraw-Hill"] = OutputData; the text for display on the LCDBug

This SEROUT command is used in lieu of the DEBUG command that was used in our previous sample PBASIC programs.

During power-up initialization (or, following a Clear Screen command), the LCDBug clears the screen, sets the default cursor style (i.e., blinking underline), and positions the cursor in the upper left corner. This initialization can result in a slight delay before any text can be displayed on the LCD. Therefore, a short pause (or, PBASIC NAP command) should be executed in your code before sending SEROUT data. For example,

```
' {$STAMP BS2}
' {$PBASIC 2.5}

NAP 5                                   ' Sleep mode
```

```
SEROUT 7, 84, [12]                                  ' Clear Screen
SEROUT 7, 84, [28]                                  ' Show Revision
PAUSE 1000                                          ' Pause Program
SEROUT 7, 84, [12]                                  ' Clear Screen
SEROUT 7, 84, ["McGraw-Hill", 10, 13, "Books"]      ' Display text
```

You can include this subroutine at the beginning of all of your Boe-Bot programs. In fact, this subroutine is a great replacement for the Start/Reset Indicator Circuit program described in Robotics with the Boe-Bot; Activity #3.

OK, let's do something practical with the LCDBug and an RFID tag. In this demonstration, rather than printing tag IDs on a tethered computer via DEBUG, I will use the LCDBug for displaying these data. Similarly, in your own programs, you can replace DEBUG commands with a series of SEROUT commands. This simple replacement will provision Boe-Bot with a remote message system for displaying sensor readings, status reports, and program variable values—all without the need for a tethered computer. In our case, however, I am going to display navigation information.

```
' {$STAMP BS2}
' {$PBASIC 2.5}

'——-[ Definitions ]——————————————————————————-

Enable        PIN 0
RX            PIN 1
Buffer        VAR Byte(10)
Tag1          DATA "0F01DD404F"
pulse_count   VAR Byte
counter       VAR Byte
distance      VAR Word

'——-[ Program Code ]——————————————————————————

Init:
  NAP 5
  SEROUT 7, 84, [12]
  SEROUT 7, 84, [28]
  PAUSE 1000
  SEROUT 7, 84, [12]
  SEROUT 7, 84, ["McGraw-Hill", 10, 13, "Books"]
```

YOU CAN'T PUT A VIRUS IN THAT THING

On March 15, 2006 researchers from the Department of Computer Science at Vrije Universiteit Amsterdam, Netherlands lit a firestorm of controversy when they released a paper titled, IS YOUR CAT INFECTED WITH A COMPUTER VIRUS? The authors of this paper, Melanie R. Rieback, Patrick N. D. Simpson, Bruno Crispo, and Andrew S. Tanenbaum, presented research work on RFID Malware: specifically RFID exploits, RFID worms, and RFID viruses.

The RFID industry was quick to pooh-pooh the notion of RFID viruses. In a BusinessWeek Online article by Olga Kharif, Larry Blue, vice-president and general manager at RFID equipment maker Symbol Technologies chided RFID viruses as a "theory" lacking in reality.

Regardless of the outcome of this research and its accompanying debate, exploring the possibility of RFID malware sounds like a worthy diversion for the weekend hacker to me.

```
Main:
    LOW Enable
    SERIN RX, T2400, [WAIT($0A), STR buffer\10]
    HIGH Enable

Check_ID:
    FOR idx = 0 TO 9
      READ (tagNum - 1 * 10 + idx), char
        IF (char <> buffer(idx)) THEN Bad_Char
    NEXT
    GOTO Tag_Found

Bad_Char:
    END

Tag_Found:
    GOSUB Show_Path
    GOTO Main

END

' ——-[ Subroutines ]———————————————————————————-

Show_Path:
    SEROUT 7, 84, ["Tag ID: ", buffer]
    PAUSE 1000
    SEROUT 7, 84, ["Right Turn"]
      Right_Turn:
      FOR pulse_count = 1 to 25
        PULSOUT 12, 1000
        PULSOUT 13, 1000
        PAUSE 20
      NEXT
RETURN
```

NOTE: Only one RFID tag is used in this sample program.

HOW TO BUILD AN RFID GLOVE AND RFID-CONTROLLED ROBOT

1 You can buy RFID tags like these from Parallax.

2 Parallax also sells an RFID Card Reader.

3 A standalone RFID system can be built around the Parallax BASIC Stamp Super Carrier Board.

4 The Parallax BASIC Stamp 2 module can process all of the data generated by a simple RFID system.

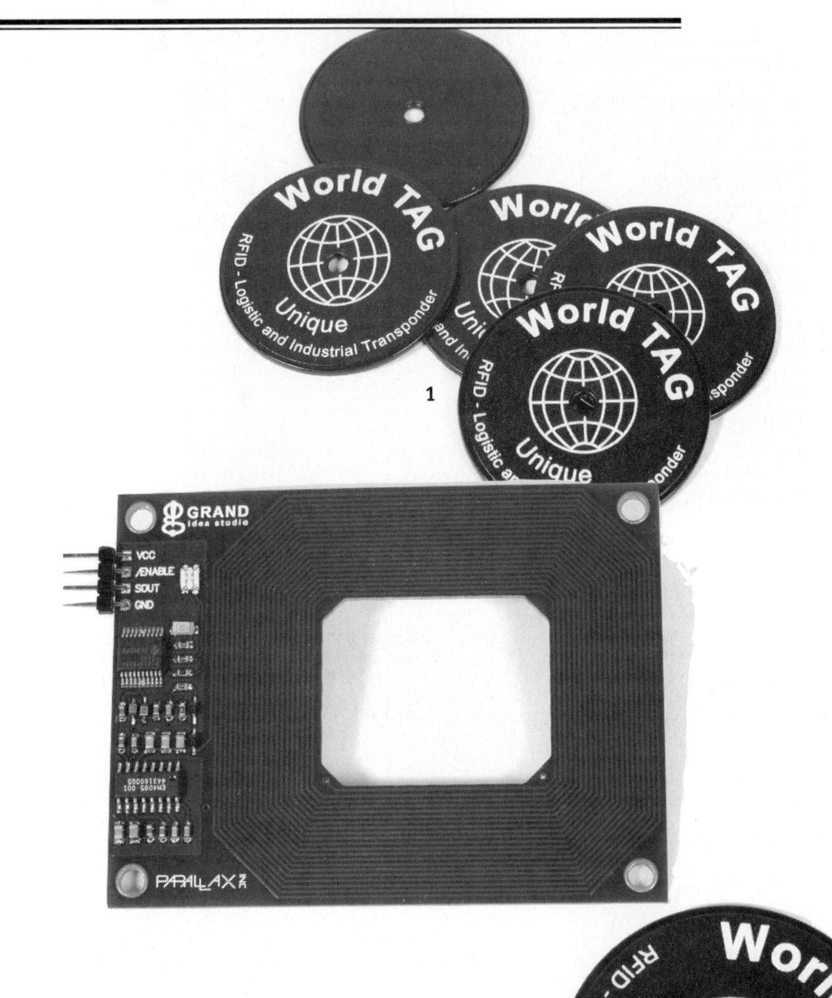

1

2

9 V Battery
Vss
Vin
J3
C3
C4
J2
Vin
Vss
PC0
PCI
Vdd
Rst
P0
P1
P2
P3
P4
P5
P6
P7
Reset
Rockln, CA — USA
PARALLAX INC™
BASIC Stamp Super Carrier
www.parallax.com (916)624-8333 (c)1999
Vdd
Vss
P15
P14
P13
P12
P11
P10
P9
P8
BS2

3

BS2
Rev C
20.0M

4

5 Install the BASIC Stamp 2 on the BASIC Stamp Super Carrier Board.

6 Hook the RFID Card Reader up to the BASIC Stamp Super Carrier Board. Connect the carrier board to a PC, and write a program for reading and responding to RFID tags.

7 When a tag comes within a couple of inches of the RFID Card Reader, the BASIC Stamp Super Carrier Board will act on your program.

8 These gloves were knitted with an RFID tag embedded inside the palm of the left-hand glove.

5

6

GRAND
9 V Battery
BS2
PARALLAX
Reset
BASIC Stamp Super Carrier
Rocklin, CA — USA

7

World TAG
Unique
RFID - Logistic and Industrial Transponder

8

9 This visible tag rests on top of the RFID tag that has been embedded in the glove's palm.

10 Bring the palm of the glove within a couple of inches of the RFID Card Reader and the BASIC Stamp 2 code can control a relay. For example, you could install this relay inside your car for operating the electric door lock on one door. Then just swipe your gloved hand near the card reader and the door's lock will be disengaged.

11 The venerable Parallax Boe-Bot™ can be controlled with RFID tags, too.

12 First, install an LCD screen on the Boe-Bot so that you won't need to be tethered to a PC for understanding your robot's actions.

13 Attach the card reader to the Boe-Bot breadboard.

9

10

11

12

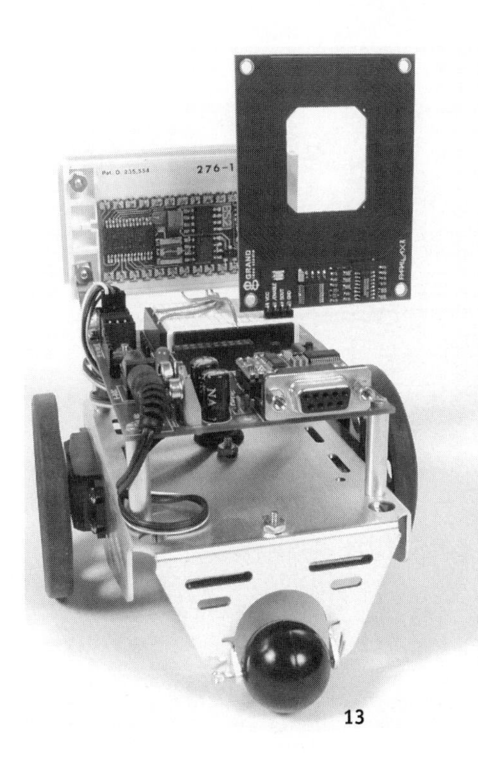

13

14 This arrangement can be used for reading tags held within a couple of inches of the reader.

15 You can enable an RFID-controlled navigation system with Boe-Bot by installing the RFID Card Reader horizontally in front of (or, behind) the robot.

16 Now Boe-Bot can react to the drive commands that have been assigned to each RFID tag.

17 The Boe-Bot GUI.

14

15

World TAG
Unique
RFID - Logistic and Industrial Transponder
PARALLAX
www.parallax.com
Actions:
Actions to be Performed:
Step #
Actions
1
Help Enabled
PC Serial Port
Auto
COM1
COM2
COM3
COM4
Sensors Enabled
STOP
Delete
Copy
Cut
Paste
Undo
Go
Clear
Test
Load
Save
Quit

16

17

18 The Boe-Bot GUI is very limited for assigning action to RFID tags.

19 The PBASIC programming editor is a better tool for writing a Boe-Bot RFID-controlled navigation system.

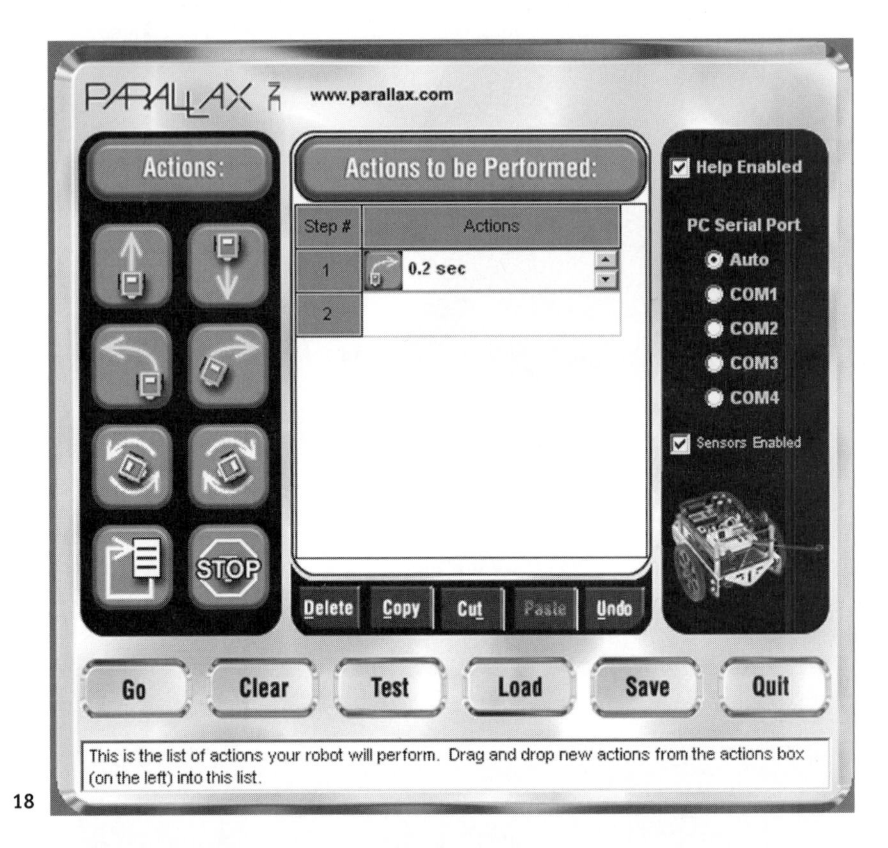

18

19

PART 2

CASE STUDIES

CASE STUDY 1

iRobot® Roomba®

A CLEAN ROOM MAKES A DIRTY MIND, ESPECIALLY WITH MIND CONTROL

OK, who doesn't wish that some sort of robot could be built for faithfully and obediently performing most, if not all, of our daily household tasks? You know, remove the drudgery of cleaning floors, rugs, and furniture from our daily lives. A "Rosie the Robot" maid from *The Jetsons* 1962 cartoon series, for example, would be the ideal incarnation, right? Well, maybe not.

Large humanoid-like robots wandering around our homes might be too much clutter and clamor (not to mention just plain creepy, especially at night) just for the sake of a cleaner house. A better household solution might be a low-profile, silent servant that inconspicuously moves around performing a specific cleaning task.

Enter iRobot Roomba (see Fig. CS1-1).

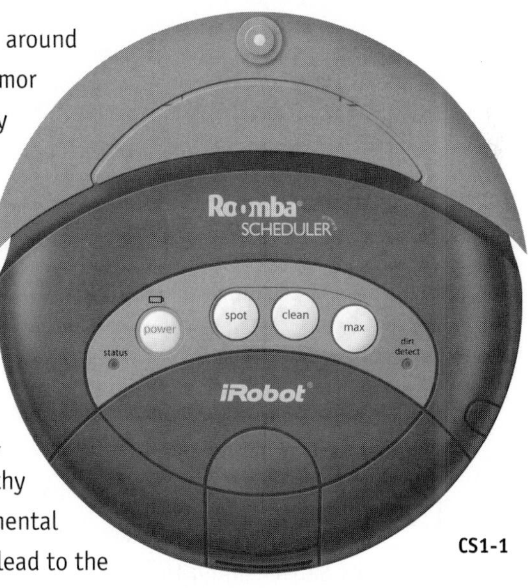

CS1-1

CS1-1 The iRobot Corporation Roomba Scheduler Robotic Floor Vac 2.1.

Founded in 1990 by three Massachusetts Institute of Technology overachievers, iRobot launched a long series of noteworthy robotic designs beginning with an experimental robot called *Genghis*. This hexapod crawler lead to the explosive ordnance-detecting crab-like *Ariel* robot in 1996. Oddly enough a doll named *My Real Baby*™ was iRobot's initial entry into consumer robotics in 2000.

In 2002, a floor-cleaning vacuum system, Roomba, had been designed. This revolutionary robot was not only affordable, but it really worked. Yes, the consumer was somewhat skeptical about a household robot that would clean floors, rugs, and carpets. This attitude changed, however, and in a big way.

Slowly reports began to emerge that Roomba really did work as advertised and, in many cases, this little robot did a better job cleaning floors than a human being (see Fig. CS1-2). In just a scant two years, and worldwide sales of Roomba had topped 1 million units sold.

CS1-2

CS1-2 **A Roomba can clean your floors for you.**

CS1-3 **A Roomba Scheduler will clean your floors at a predetermined time. Also, a Roomba is great for cleaning a garage or workspace when you're gone.**

CS1-4 **A Virtual Wall keeps a Roomba Discovery out of an area that shouldn't be vacuumed. [NOTE: The shading is a pictorial representation of the IR beam emitted by the Virtual Wall.]** (*Photograph courtesy of iRobot.*)

CS1-5 **You can corral your Roomba Discovery into a user-defined area with the IR beam-emitting Virtual Wall.** (*Photograph courtesy of iRobot.*)

Dirty floors became a historical footnote to the twentieth century. Roomba would actually sense dirt, seek it out, suck it up, and automatically return to a self-charging docking station for a quick sip of electricity for recharging its battery (see Fig. CS1-3). Unlike other household robots, Roomba wouldn't fall down stairs, could be blocked from entering certain rooms of your house, and could be programmed to clean on *your* schedule and *not* when the robot felt like vacuuming.

By using a sophisticated battery of sensors, along with a generous dose of artificial intelligence (iRobot calls this feature AWARE™), Roomba is able to locate its Self-Charging Home Base™ while avoiding pitfalls like stairs. If you want to restrict Roomba from an area, you can place a special infrared transmitter called a Virtual Wall® in its path (see Fig. CS1-4). When the robo-vac senses the signal from the Virtual Wall it reacts just like it bounced into a physical wall or chair (see Fig. CS1-5).

What Roomba did for the vacuum cleaner, Scooba™ did for the mop. Released to the public in 2005, Scooba was a floor-washing robot that also marked a unique collaboration between iRobot and The Clorox Corporation.

This joint development was distinctly different from Roomba.

A special, no-bleach, cleaning solution developed by The Clorox Corporation enabled Scooba to wash, scrub, and dry hardwood, linoleum, tile, and marble floors better than most bipedal life forms could do with a mop, bucket, and cleaner.

A new era in cleaning our household floors was born; coincidentally, right at the dawn of the twenty-first century. Lucky us. And all it took was two little robots and iRobot to save the average homeowner about 26 hours per year in cleaning, vacuuming, and mopping chores. Thanks iRobot. Now how about getting to work on that clothes ironing robot?

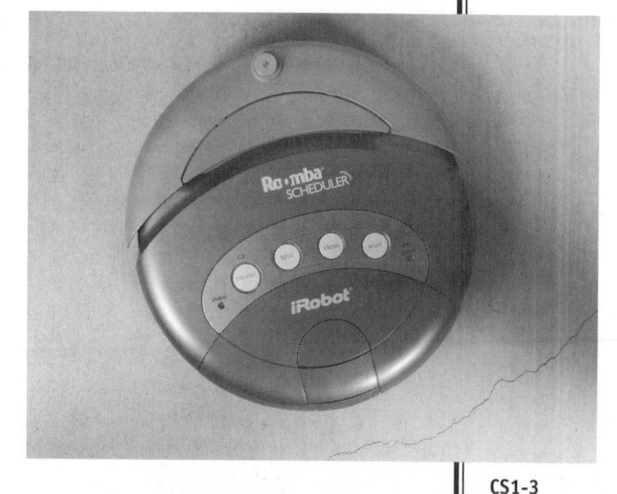

CS1-3

CS1-4

I Spy, SCI

All Roombas manufactured after October 2005 featured a special interface that enabled users to control this floor vacuum system's behavior and monitor the robot's sensors. This interface is the iRobot Roomba Serial Command Interface (SCI) (see Fig. CS1-6).

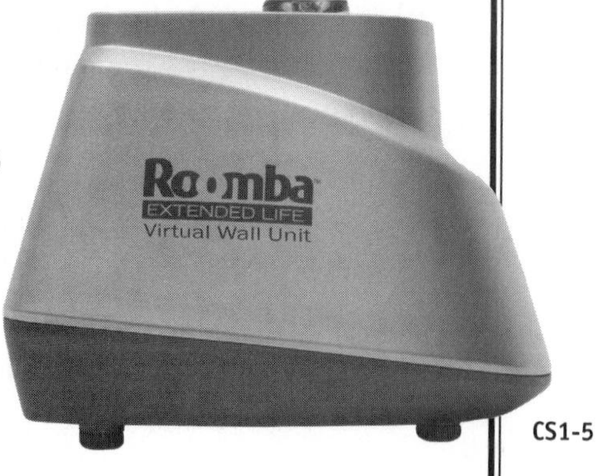

CS1-5

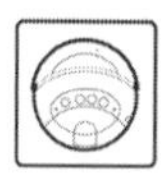

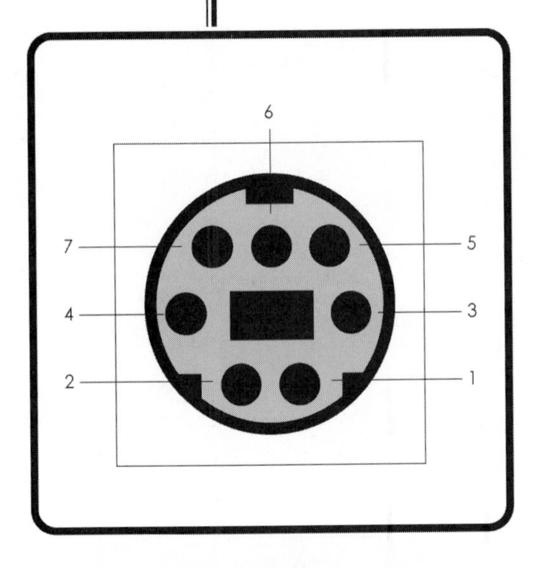

Pin	Name	Description
1	Vpwr	battery + (unregulated)
2	Vpwr	battery + (unregulated)
3	RXD	0 – 5V Serial input
4	TXD	0 – 5V Serial output
5	DD	Device Detect input (active low)
6	GND	battery ground
7	GND	battery ground

CS1-6 The Roomba SCI.

The Roomba SCI is a serial protocol mini-DIN connector with a built-in software command structure that can control all of Roomba's actuators (motors, LEDs, and speaker), as well as poll sensor data from Roomba's sensors. By using the SCI, users can modify the functionality of a normal Roomba and create a new set of operating instructions.

Oh, your Roomba doesn't have the SCI? A special hardware update called OSMO/Hacker (versions 1 and 2; based on the age/model of your Roomba) can be ordered from iRobot.

There are two distinctly different methods for achieving this control over the Roomba: hard and easy. In the hard method, users must build a serial communication hardware interface with the Roomba SCI. Furthermore, users who select this hard method for controlling a Roomba must also write machine-level software for gaining access to the Roomba SCI command structure. While achieving successful results with this hard method are possible, it does require mastering both hardware circuit design and sophisticated software engineering. Doable, but yuck.

In the easy method, users will only have to acquire a special Roomba controller called Mind Control. Mind Control is manufactured by Element Products and consists of an elegant lipstick-sized, removable mini-DIN controller module and a special Mind Control programming board.

Just plug this programming board into your computer's USB port, write your program in a simple C or C++ programming language environment, send your program to the Mind Control module, plug the module into the Roomba SCI, and run your program. That's it. There's nothing to build and no command structures to master.

Using Mind Control with an SCI-equipped Roomba is the surest method for building an autonomous, programmable robot that can also vacuum your floors for you.

Regardless of which method you opt for, the Roomba SCI physical connections remain the same:

PIN	NAME	DESCRIPTION
1	Vpwr	Roomba battery + (unregulated)
2	Vpwr	Roomba battery + (unregulated)
3	RXD	0 to 5-V serial input
4	TXD	0 to 5-V serial output
5	DD	Device detect input (active low)—wake up Roomba from sleep
6	GND	Roomba battery ground
7	GND	Roomba battery ground

NOTE: If you're thinking about building your own SCI-to-PC control circuit, please note that the RXD, TXD, and Device Detect pins use 0 to 5-V logic. Therefore, you will need to employ some sort of level shifter (e.g., MAX232) for enabling the Roomba to communicate with a PC using RS-232 serial communication voltage levels.

Don't Be So Mode(y)

There are four modes of operation for the Roomba SCI: off, passive, safe, and full. OK, "off" mode is pretty easy to figure out; no power and the SCI is off. Ta-dah. When in the off mode, the SCI waits for the Start command. If the Start command is received, the Roomba enters passive mode.

In passive mode, users can do the following:

- Request and receive sensor data using the Sensors command
- Execute Power, Spot, Clean, and Max commands
- Define a song, but not play one
- Set force-seeking-dock mode

NOTE: Users cannot control any of Roomba's actuators when in passive mode.

When in passive mode, sending the Control command to Roomba will force it into safe mode. In safe mode, users will have full control of the robot, except for affecting the following conditions:

- Detection of a cliff while moving
- Detection of wheel drop
- Charger plugged in and powered

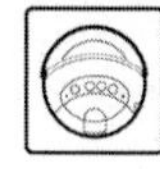

Five Questions with Ben Wirz, President, Element Products, Inc.

QUESTION ONE: You are the co-author of two books, "SUMO BOT: Build Your Own Remote-Controlled Programmable Sumo-Bot" (McGraw-Hill, 2002) and "TAB Electronics Build Your Own Robot Kit" (TAB Books, 2001). Each of these books is noteworthy in that they actually contain a robot kit that is wrapped around a "book" concept. Where did this unique concept originate?

Scott Grillo of McGraw-Hill spearheaded the original concept focusing on the marketing and business aspects. Myke Predko and I were brought in as the "authors" to perform the design work and write the text. Myke's focus was on the microcontroller firmware and documentation. I handled the electrical and mechanical engineering and managed the manufacturing effort. We each had a unique and complimentary set of skills that made us an effective team. I've always felt a certain sense of pride for being part of a project which brought an educational robot kit to the mass market. The kit wouldn't have been possible without a marketing partner like McGraw-Hill. Scott deserves credit for convincing McGraw-Hill to break out of their book publishing model and take a chance on an unproven product concept. They exposed many new people to the exciting world of hobby robotics.

FOLLOW-UP QUESTION: Your co-author with these robot books was the legendary "Mr. PIC," Myke Predko. While Predko is well known for his patent-pending design for automated testing of PC motherboards, you are also a patent holder of some robotics technologies. Once you have successfully received a patent, do you consciously attempt to apply that technology in all of your work?

Patents are mostly the domain of large corporations and I've always worked for smaller companies. The patents in which I have been involved were designed to protect a particular product which was a unique collection of existing technologies rather than a new technology itself. So I generally don't apply patented technology to new product designs.

FOLLOW-UP QUESTION: Do you subscribe to the notion that some products are "designed to be hacked"? If so, does "hacking" extend your patent to a greater user base or infringe on your patent's definition?

I think it is the rare case that a product is intentionally designed to be hacked. The hacker market tends to be a small fraction of the total market share for most consumer products. The additional sales generated from making a product hacker friendly wouldn't justify the engineering expenditure. Consumer product companies also have to worry about tech support and liability issues. I suspect that companies actually often work to make a product more difficult to hack. When you see products with hacker features built in, it is likely a labor of love from an engineer done behind the scenes rather than something intentionally included. I imagine hackers probably prefer it this way. Hacking wouldn't be nearly as much fun if manufacturers made it too easy.

QUESTION TWO: Prior to starting Element Products, you worked for iRobot Corporation. Well known for their wildly popular Roomba robotic vacuuming system, iRobot is also involved in the design, testing, and manufacturing of commercial robots. Many of these commercial robots are "employed" in Department of Defense (DoD) military applications. What was the scope of your job involvement with iRobot?

I was initially hired by iRobot for an industrial robotic cleaning program. It was the company's first attempt at a nonresearch product so there was a lot of excitement surrounding the project. The team designed a massive 500-pound robot that was highly effective at wet scrubbing, waxing, and burnishing floors in a retail setting. Following the cleaning program, I joined iRobot's R&D division which primarily focused on DARPA and DoD government-sponsored research projects. I designed some of the control electronics for the crab-inspired robot Ariel which detected and removed mines in the surf zone along a beach. Ariel's unique leg configuration allowed it to scramble over obstacles and right itself when flipped over by a wave. I then moved onto the Swarm project which was one of the most interesting projects that I worked on at the company. The goal of the Swarm project was to develop distributed algorithms for robotic swarms composed of hundreds of individual robots. We built a small but sensor-rich and computationally powerful robotic platform for the researchers. Finally, I joined iRobot's Packbot Team. The Packbot is a highly mobile tracked robot that was initially designed to perform military and police reconnaissance missions. The Packbot platform has been highly successfully and now forms the backbone of iRobot's Government and Industrial division.

FOLLOW-UP QUESTION: It has been suggested that Alfred Nobel, the inventor of dynamite and the founder of the Nobel Foundation which awards the Nobel Peace Prize to individuals whose work has "been of the greatest benefit to mankind," was so appalled with the "employment" of dynamite in military actions that he established his annual prize as a penitence for this unforeseen misapplication of his discovery. How do you feel about the usage of robots in warfare?

War history has always been an area of interest for me so I find this topic particularly intriguing. Technology has proven a significant deciding factor in every war fought in modern history. The appeal of a combat robot is strong. Why risk a solider when the military can send a replaceable machine to do the fighting? iRobot's Packbot and Foster Miller's TALON robots are saving U.S. soldiers' lives in Iraq by performing dangerous reconnaissance and explosive disposal missions today. A combat robot could also be much more precise than air bombing which has historically been one of the largest sources of civilian causalities.

The potential advantages of a combat robot seem promising but what about the moral implications of having a machine that can kill remotely?

Many conflicts in recent history have been concluded not by one side surrendering but instead from the domestic political pressure that results from a country losing its own soldiers. I worry that a world power that wields the capability to fight with minimal risk to its own soldiers will enter combat more quickly and sustain combat for longer periods of time.

QUESTION THREE: Following your departure from iRobot, you founded Element Products. While many of us might contemplate leaving the safety of a corporate job and attempt to do our own thing, few of us ever do it. But you did. Some have said that you are a true entrepreneur—a dreamer who follows a different drum beat. What dream do you hope that Element Products will fulfill?

I think many business owners have the same dream of financial independence but you really need something more to sustain yourself through the inevitable trials and tribulations. My dream for Element Products revolves around my desire to have a job that makes me excited to come into work in the morning. I have always loved to design and build things since I was child. I have a particular interest in designing automated machines with sensors and actuators. Starting a small robotics company with my business partner seemed like a natural path to fulfilling this dream.

FOLLOW-UP QUESTION: Running your own business is never like collecting a biweekly pay check. Suddenly you are responsible for everything: payroll, health care, retirement plans, marketing, office management, and even janitorial duties. Oh, and then you have to tiptoe through the client minefield—placate unreasonable clients, strong-arm deadbeat clients, and solicit new clients. Those are a lot of different hats for one person to wear. What is your favorite aspect of owning your own business?

Working to start a new company on a shoestring budget has certainly been one of the biggest challenges of my life but I have really loved it.

There seems to be something about my personality that thrives on multitasking to a degree that would drive a normal person crazy. I find it is interesting that this same trait which made me feel stifled in the corporate world has proven a real asset in working at a start-up. My three favorite aspects are having a greater sense of control, the thrill of working to build something with a small team, and a flexible work schedule.

QUESTION FOUR: By some accounts, the twenty-first century could be dubbed the "dawn of the robotic age." Many of our more popular toys are robots (e.g., WowWee Robotics Robosapien), we coexist with robots in our homes (e.g., iRobot Roomba), and we trust robots to build our automobiles (e.g., industrial robots). This type of pervasive encroachment into our daily lives would have been unthinkable 100 years ago. Indeed, I keep waiting for some "futurist" to marry a robot someday. How do you foresee robots changing our lives in this century?

I think we are decades away from robots truly changing our lives in significant ways. Even the simplest robotic task requires a massive engineering effort to solve, and then the solutions are often uneconomical. Dish and clothes washing have been automated for many years but most people don't think of these appliances as robots. More recently carpet vacuuming, floor scrubbing, and grass mowing have also been automated by robots. You only have to think about the other chores around your home to see more potential robotic applications. I believe that we will continue to see many smaller changes in our lives as single-purpose robots become appliances in our homes. The multipurpose Rosie the Robot is still decades away though, and even when it becomes technically feasible it will still have to compete economically with a minimum wage worker. That's a tall order!

FOLLOW-UP QUESTION: My three daughters take robots in their daily lives for granted. At times, they even treat robots as if they possess a form of innate intelligence. Conversely, most adults completely lack

any appreciation for robots and typically snort a retort of contempt when offered a robotic alternative to a common task or chore... like vacuuming the floor. Why are children always receptive to new technology while adults are generally dragged "kicking and screaming" into a new era?

That's an interesting question. It probably partially comes from children's lack of preconceived ideas about how things should work. They are still discovering the world so everything is new to them. I've had a chance to observe a number of different reactions to the Roomba robot vacuum cleaner. Children generally love it and often think of it as a pet or even a friend. They see personality traits and quirks in what I know are just lines of code responding to the robot's sensors in a deterministic way. I've heard some adults say, "Only a lazy person would own a Roomba" while others love the idea. I don't think my mom has ever taken her Roomba out of the box but it has improved the lives for some elderly people who aren't physically able to vacuum anymore.

QUESTION FIVE: I once heard that a screw was invented by a corporate lawyer, while the screwdriver was invented by a hacker. The hilarious double entendre of these definitions notwithstanding, the term "hacker" is slowly regaining respect in public opinion. As an engineer, a patent holder, and an entrepreneur, how do you view a hacker?

I like to think of hackers as curious technical people with extra time on their hands. Many people have a real love of technology. They eat and breathe technology. If their job doesn't satisfy that itch then they may wind up tinkering in their free time. I think hackers have a positive influence on technology. They are out on the fringe using technology in new ways spurring innovation and new products ideas.

FOLLOW-UP QUESTION: Are you a hacker?

Of course! Most engineers are to some degree.

NOTE: If any of these safe mode conditions occurs, Roomba will stop and enter passive mode.

Finally, if Roomba is in safe mode, you can enter full mode by sending the Full command. Full mode disables the two detection conditions of the safe mode while retaining the powered charger condition. On the other hand, full mode enables unrestricted control of Roomba's actuators.

NOTE: In order to return to safe mode, just send the Safe command to the SCI. Additionally, a 20-millisecond pause must be issued between all mode-changing commands.

Roomba SCI Commands

These are the commands that can be sent to the SCI for controlling a Roomba. Remember, just like your own kids, Roomba will not respond to any command while it is asleep. Wake it up by toggling the Device Detect pin of the mini-DIN to a low state for 500 milliseconds.

Start command opcode: 128
Number of data bytes: 0
Starts the SCI.
Serial sequence: [128]

Baud command opcode: 129
Number of data bytes: 1
Sets the baud rate in bits per second (bps) for the SCI
Serial sequence: [129] [Baud Code]

Control command opcode: 130
Number of data bytes: 0
Enables user control of Roomba.
Serial sequence: [130]

Safe command opcode: 131
Number of data bytes: 0
This command puts the SCI in safe mode.
Serial sequence: [131]

Full command opcode: 132
Number of data bytes: 0
Enables full mode control of Roomba.
Serial sequence: [132]

Power command opcode: 133
Number of data bytes: 0
Puts Roomba to sleep.
Serial sequence: [133]

Spot command opcode: 134
Number of data bytes: 0
Starts a spot cleaning cycle.
Serial sequence: [134]

Clean command opcode: 135
Number of data bytes: 0
Starts a normal cleaning cycle.
Serial sequence: [135]

Max command opcode: 136

Number of data bytes: 0

Starts a maximum time cleaning cycle.

Serial sequence: [136]

Drive command opcode: 137

Number of data bytes: 4

Controls Roomba's drive wheels.

Serial sequence: [137]
[Velocity high byte]
[Velocity low byte]
[Radius high byte]
[Radius low byte]

Motors command opcode: 138

Number of data bytes: 1

Controls Roomba's cleaning motors.

Serial sequence: [138] [Motor Bits]

LEDs command opcode: 139

Number of data bytes: 3

Controls Roomba's LEDs.

Serial sequence: [139] [LED bits]
[Power color] [Power intensity]

Song command opcode: 140

Number of data bytes: 2N + 2, where N is the number of notes in the song

Specifies a song to the SCI to be played later.

Serial sequence: [140]
[Song number] [Song length]
[Note number 1] [Note duration 1]
[Note number 2] [Note duration 2]
...

Play command opcode: 141

Number of data bytes: 1

Plays one of 16 songs, as specified by a Song command.

Serial sequence: [141]
[Song number]

Sensors command opcode: 142

Number of data bytes: 1

Requests the SCI to send a packet of sensor data bytes.

Serial sequence: [142]
[Packet code]

A RooTooth Ache

Roomba Dev Tools, a division of Robo Dynamics, created a Bluetooth connection for Roomba. Called RooTooth, this device enables you to communicate with your Roomba via any Bluetooth device. Using the Bluetooth Serial Port Profile, RooTooth can give your vacuum a wireless connection to your Bluetooth cell phone for letting you clean that apartment before you arrive home with the boss.

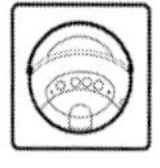

Roomba SCI Sensor Packets

Roomba will send back one of four different sensor data packets (e.g., packet subsets 0 through 3) in response to a Sensor command.

PACKET SUBSET 0

All Sensor Data Bytes

Packet subset: 0

Packet size: 26 bytes

PACKET SUBSET 1

Bumps Wheel Drops

Packet subset: 1
Range: 0 to 31

The state of the bump and wheel drop sensors.

Wall

Packet subset: 1
Range: 0 to 1

The state of the wall sensor.

Cliff Left

Packet subset: 1
Range: 0 to 1

The state of the cliff sensor on the left side of Roomba.

Cliff Front Left

Packet subset: 1
Range: 0 to 1

The state of the cliff sensor on the front left side of Roomba.

Cliff Front Right

Packet subset: 1
Range: 0 to 1

The state of the cliff sensor on the front right side of Roomba.

Cliff Right

Packet subset: 1
Range: 0 to 1

The state of the cliff sensor on the right side of Roomba.

Virtual Wall

Packet subset: 1
Range: 0 to 1

The state of the virtual wall detector.

Motor Overcurrents

Packet subset: 1
Range: 0 to 31

The state of the five motor overcurrent sensors.

Dirt Detector Left

Packet subset: 1
Range: 0 to 255

The current dirt-detection level of the left side dirt detector.

Dirt Detector Right

Packet subset: 1
Range: 0 to 255

The current dirt-detection level of the right side dirt detector.

PACKET SUBSET 2

Remote Control Command

Packet subset: 2
Range: 0 to 255
(with some values unused)

The command number of the remote control command currently being received by Roomba.

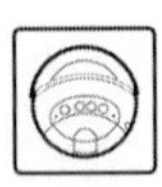

Buttons

Packet subset: 2
Range: 0 to 15

The state of the four Roomba buttons.

Distance

Packet subset: 2
Range: –32768 to 32767

The distance that Roomba has traveled in millimeters since the distance was last requested.

Angle

Packet subset: 2
Range: –32768 to 32767

The angle that Roomba has turned through since the angle was last requested.

PACKET SUBSET 3

Charging State

Packet subset: 3
Range: 0 to 5

A code indicating the current charging state of Roomba.

Voltage

Packet subset: 3
Range: 0 to 65535

The voltage of Roomba's battery in millivolts (mV).

Current

Packet subset: 3
Range: –32768 to 32767

The current in milliamps (mA) of Roomba's battery.

Temperature

Packet subset: 3
Range: –128 to 127

The temperature of Roomba's battery in degrees Celsius.

Charge

Packet subset: 3
Range: 0 to 65535

The current charge of Roomba's battery in milliamp-hours (mAh).

Capacity

Packet subset: 3
Range: 0 to 65535

The estimated charge capacity of Roomba's battery.

End of Roomba SCI Sensor Packets

Force-Seeking-Dock command opcode: 143

Number of data bytes: 0

Turns on force-seeking-dock mode.

Serial sequence: [143]

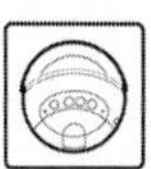

Five Questions with Phil Mass, Vice President, Element Products, Inc.

QUESTION ONE: You graduated from the University of Washington with a degree in sculpture, as well as one in physics. During the preparation of a book that I wrote about Robosapien (THE OFFICIAL ROBOSAPIEN HACKER'S GUIDE, McGraw-Hill, 2006), I interviewed Mark W. Tilden. We all know Tilden as the inventor of Robosapien and the father of BEAM (Biology, Electronics, Aesthetics, Mechanics) robotics, but I also noted that he was a gifted artist. When I approached this subject he bristled with an indifference toward his sensitive artistic side. How do you balance these distinctly "right brain," "left brain" achievements in your life?

I don't see it as a contradiction, but rather two different viewpoints on a subject. To me, science and art are closely related. They both satisfy my same need to create things. You often hear people who write software refer to beautiful code. I think that they're referring to beauty in a different context, but the meaning is the same. Aesthetics run through most human endeavors and they are a real force in good design. So, I'd say that, if anything, art is more important to me than the technical aspects and the technical work is, hopefully, an expression of good aesthetics. The processes of art and engineering can also be similar. For me, sculpting and writing code are both iterative processes. First, I get an idea of the project in my mind, and then put a first draft down on paper or clay. From there, I slowly change things, growing and shaping the design until it works in the way it needs to. And with both, I often come to a point partway into the design where I need to throw it all out and start from scratch again because after building it once I see how it should really be put together.

FOLLOW-UP QUESTION: Good industrial design practice is one element that is sadly delinquent in many of today's robots. For example, I shudder when I see a WowWee Robotics Robopet. What were they thinking? A "pet" without touch sensors that gimps along like some

hobbled Soviet-era cyborg. One thing that Apple Computer has clearly proven, however, is that good industrial design sells. One robot which does successfully blend its function into an appropriate form factor is the iRobot Roomba. Prior to joining Element Products, you worked at iRobot Corporation on the Roomba robotic vacuuming system project. Did you assert any of your sculptor sensibilities into the design of the Roomba?

Being part of the Roomba design team at iRobot was one of the best work experiences I've ever had. There is a lot of talk at companies and in business books about teams and teamwork, but I think that there are very few real teams out there. For me, it was the first time in my life that I felt like I was part of a real team, and I was amazed at how powerful it was. It wasn't like we didn't have plenty of disagreements over all sorts of issues, but behind it all we were working toward the same goal of making a simple robot that did real work. It was a fundamentally different, and much more enjoyable, work experience than any I'd had up to that point. So, there ended up being contributions from each of the team members in all corners of the design, just as there were ideas big and small from all of the team members in the software I wrote. But to answer your question, I wasn't involved in the industrial design of the Roomba directly, except for the user interface. I was responsible for creating how the end user would control the Roomba, following our team philosophy of keeping it as simple as possible. Looking back at it now, I wish I had just used one button instead of three, but you always want to improve on your designs. If you indulge that urge too much, though, products never reach the shelves.

FOLLOW-UP QUESTION: Ironically, your major contribution to Roomba was through programming. Can programs be sculpted into good designs?

Definitely, even literally. After the original Roomba was finally released to production after years of work, I had the urge to make a large interactive sculpture to illustrate the working of the code. I had been so involved with the code for a long time and knew what was contained in every byte of memory, so I wanted to make it physical and visual. The idea was a metal sculpture maybe ten feet tall with running water, levers, and rotating dials. Water would fill reservoirs as variables were integrated, spilling over when the value reached its threshold. And people would be able to interact with it, to press the giant bumper to see what would happen inside. I'm not sure if it would actually work, but it was fun to think through. Sometimes I do wish there were better tools for visualizing software. I think that if there were a program that would translate code into a graphical representation, it would help coders to see the flaws. People tend to be very good at visual pattern matching. Maybe it could also interest other more artistic and visual types of people in coding.

QUESTION TWO: The theatrics of Isaac Asimov's Three Laws of Robots notwithstanding, (e.g., 1. A robot may not injure a human being, or, through inaction, allow a human being to come to harm. 2. A robot must obey the orders given it by human beings except where such orders would conflict with the First Law. 3. A robot must protect its own existence as long as such protection does not conflict with the First or Second Law.) how do you account for "human variables" in programming robots?

I read Isaac Asimov for the first time a couple of years ago and I have to admit I was pretty disappointed. There wasn't much of what I find compelling about robotics in iRobot. It was so clean and theoretical, more like the philosophy of psychology than robotics. I don't think we'll ever need the three laws. I think that problems with robots will be much more like industrial accidents than the moral quandaries of the three laws, like a robot falling over due to a burnt out motor and breaking someone's leg. A friend of mine from iRobot, Clara Vu, had rules she'd tell friends to follow if a robot ever threatened them. Things like "Go upstairs," "Pour water on it," or "Just wait a little while and its battery will run down." This will protect you from most robots we can create now and in the near future. I've always been interested in robots as creatures, not human-like companions. The work of Rodney Brooks got me interested in robotics in the first place, and his idea was to build robots from the bottom up, starting by emulating relatively simple creatures like insects. I do think it's great that many people, including my mom, think of their Roombas as pets. This may be a good model of how we will interact with many robots in the future. It may also make us more forgiving of their shortcomings. The ultimate goal is to design robots that people interact with naturally without ever having to read a manual.

FOLLOW-UP QUESTION: One of the most significant third-party products to emerge for the budding Roomba accessory market is Mind Control. Mind Control was designed (and is being manufactured and sold) by your company, Element Products. This product enables anyone to access and control all of the functions of a Roomba. Did you program Roomba with an eye toward designing a product like Mind Control which would allow users to "program" the robot themselves?

Actually, no. The only thing we were focused on at the time was making a product that was easy to use and cleaned floors well without getting stuck. At first, the Roomba was not even marketed as a robot because we didn't know how people would react to the idea of a household robot. But, now with the success of the product, I think that there is a great opportunity to expand the capabilities of the Roomba with products like the Mind Control. There is a large base of Roomba owners that can now take control of their robot and personalize it in any way they choose. And it's not an either/or proposition. With the Mind Control, you can use your Roomba as a robotics experimental platform at night, and during the day still have it clean your floors. Because of the low cost, we envision that Roombas and Mind Controls will also be used in the classroom to teach robotics. Instead of one robot for an entire class with everyone having to wait their turn, now every student can have their own robot to program. I can't wait to see a classroom full of a couple dozen Roombas, each dancing to a different student's beat.

QUESTION THREE: You've had the unique opportunity to work for two of the biggest players in the home consumer robotics marketplace: iRobot Corporation and Parallax, Inc. In the case of Parallax, your contribution was the design of the Scribbler™ Robot. This low-cost introductory robot was targeted at the classroom environment as a platform for teaching rudimentary programming. A cursory examination of local schools, however, would seem to indicate that the Scribbler Robot has been a marketing flop. My quick poll of six Mississippi school districts located only one of these introductory robots...and that robot was mine! Since the Scribbler Robot was your "baby," do you feel a personal loss when one of your robots is not accepted by the public?

I don't consider the Scribbler to be a flop at all. The difficult truth we learned with the Scribbler is how hard it is to get the word out about a product as a small company. Without large advertising and marketing budgets, you have to rely primarily on word of mouth. While very effective, it is a slow method. But the Scribbler keeps getting good reviews in a wider and wider circle. I think one of our primary Scribbler customers will be parents who want to give their children an educational toy that will instantly engage them, but also keep their interest for a long time. This market is hard to reach, but the Scribbler's presence is growing slowly, but surely, and I think that once all of those parents know what a Scribbler is, it will be a success. Our other primary customers, educational systems, are slow to adopt new technologies for several good reasons. They are on a yearly cycle, so they can't easily add something new in the middle of the year and they also need to be sure that what they are adding has real educational value for the students. But Scribblers have already been used in a computer science class at Princeton and we've gotten great responses from teachers who have seen the product. But as a small company, patience is indeed a virtue.

FOLLOW-UP QUESTION: **Actually, I have a few gripes with the Scribbler Robot. The failure to bundle batteries and a suitable writing instrument in the package is irksome. How many of us have 6 AA batteries and a Sharpie® pen laying around for use with a robot? Furthermore, selling a product for the educational market without addressing software support for the Mac OS is not only ignorant, but it is dooming the product for failure. As a programmer, I'm sure that you can appreciate the need for matching a product to its intended market. When dealing with a client, how do you lobby for adding a feature or capability to a product that is clearly against the client's specifications?**

Deciding on the right features for a new product is always a difficult process. On the Scribbler, Roomba, and other products I've been involved with, we tried to give consumers the most for their money. There are never easy answers. You try to guess what your customers want by talking to as many potential users as possible. But you also need to be aware of how much a feature will add to the final cost of the product. So, you gather the input of the customers, design team, and management, and slowly iterate toward the best product. In terms of the specific issue of Scribbler Mac support, you can currently program your Scribbler on a Mac using the PBASIC editor. And the graphical programming software is open source Perl code that can be compiled and run on a Mac, but only under X Windows for now until the Tk/Tcl libraries are compiled natively for OS X. In other words, we're working on it.

QUESTION FOUR: **You seem to be very passionate about your designs. For example, in the November 2005 issue of SERVO MAGAZINE**

(Vol. 3, No. 11) you wrote a tutorial article about programming the Scribbler Robot ("Extending the Scribbler Robot" by Phil Mass). This type of "guerrilla" support for a product is typically a service performed by a marketing or public relations agency...not the designer. Do you feel a personal attachment to your designs?

Yes, of course. I think that it's the only way to sustain your effort on long-term projects. It would be hard to work on anything for many months or years if you didn't feel that connection with it. The more you enjoy something, the easier it is to do.

FOLLOW-UP QUESTION: **Should more designers be this passionate about their creations?**

I think that most designers are inherently passionate about what they create. Unfortunately, I think that passion can get lost in a bad work environment. Collaboration, conflict, and compromise are important parts of the creative process, especially when designing with other people. But politics and ego can create a dysfunctional environment that breeds bad design.

QUESTION FIVE: **If you are like a protective father with your designs, how do you relate to hackers who alter the function of one of your "children"?**

I think it's great. It means that the product is still alive and changing. I work hard to have my designs fulfill their intended purpose, but that doesn't mean it should stop there. I think that one of the extra benefits of consumer robots is that they can be mass-produced at a lower cost, thereby supplying hackers with less expensive raw materials for their creations. In a way, I get to interact with the hackers when they alter something I've helped design. I'm really looking forward to seeing what people create with the Mind Control and their Roombas.

FOLLOW-UP QUESTION: **Are you a hacker?**

I can't claim to be a hacker. I've always been fascinated with how things work, from thunderstorms to botany and DNA. And I've always loved building things, but I'm lucky enough to get most of my need to create things fulfilled in my work. My grandfather was a big hacker from a different era. He worked in a Northern Pacific Railway machine shop for over 50 years and brought home anything he could to machine or weld together into something new. You should have seen his garage and workshop; every surface was piled high with parts. I do write some software in my spare time and would like to get involved with an open source software project, since I think that open source software will have a very positive influence on the future of technology. But mostly, after building things all day, I appreciate some of the other joys of life, like good conversation, music, and baseball.

Hey, Houdini, How About Some Mind Control

If you're looking for the easiest, least painful method for controlling Roomba, then look no farther than Mind Control. Mind Control lets you take control of Roomba through conventional C or C++ programming. Furthermore, Mind Control comes with everything that you need for making your Roomba, well, *your* Roomba.

While the "official" statement from Element Products states that the Atmel ATmega168 microcontroller inside Mind Control should be programmed with the Atmel WinAVR programming tool on a Windows XP PC, there is an option for Roomba users who wish to use Mac OS X.

AvrFlasher by Henri-Pierre Garnir is a Mac OS X application originally developed for programming the AVR Butterfly board, but it can be applied to other AVR microcontrollers.

Regardless of your programming environment, a sample program for Mind Control would look like this:

```
// Included files
#include <avr/interrupt.h>
#include <avr/io.h>
#include <avr/delay.h>
#include "sci.h"

// Functions
DEFINE YOUR FUNCTIONS HERE

int main (void)
{
  uint8_t buttons = 0;

  // Initialize the AVR
  initialize();

  // Wake the Roomba using the DD pin on the mini-DIN
  wake();
```


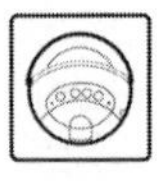

```
// Start the interface
byteTx(CmdStart);

// Change to 28800 baud
baud28k();

// Take full control of the Roomba
byteTx(CmdControl);
byteTx(CmdFull);

// Turn on the spot and clean leds
byteTx(CmdLeds);
byteTx(0x0C);
byteTx(0);
byteTx(0);

// Get rid of unwanted data in the serial port receiver
flushRx();

for(;;)
{
 // Request the 6 middle sensor bytes
 byteTx(CmdSensors);
 byteTx(2);

 // Read the 6 bytes, only keep the buttons data byte
 byteRx();
 buttons = byteRx();
 byteRx();
 byteRx();
 byteRx();
 byteRx();

 // If the clean button is pressed
 if(buttons & 0x02)
 {
  // Turn on the vacuum motor
```


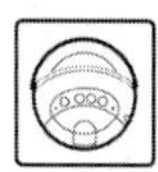

```
      byteTx(CmdMotors);
      byteTx(0x02);
    }
    // Else, if the spot button is pressed
    else if(buttons & 0x04)
    {
      // Turn off the vacuum motor
      byteTx(CmdMotors);
      byteTx(0x00);
    }
  }
}
```

DEFINE THE REST OF YOUR SUBROUTINES LIKE:

```
uint8_t byteRx(void)
{
  // Receive a byte over the serial port (UART)
  while(!(UCSR0A & _BV(RXC0))) ;
  return UDR0;
}
```

NOTE: This code snippet is from a real Mind Control program that was provided by Element Products.

Jacking into Roomba

Once you have a handle on programming the SCI with Mind Control, get ready for prime time—removing the Roomba main printed circuit board (PCB) and repurposing it in another robot design. Unfortunately, the Roomba PCB is not annotated with functional descriptions for each of its connectors. In other words, it's very difficult to determine which connector controls which actuator or sensor just by looking at the PCB. Now you might begin to appreciate all of the verbose labels that WowWee Robotics and Mark W. Tilden added to the Robosapien PCB, eh?

The Roomba PCB is an extremely powerful robot controller. And it is worth the extra effort to trace each connector to its actuator and sensor. Now you can

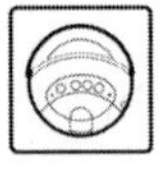

either manually follow each wiring harness from its source to the appropriate connector or you can refer to Hugo Perquin's Roomba PCB connector Web site.

Hugo Perquin's complete analysis of the Roomba connectors can be found at prj.perquin.com/roomba/pcb.php.

Three of the more important Roomba connectors for interfacing with other robots are:

- J4—left motor
- J16—right motor
- J7 red/white/black—battery

Imagine, with just the Roomba PCB, Mind Control, and a suitable platform, you could design an extremely powerful robot with very little effort. In fact, a home security system (see Chapter 4), a Roomba-controlled lawn mower (see Chapter 10), or a pet-sitter robot (see Chapter 15) are all possible hacks with a little Mind Control.

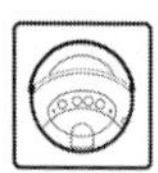

THE ANATOMY OF THE ROOMBA OSMO/HACKER

1 The OSMO/Hacker.

2 Locate the charge plug on the side of your Roomba.

3 Remove the cover above this plug.

4 Slip the OSMO/Hacker into place.

1

2

3

Roomba
SCHEDULER
power
spot
clean
max
dirt
detect
iRobot
iRobot
Roomba
OSMO//hacker

4

5 Remove the screws on bottom of the OSMO/Hacker.

6 The OSMO/Hacker PCB.

7 The underside of the OSMO/Hacker PCB.

8 You can remove the SCI Din plug from the OSMO/ Hacker for use in your own Roomba hacks.

5

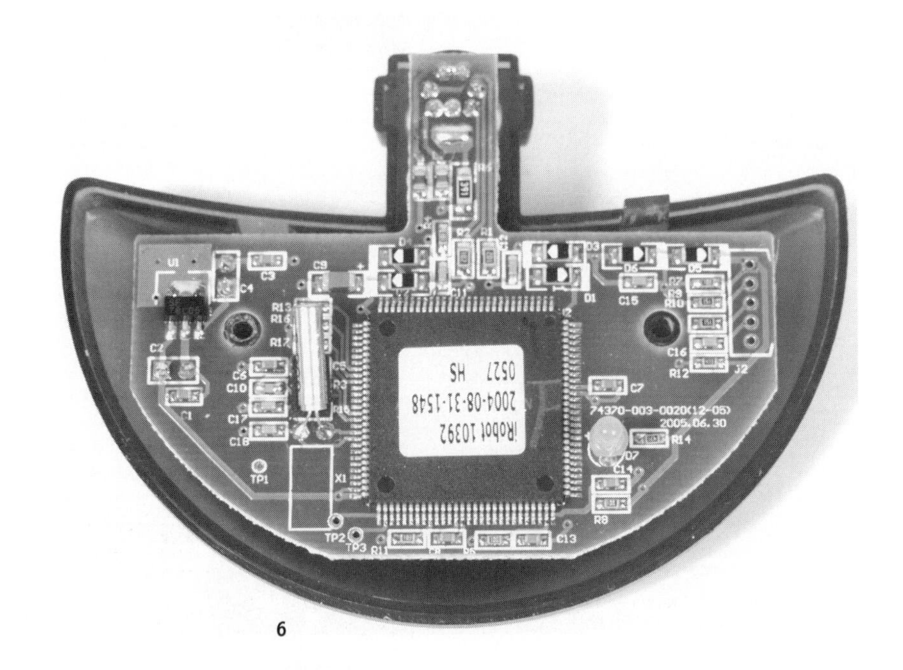

6

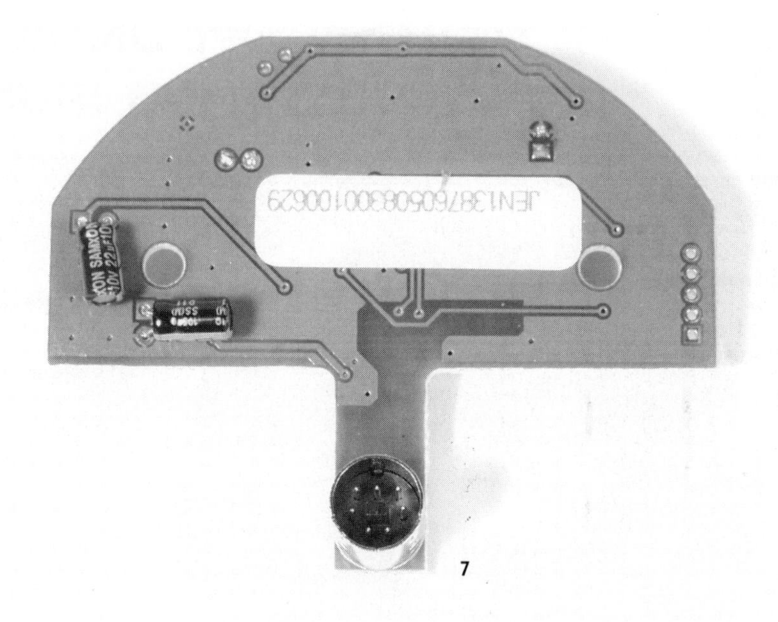
7

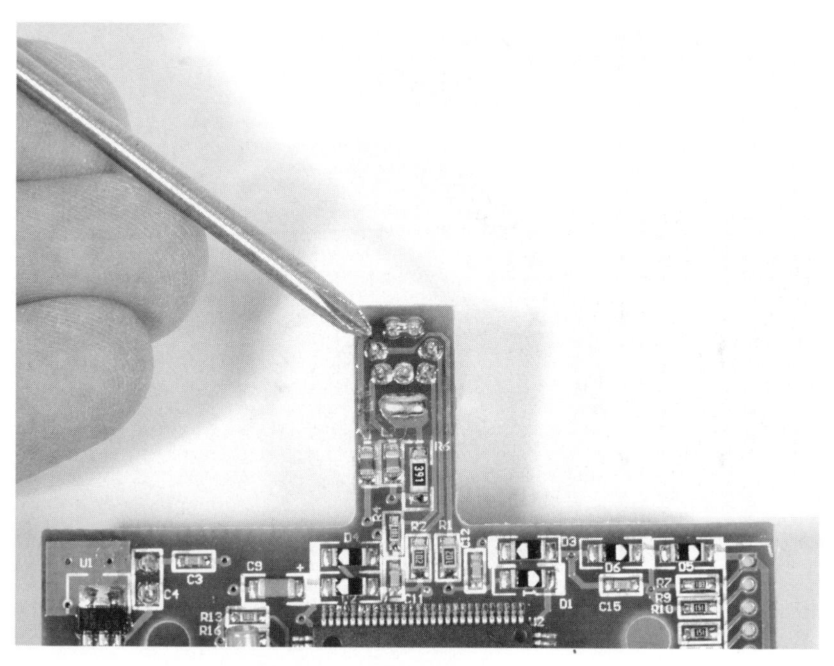
8

HOW TO GET UNDER ROOMBA'S SKIN

1 A fresh Roomba Red.

2 Turn the Roomba over and remove the battery.

3 Remove the brush assembly.

4 Remove the rear dust collection bin.

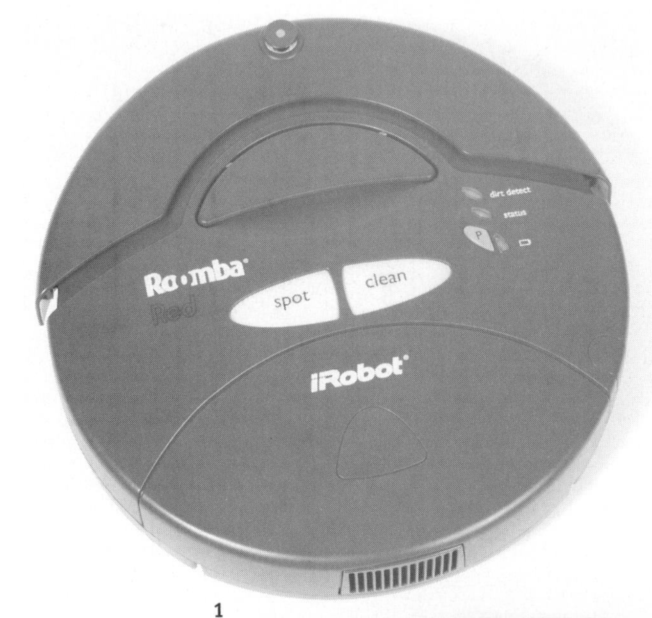

1

2

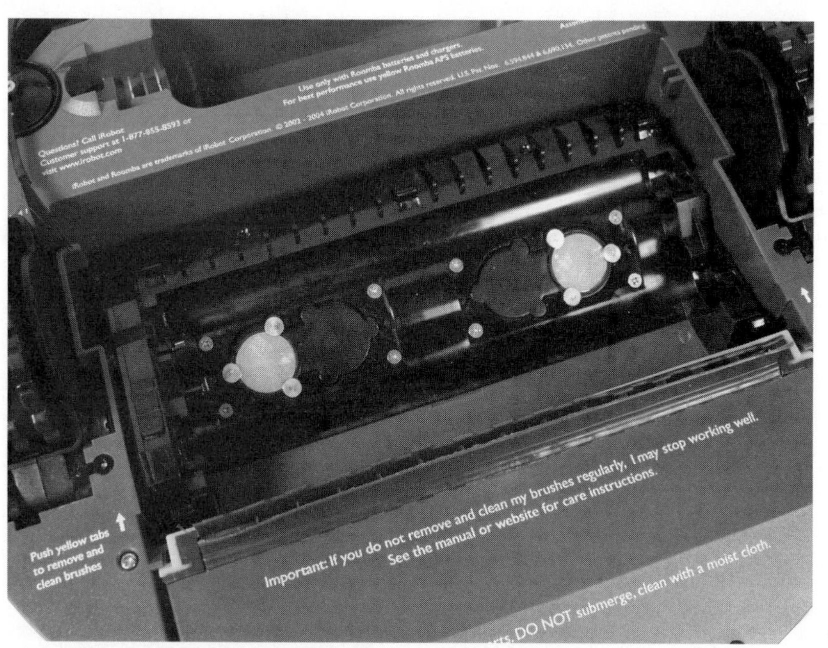
Questions? Call iRobot
Customer support at 1-877-855-8593 or
visit www.irobot.com
Use only with Roomba batteries and chargers
For best performance use yellow Roomba APS batteries.
iRobot and Roomba are trademarks of iRobot Corporation. © 2002 - 2004 iRobot Corporation. All rights reserved.
Push yellow tabs to remove and clean brushes
Important: If you do not remove and clean my brushes regularly, I may stop working well.
See the manual or website for care instructions.
DO NOT submerge, clean with a moist cloth.

3

Push yellow tabs to remove and clean brushes
Important: If you do not remove and clean my brushes regularly, I may stop working well.
See the manual or website for care instructions.
Bin contains electronic parts. DO NOT submerge, clean with a moist cloth.

4

5 Start removing screws.

6 And more screws.

7 And more screws.

8 There are even screws hiding in the battery compartment.

9 Don't forget the screws in the underside of the front bumper.

10 Disconnect the plug under the front bumper.

5

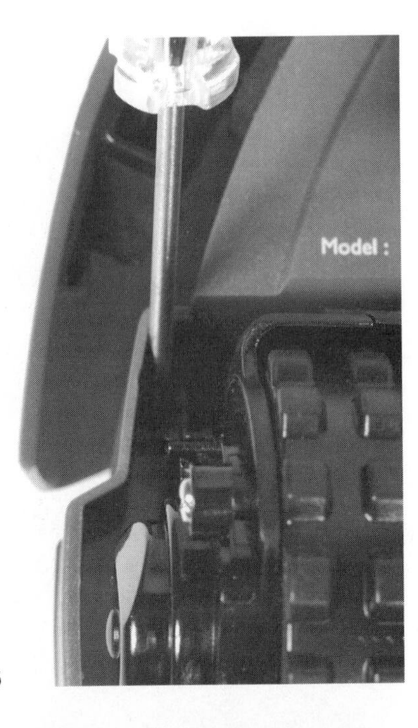

6

7

8

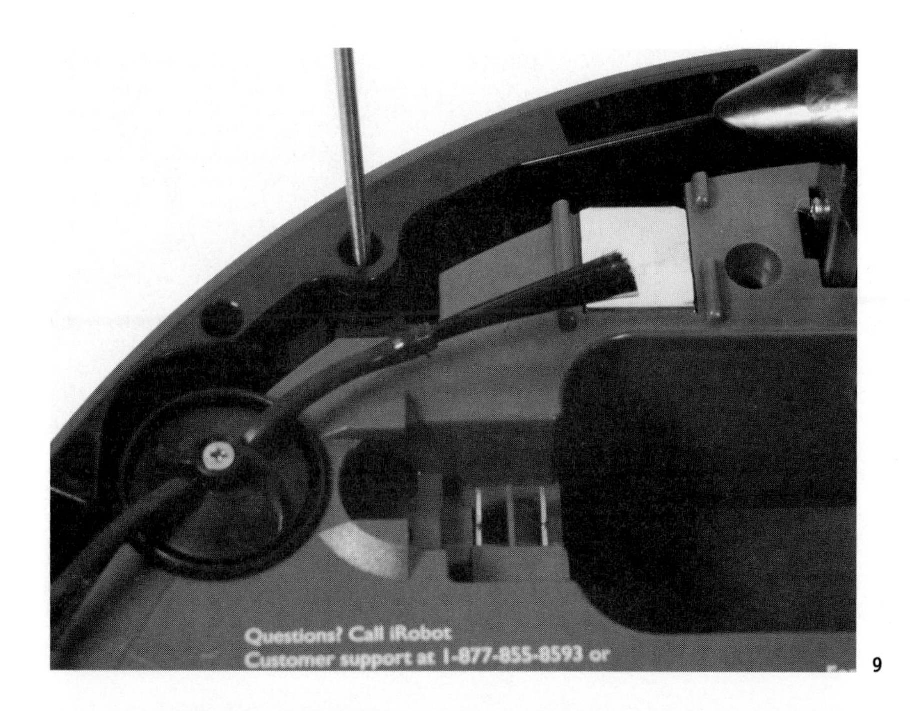

9

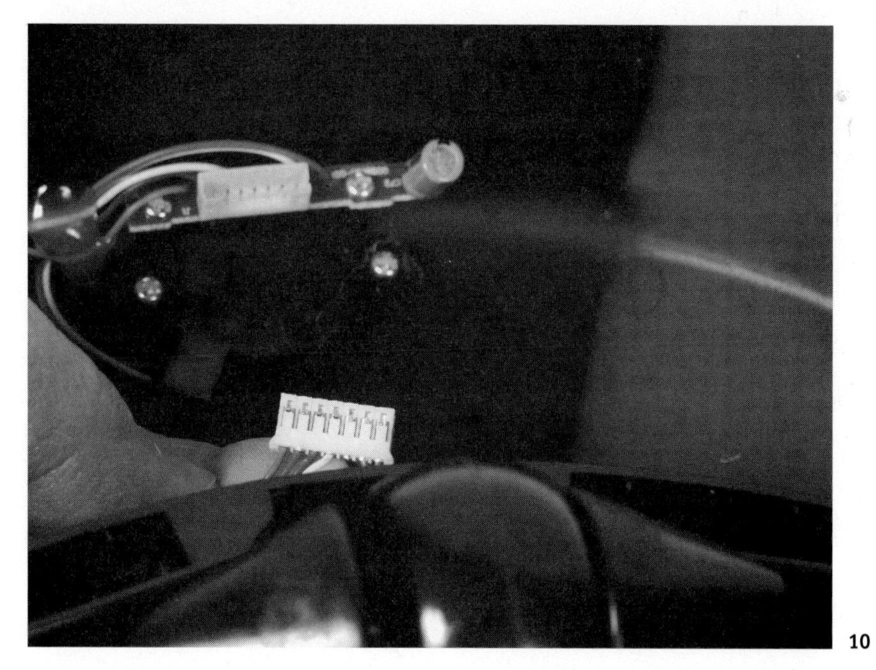
10

11 Remove the plug from the control panel that is connected to the main printed circuit board (PCB).

12 The drop wheels have hefty spring returns.

13 The wheels are driven by a pair of high-torque motors.

14 The right wheel spring return is held in place with a plastic plate.

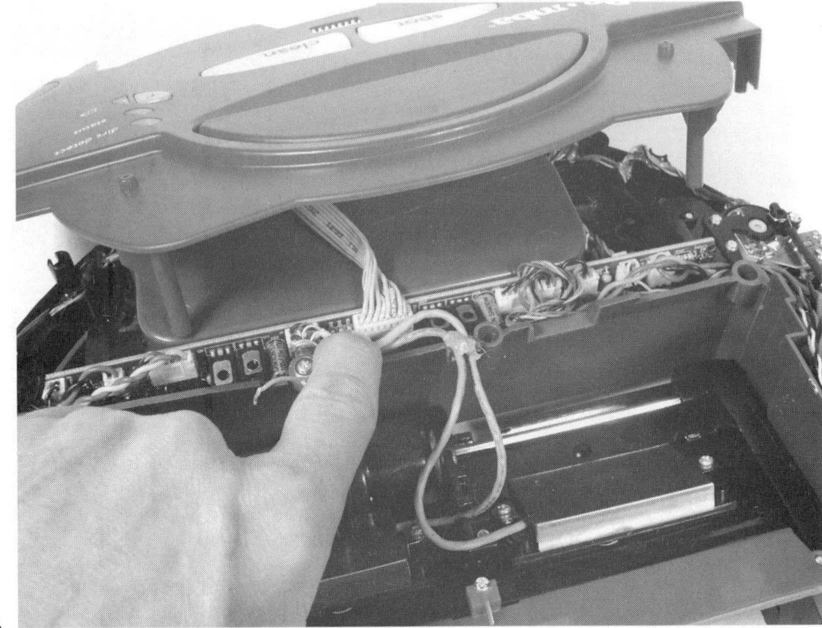

11

12

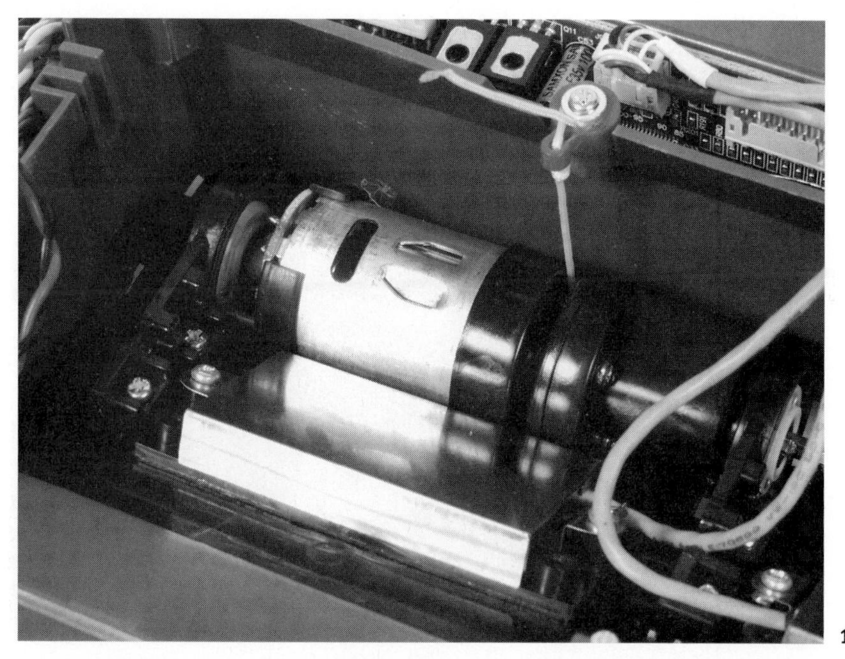

13

14

15 Disconnect all plugs from the top edge of the main PCB.

16 Trace the routes of each plug to its connector. Record this information on the Roomba PCB drawings later in this Case Study.

17 The front bumper connects to this PCB touch sensor.

18 Remove one screw from the right wheel spring plate.

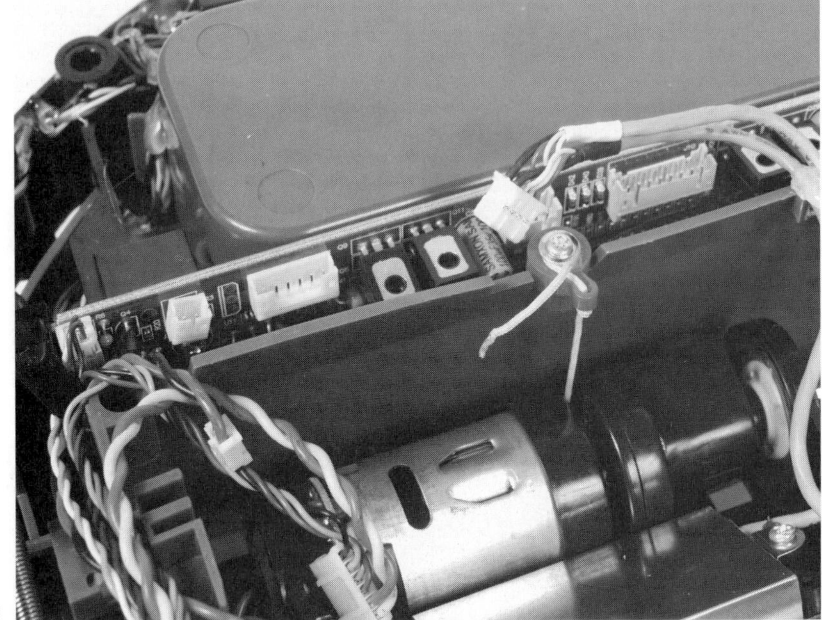

15

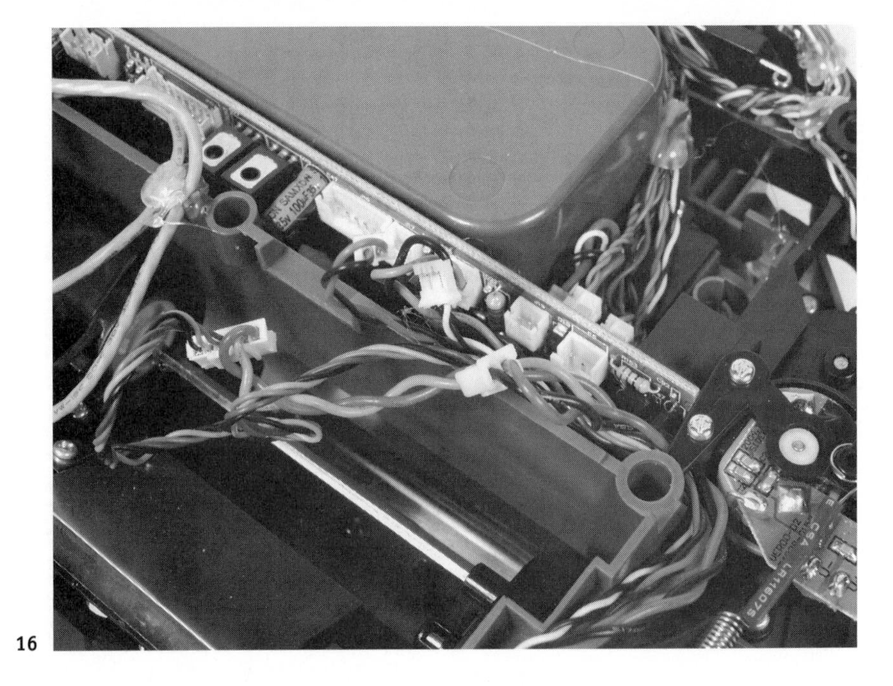

16

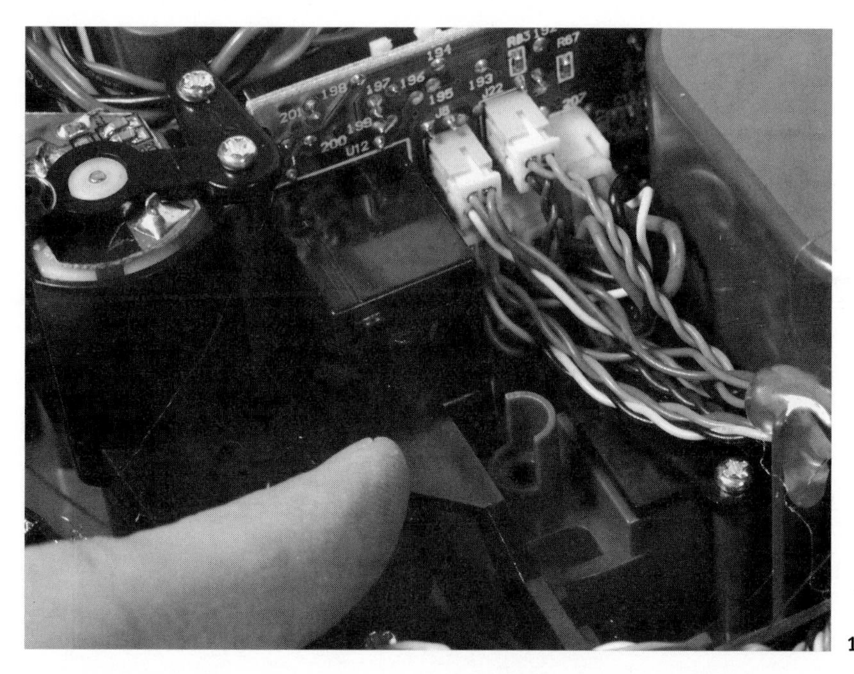
201
198
197
196
194
193
J22
R83
R67
200
U12

17

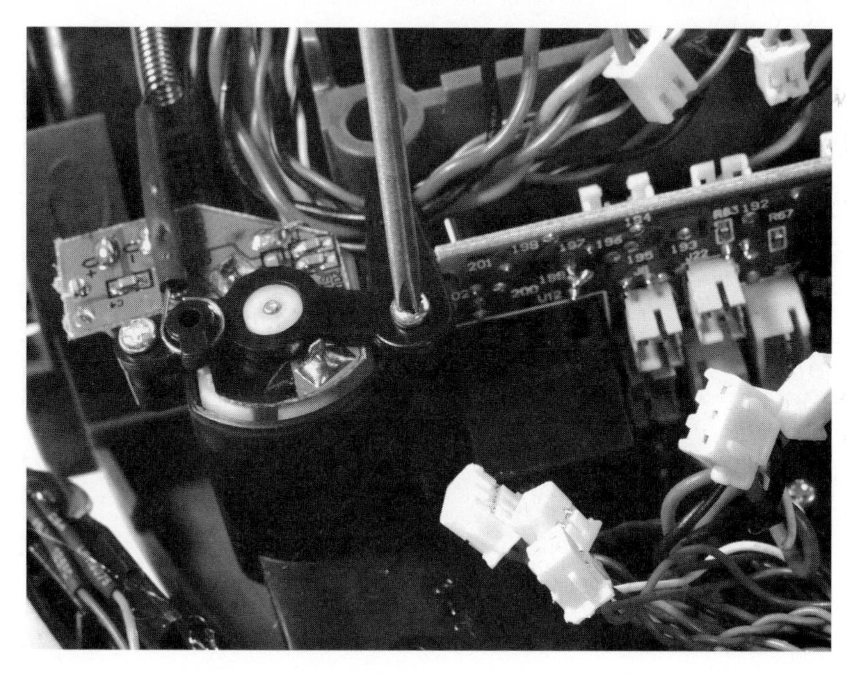
201
198
197
196
194
193
J22
R83
192
R67
200
U12

18

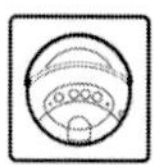

19 Remove the second screw. Watch that spring's tension when you remove the last screw.

20 Remove the screw from the left wheel spring plate.

21 The side sweeping brush motor.

22 Carefully scrape away the glue from the wiring harness in the front bumper.

19

20

21

22

23 Just about there—the twin front bumper touch sensors must be removed before the main PCB can be extracted.

24 The right touch sensor trigger has been removed from the main PCB.

25 Free at last. The main PCB can just about be removed.

26 Slip the charging board/SCI port out of the Roomba body.

23

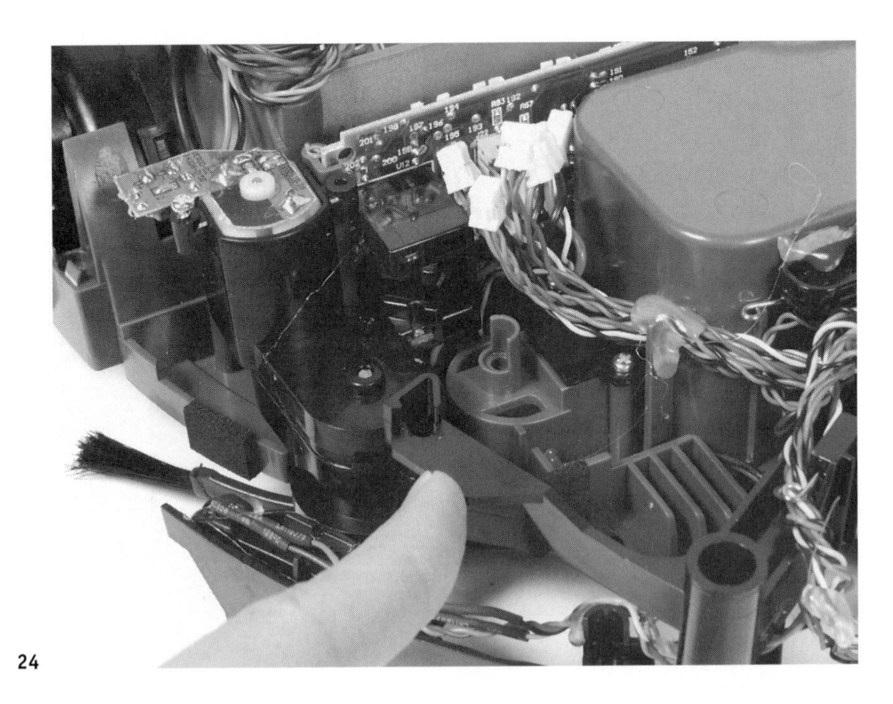

24

25

26

27 Carefully, route the charging board/SCI cable through the Roomba chassis.

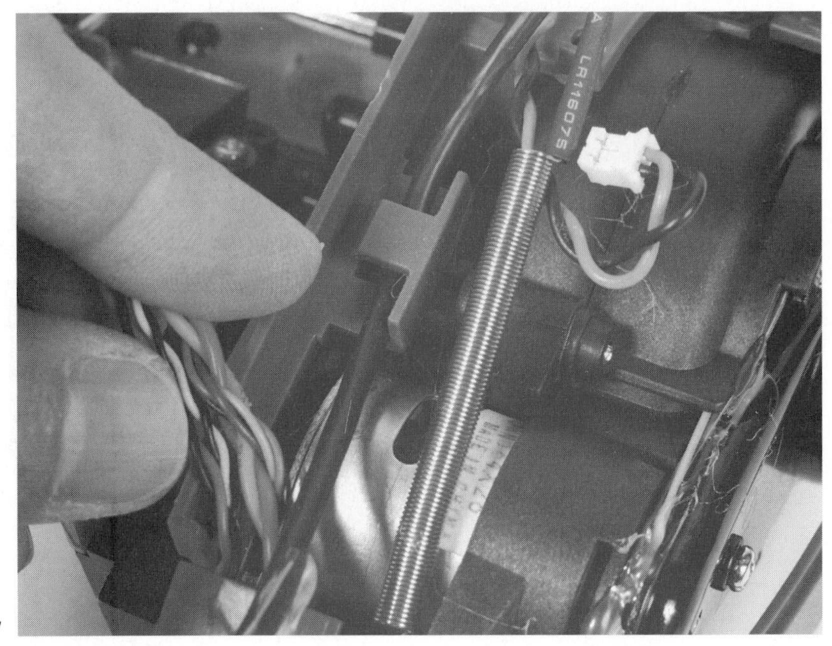

27

THE ROOMBA PCB

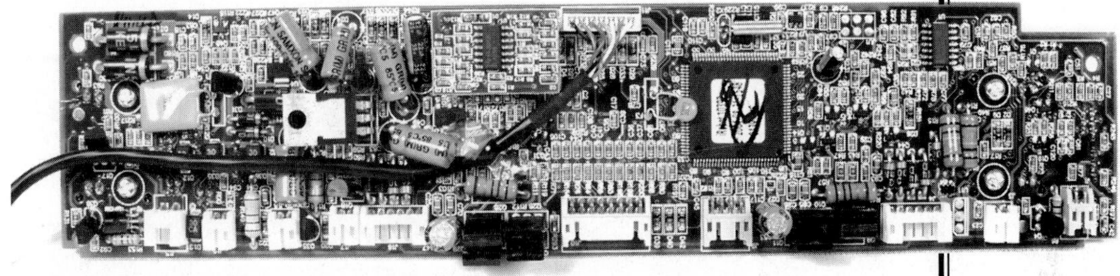

1

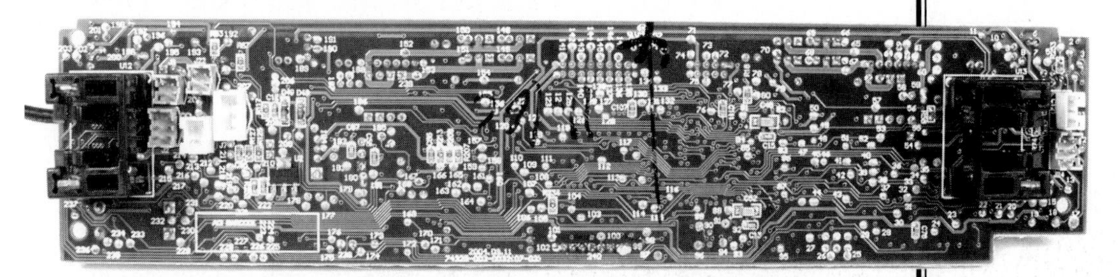

2

1 The Romba PCB front.

2 The Roomba PCB back.

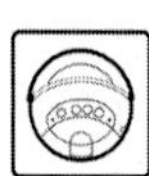

3 The right touch sensor (U12) on the PCB back.

4 The left touch sensor (U13) on the PCB back.

5 The PCB's left front.

6 The PCB's left front with the daughter card with U1.

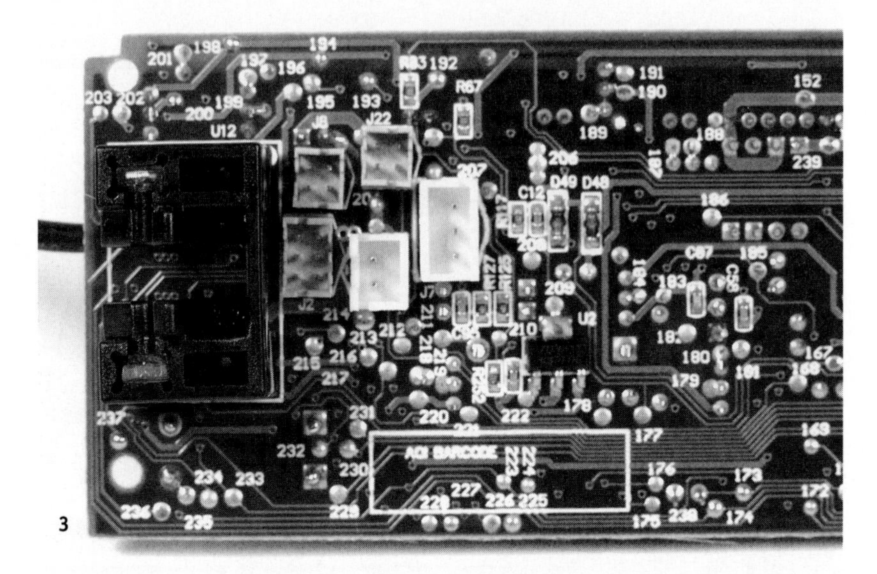

3

4

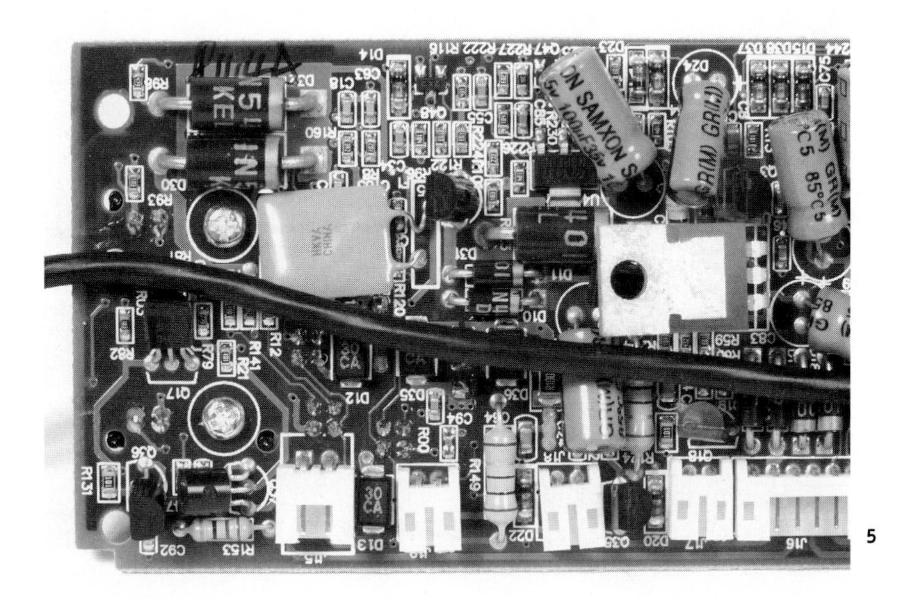

5

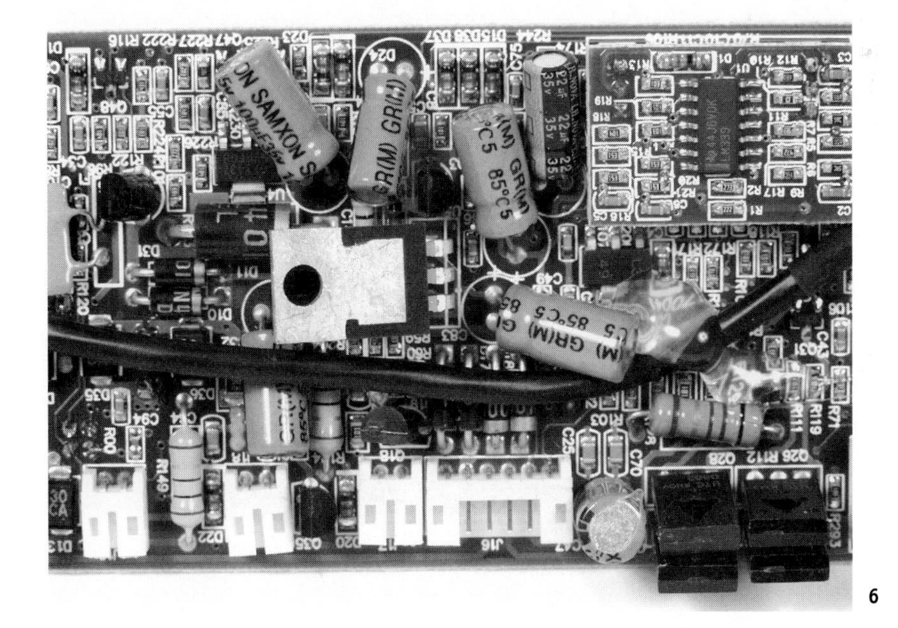

6

7 The PCB's left center front.

8 The PCB's right center front.

9 The PCB's right front.

10 The PCB's front attachment point for the left touch sensor.

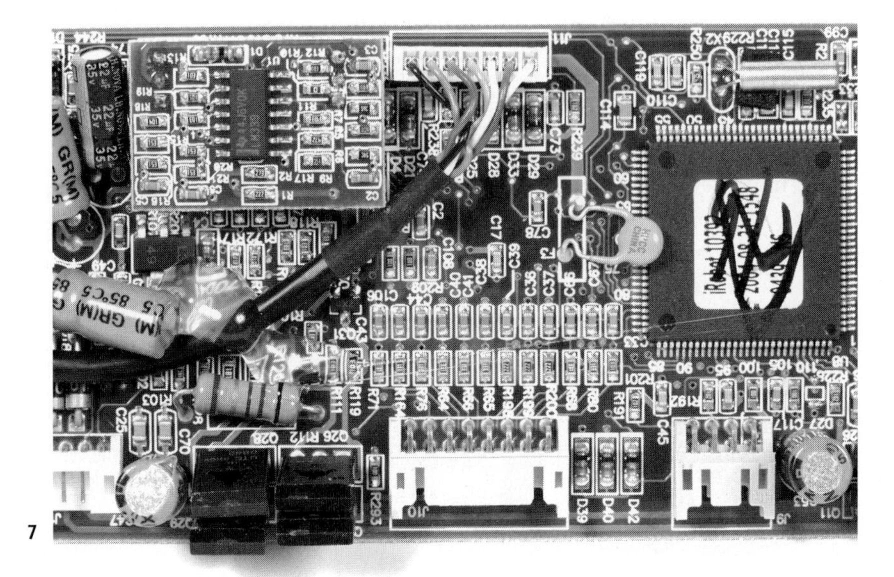

7

8

9

10

11 The connectors on the PCB front.

12 The connectors on the PCB back.

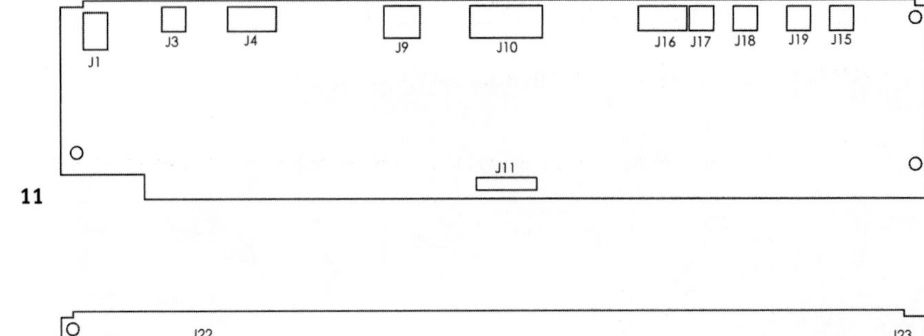

11

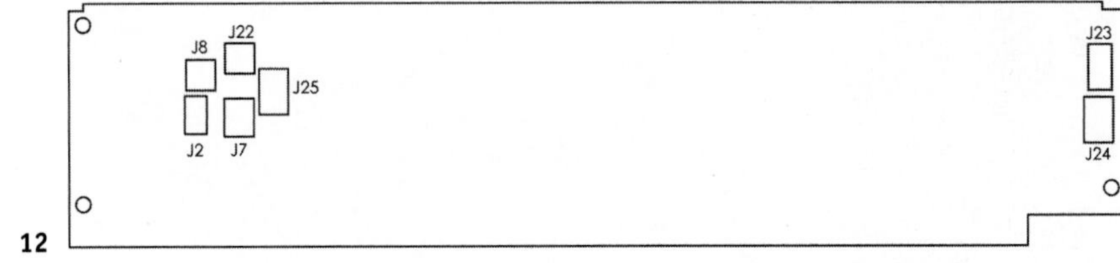

12

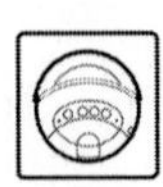

ELEMENT PRODUCTS' MIND CONTROL

1 Element Products' Mind Control.

2 This tiny DIN plug can be programmed with your own code for commanding Roomba.

3 Mind Control plugs into the Roomba SCI.

4 The slim form factor of Mind Control enables Roomba to execute your custom programming without any physical obstructions that could impede Roomba's movements.

5 An AVR programmer is included with Mind Control.

6 One connector on the AVR programmer receives Mind Control.

3

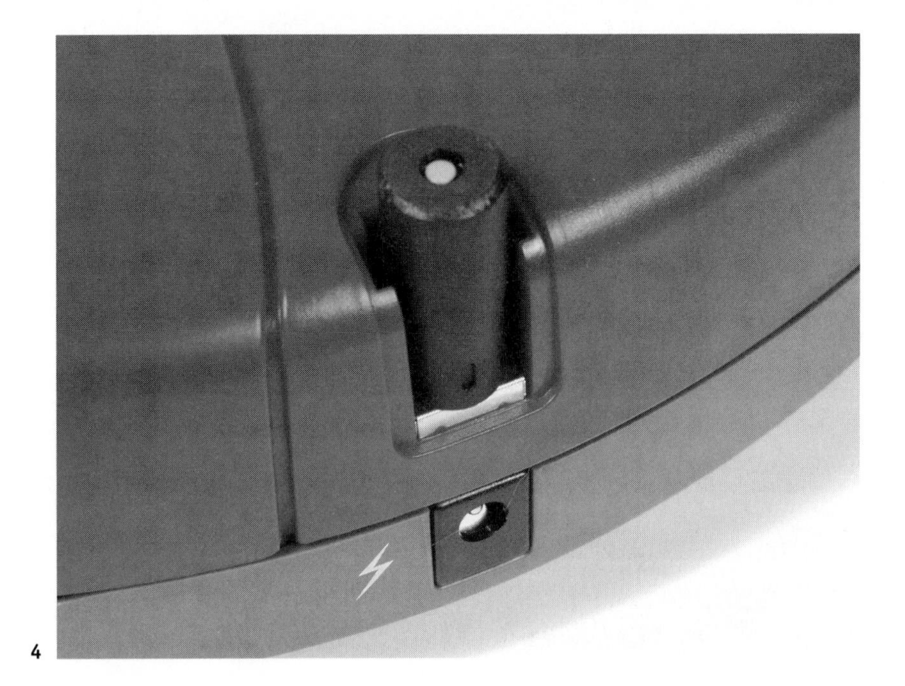

4

elementdirect.com
TX/RX
FTDI
6.00ECSH
Rev A
EP10447A
mm
10
20
30
40
50
60
70
4 8 12 16 20 24 28
Inches
1
STAINLESS STEEL
2
NO.311-me

5

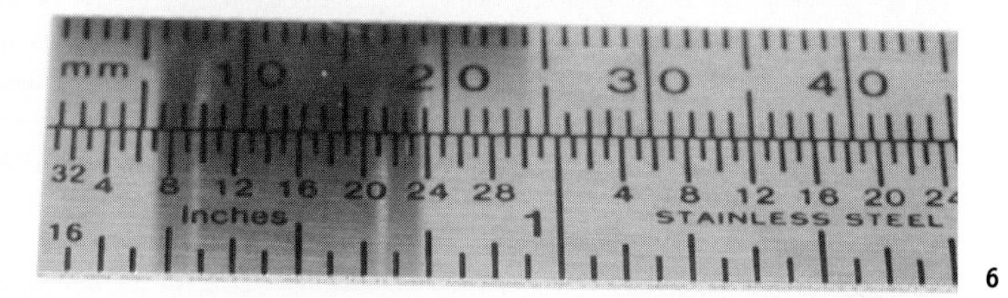
mm
10
20
30
40
Inches
1
STAINLESS STEEL

6

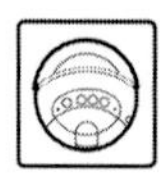

7 The other connector on the AVR programmer connects to your PC USB port.

8 You can program Mind Control on any PC with a USB port.

9 Element Products has done instrumental development work on both Roomba and the Parallax Scribbler robot.

10 Element Products has a history of developing user-programmable robots and components. Therefore, it should come as no surprise that custom control programs can be written for the Scribbler robot.

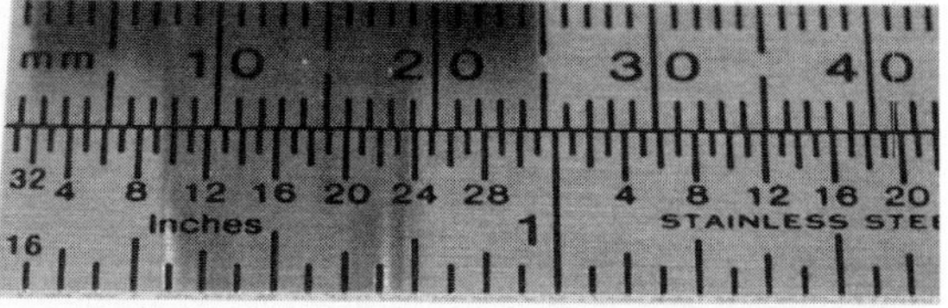

7

8

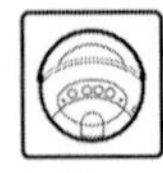

SCRIBBLER ROBOT
dirt detect
status
Roomba
spot
clean
iRobot
9

SCRIBBLER ROBOT
www.ScribblerRobot.com
PARALLAX
10

HOW TO PROGRAM MIND CONTROL WITH A MAC

1. AVRFlasher can be used for programming Atmel microcontrollers on a Mac.
2. When you start AVRFlasher you must link it to the AVR device.
3. Set up your connection speed.
4. You will need the AVR GCC compiler and library for making your programs.
5. Files like your Make File can be written inside BBEdit.

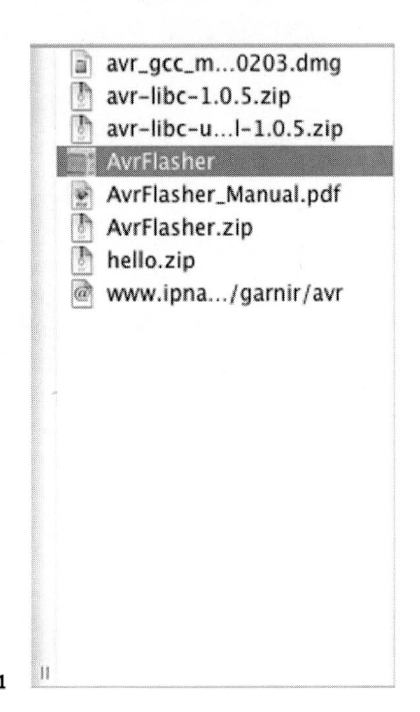

1

AvrFlasher ©ULg.(1.2 [8/12/2005])

Read from... Flash verify Save bin

Serial : no serial!

Link : AvrFlasher not linked!

File : No file

Status : Not ready!

2

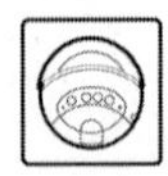

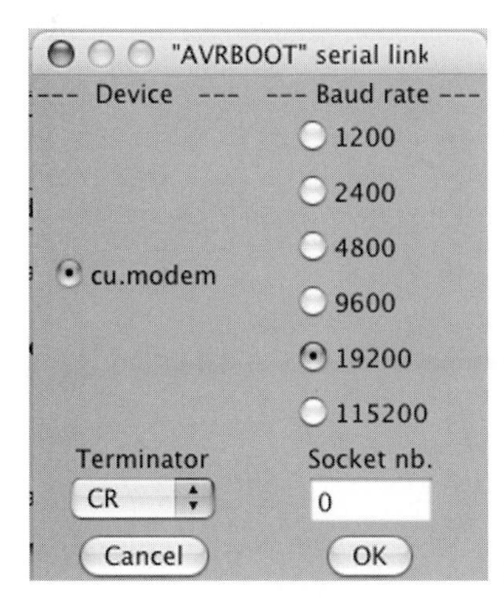

3

AVR-compiler

Set working dir...
MAKE all
MAKE clean
Load .hex
Log window
Terminal...

4

makefile

Last Saved: 03/17/06 08:23:07 AM
File Path: ~/Desktop/makefile

```
# Hey Emacs, this is a -*- makefile -*-
#----------------------------------------------------------------------------
# WinAVR Makefile Template written by Eric B. Weddington, J^rg Wunsch, et al.
#
# Released to the Public Domain
#
# Additional material for this makefile was written by:
# Peter Fleury
# Tim Henigan
# Colin O'Flynn
# Reiner Patommel
# Markus Pfaff
# Sander Pool
# Frederik Rouleau
#
#----------------------------------------------------------------------------
# On command line:
#
# make all = Make software.
#
# make clean = Clean out built project files.
#
# make coff = Convert ELF to AVR COFF.
#
# make extcoff = Convert ELF to AVR Extended COFF.
#
# make program = Download the hex file to the device, using avrdude.
#                Please customize the avrdude settings below first!
#
# make debug = Start either simulavr or avarice as specified for debugging,
#              with avr-gdb or avr-insight as the front end for debugging.
#
# make filename.s = Just compile filename.c into the assembler code only.
#
# make filename.i = Create a preprocessed source file for use in submitting
#                   bug reports to the GCC project.
#
# To rebuild project do "make clean" then "make all".
#----------------------------------------------------------------------------


# MCU name
MCU = atmega168


# Processor frequency.
#     This will define a symbol, F_CPU, in all source code files equal to the
#     processor frequency. You can then use this symbol in your source code to
#     calculate timings. Do NOT tack on a 'UL' at the end, this will be done
#     automatically to create a 32-bit value in your source code.
F_CPU = 18432000


# Output format. (can be srec, ihex, binary)
FORMAT = ihex


# Target file name (without extension).
TARGET = simple
```

5

6 **Write your C code.**

7 **Use the Mind Control header file.**

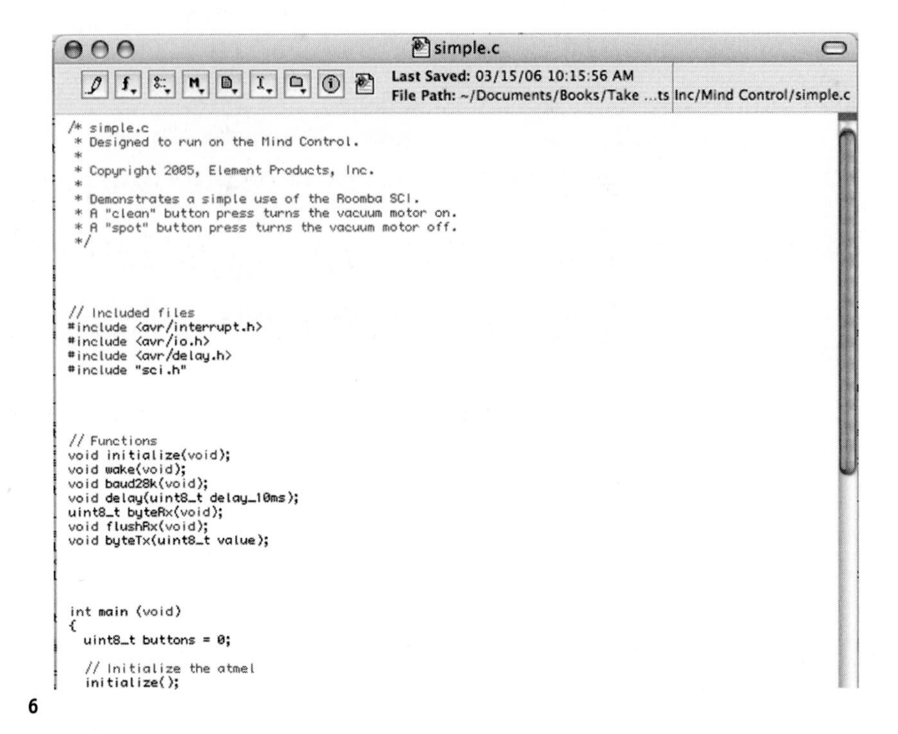

simple.c

Last Saved: 03/15/06 10:15:56 AM
File Path: ~/Documents/Books/Take ...ts Inc/Mind Control/simple.c

```
/* simple.c
 * Designed to run on the Mind Control.
 *
 * Copyright 2005, Element Products, Inc.
 *
 * Demonstrates a simple use of the Roomba SCI.
 * A "clean" button press turns the vacuum motor on.
 * A "spot" button press turns the vacuum motor off.
 */



// Included files
#include <avr/interrupt.h>
#include <avr/io.h>
#include <avr/delay.h>
#include "sci.h"



// Functions
void initialize(void);
void wake(void);
void baud28k(void);
void delay(uint8_t delay_10ms);
uint8_t byteRx(void);
void flushRx(void);
void byteTx(uint8_t value);



int main (void)
{
  uint8_t buttons = 0;

  // Initialize the atmel
  initialize();
```

6

sci.h

Last Saved: 03/17/06 08:23:12 AM
File Path: ~/Desktop/sci.h

```
/* sci.h
 * Designed to run on the Mind Control.
 *
 * Copyright 2005, Element Products, Inc.
 *
 * Definitions for the Roomba Serial Command Interface (SCI)
 */


// Command values
#define CmdStart        128
#define CmdBaud         129
#define CmdControl      130
#define CmdSafe         131
#define CmdFull         132
#define CmdPower        133
#define CmdSpot         134
#define CmdClean        135
#define CmdMax          136
#define CmdDrive        137
#define CmdMotors       138
#define CmdLeds         139
#define CmdSong         140
#define CmdPlay         141
#define CmdSensors      142


// Sensor byte indices
#define SenBumpDrop     0
#define SenWall         1
#define SenCliffL       2
#define SenCliffFL      3
#define SenCliffFR      4
#define SenCliffR       5
#define SenVWall        6
#define SenOverC        7
#define SenDirtL        8
#define SenDirtR        9
#define SenRemote       10
#define SenButton       11
#define SenDist1        12
#define SenDist0        13
```

7

CASE STUDY 2

WowWee Robotics Robosapien and Friends

THE WOWWEE FACTOR

Want to "jump-start" your Robosapien? Here are two "hacks" for turning your Robosapien into a robot version of the *Six Million Dollar Man*—minus Lee Majors, of course.

In the Beginning

Arguably, one of the hottest-selling toys from the 2004 Holiday season was Robosapien from WowWee Ltd. (this manufacturer is now doing business as WowWee Robotics). With over 1.5 million units sold, Robosapien was the ideal crossover toy for bringing robotics to the masses—masses of kids and masses of adults.

But that's only part of the story. Robosapien was one of those few toys that you absolutely had to take apart, for three very important reasons. First, as you disassembled Robosapien you were truly amazed at the simple beauty of its design. Counterbalancing springs, integrated plastic strain reliefs, and intricately geared servo motors delighted even the most jaded toy buyer. This robot "weren't no bucket of bolts."

The second major reason for opening up Robosapien was learning the basics in robot and toy design. Yes, the insides of this robot were well documented, ensuring a good, competent education in robotics; that is, if you were willing to pick up a screwdriver and open it up.

Finally, you had to open up Robosapien if you wanted to become one with this robot's greatest inner strength—its design to be hacked. Likewise, you became a seasoned robot hacker from the experience.

All of the finer points about hacking Robosapien were collected into my landmark book, *The Official Robosapien Hacker's Guide* (McGraw-Hill, 2006). This book detailed over one dozen great modifications, construction projects, and hacks that can be performed on this remarkable robot. Regrettably, there were two fantastic hacks that I couldn't adequately discuss in that book. And you are about to benefit from these two omissions.

Robosapien, Can you Hear Me?

The easiest of these two Robosapien hacks is the replacement of the main processor crystal. Labeled "Y1" on the Robosapien main circuit board, this crystal is actually a ceramic resonator which acts like a monolithic capacitor. As such, it doesn't require a big leap of imagination to think that switching this resonator's crystal value could result in a different Robosapien "personality."

And that's exactly what will happen: The frequency of the robot's operation can be slowed down or sped up by almost 50 percent just by using different-sized ceramic resonators. Be forewarned, however, that altering the frequency of the circuit board crystal will cause the Robosapien IR remote control to not function properly. So, this hack allows you to vary the speed of motor actions for either fast/lightweight or slow/powerful designs, but you might not be able to control this resulting "speed demon."

The robot's startup routine *will* be significantly altered by this crystal frequency change. So you will be able to clearly see the fruits of your labors.

SCALPEL, CLAMP, SCREWDRIVER

Before you begin this hack, beware that, if you don't know exactly what you're doing, you could damage your Robosapien. While these instructions make every effort at holding your hand through these processes, one errant soldering mistake could render your robot into a gigantic paperweight.

The first step (and one of the hardest in this hack) is to get inside Robosapien to the main circuit board. All that you will need for this portion of the hack is a No. 0 Phillips screwdriver.

Before you begin any hacking surgery, however, make sure that you remove the four D-cell batteries from the Robosapien feet. With your patient now suitably anesthetized, there are four screws that hold the back plate to the Robosapien body—one in each shoulder and two in the waist.

Once you remove these screws, the back plate will lift off. Be careful, how-

ever; the power switch wiring harness (this also holds the speaker wiring) is attached to the main circuit board. Just disconnect the main circuit board plug from this power switch harness (a pair of needle-nose pliers might be needed for this step) and the back plate can be removed. Set both the front and back plates aside.

RESONATING WITH POWER, MOO-HA-HA

The main circuit board is located on the back of Robosapien. Take a moment to study all of the lovingly applied labeling that WowWee Ltd. added to the main circuit board—all of these silkscreen labels were done for helping you, the hacker.

Locate the ceramic resonator crystal. It is to the left of the IC U3 and labeled "Y1." Although this crystal "looks" like a capacitor, it is actually a ceramic resonator. Take a pair of diagonal cutters, snip off the Y1 crystal, and remove it from the circuit board. You will now solder a new ceramic resonator in its place. You can add virtually any frequency value of inexpensive ceramic resonator (e.g., DigiKey; www.digikey.com). Your beginning frequency for the resonator installed inside Robosapien is 4.00 MHz.

Typically, resonators with higher frequencies, 6.00 to 12.00 MHz for example, will result in a "faster" Robosapien. While lower frequency resonators, 3.58 to 2.00 MHz , will make the robot behave more slowly. Just solder this replacement resonator crystal to the decapitated leads from the old crystal.

Remember to hold onto the original Y1 crystal, so that you can return your robot to its factory state of mind.

If you really want to experiment with a wide variety of ceramic resonator crystals, you might wish to solder two header pins to the Y1 crystal pass-through pads. Then you can just temporarily attach your resonators to these header pins, until you find the perfect hack.

Racing stripe is optional.

Selective IR Remote Control of Multiple Robosapien Robots

BY DANIEL ALBERT

[**NOTE:** I met Daniel at RoboNexus 2005 in San Jose, California. He showed me his marvelous Robosapien hack and I lamented that I hadn't included it in *The Official Robosapien Hacker's Guide*. Then *Take This Stuff and Hack It!* came

along and I asked Daniel if he wouldn't mind contributing his hack to this book. His answer was an enthusiastic "yes" and you are about to benefit from his generous contribution.]

HACK ABSTRACT: Robosapien Transmitter Board IRTXv1. Program from 1 to 256 unique robot ID codes into module. Add up to four additional analog/digital inputs/outputs. Easily interfaces to IR output LED and Sharp IR distance module. No soldering or wiring on Robosapien required—simply plug in receiver module inside Robosapien without modifications.

Remote transmitter module requires four simple solder connections. For those who want plug and play without soldering, you can use a programmable handheld device to control individual or multiple bots.

HACK BACKGROUND: When I first saw the Robosapien I was floored. I couldn't believe you could build a bipedal robot that could sell in the sub-one-hundred-dollar market. Yet there it was. So I bought two. I needed one to experiment with and one that was static. It's a technique they teach in high school science class called a control group.

So there I was with my robots wondering what to do with them. I got a friend of mine and we loaded them with batteries and started to play. And then we realized the awful truth. Each remote controlled both robots. After researching on the Web I found some clever person put a toilet paper roll on the head for IR directional control. It worked but it was not my style. So we brainstormed and came up with a great idea.

PLAN A. CHANGE THE CARRIER FREQUENCY.

What if we changed the 40-MHz carrier frequency to some other frequency? So we bought an IR receiver that was modulated at 36.7 kHz. We would either change the IR receiver in the head or just unplug the old one and plug in our replacement. Without opening up the head we could just mount the new receiver on the top of the head with sticky tape. To change the remote carrier frequency we could put a PIC microcontroller chip inside to send the codes at that different frequency. Seemed simple enough, so we did it. We made up some printed circuit boards (PCBs) using ExpressPCB and put a PIC12c509 processor on it.

Along with some programming and some hardwiring we were ready. Now for the moment of truth—we tested out the robot and everything functioned correctly just as before the modification. But then came our control group test.

We put both bots together in the same room and to our dismay when the modified remote was close to the unmodified bot, it still accepted commands. Uh-oh, something went wrong. The carrier frequencies were too close and the receiver picked up some of the signal. Well it seemed obvious that this was not a good solution so we went with Plan B.

PLAN B. (THE ONE THAT WORKED.)

This hack required two mods. One PIC microcontroller chip would go in the remote and one in the robot. We decided to alter the IR code itself.

First let me say, there isn't much programming space inside the PIC microcontroller. You're not going to do this hack in BASIC programming language. Even C can be too large when you only have 1K. Yea, that's a thousand processor instructions. Assembly code is the only good choice for this type of embedded processing. We worked very hard on this code and at first thought we would keep it a secret. There isn't a lot of money to be made with this hack so we decided to try to win the five hundred bucks being offered by the *SERVO Magazine* "Hack-a-Sapien" contest. After all, fame is better than fortune.

The transmitter chip was tackled first. We captured the IR signal just prior to the output transistor that is attached to the remote's LED (see Fig. CS2-1).

CS2-1 The "before" IR waveform (left) and the "after" IR waveform (right).

After sending the AGC pulse and gap we sent an 8-bit ID preamble with the original 7-bit code concatenated. This is then followed by the end bit. If you look at the before and after signals on a logic analyzer you can see the difference.

I have omitted the 40-kHz carrier part of the waveform. You would need to zoom in to see the 25-μS pulses—12.5 μS high and 12.5 μS low. If you want to do the math it's 1/40,000. The ID code we used in this example is a 0x0f. See if you can spot the 00001111 bits that are inserted into the pulse train.

Then we set out to program the receiver's PIC microcontroller chip to convert the signal back to something the Robosapien understands. This turned out to be easier than the transmitter. At the point where we capture the signal, after the IR receiver, we no longer had to worry about the 40-kHz carrier frequency. Timing was less of an issue. Most of the code was reused with the carrier code taken out. The waveform looked like the "after" signal mentioned above.

It was simple to retrieve the first 8 bits of data and store that as the ID byte received. A simple compare against the ID preprogrammed in the receiver

and you could determine if this command was valid for this bot. If not, we send the processor to sleep and wait for the next command. If the command is valid, we parse the rest of the command and send it out to the bot's processor board.

A test with both our modified bot and our control bot proved successful this time. Each bot responded to its, and only its, remote (see Fig. CS2-2).

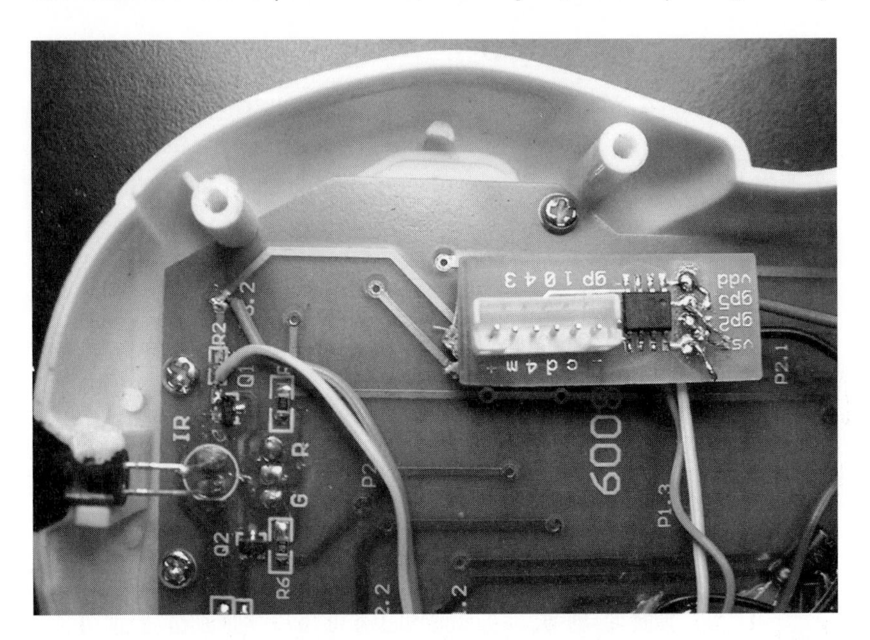

CS2-2 **IR remote control with PIC microcontroller.**

Putting this hack in two bots and two remotes, both with separate ID codes, also proved successful so we felt confident this plan would work.

Well step one was now complete. We had a working hack. But what good is this kind of hack if it requires delicate soldering to surface-mounted components? So we set out to make the mod "plug and play."

PLAN A. A CUSTOM BOARD MOUNTED UNDER THE CHEST PLATE OF ROBOSAPIEN.

This is where most of the free room is inside the bot. We designed a board and started to put together a list of required parts. Here's what we needed:

- One 10-pin connector with pins and wires
- One board-mounted header socket for existing connector
- One custom PCB
- One PIC microcontroller processor
- One 6-pin header socket for flashing the PIC

Wow that adds up. To make matters worse, DigiKey (the vendor we used for supplying our parts) had "good news/bad news" concerning the matching connectors. The good news was each connector was under 50 cents each. The bad news was a required minimum order purchase quantity of 19,000. I'm not kidding! This pretty much nixed this plan and so once again we went to Plan B.

PLAN B. A BOARD THAT HAS NO WIRES AND PLUGS INTO EXISTING CONNECTOR.

So how much room is available on the back of Robosapien where the IR connector is located? We needed to find out if there was clearance for the board (see Fig. CS2-3). I said, "If the speaker had been put somewhere else this

CS2-3 Robosapien embedded controller implant.

wouldn't be such a problem." It's easy to be a Monday morning quarterback, so I stopped complaining and we set out to make it work. We used some "biofoam" to make an impression of the existing room in the back plate cavity.

Biofoam is really neat stuff that a podiatrist friend of mine uses for foot impressions. It's soft as foam and if you push your hand in it you get a great impression that stays put. I inserted a small piece of biofoam cut to the approximate inside shape of the Robosapien cavity and placed it on the processor board. The result was a perfect model of the space inside the robot after the cover was closed tight. As expected there wasn't much room, but there was enough for a small board.

Working with the technical support people at DigiKey we were able to find less costly replacements for the connectors. These replacement connectors don't look as nice as the previous Plan A versions, but they are readily available in small quantities. Since we were redesigning the board anyway, it seemed logical to add more features. This is called "feature creep." The term is defined as "an uncontrollable urge to continually add many new features where the design soon becomes too complicated and nothing works." But we kept feature creep to a minimum and the 12509 was switched out for the PIC 12F675 and/or 12F683.

These chips are pin compatible and gave us an additional 4 GPIO (general purpose input output) pins. These can be digital or analog. It's really up to the hacker as to what to do with these pins but we decided to add IR output from the bot and a Sharp IR range detector. This leaves two available pins for...well...whatever you want. The board seen here is the third concept board and we are planning some more mods to it for making it fit better inside the limited space.

The hacker has two options for the transmitter. The first is not for the faint of heart. It requires opening up the remote control and de-soldering the surface mounted resistor prior to the output transistor and soldering a wire in its place. Three other wires need to be soldered also to complete the hack.

The second option, which we are working on now, does not involve any soldering. Others have done this so we are reinventing the wheel with a new twist. We have a handheld Pocket PC PDA with IR I/O that is programmable.

We are writing an interface to transmit and receive the IR codes. This application will be ready for our RoboPraxis event in December. One of the competitions at this event is a Robosapien relay race—avoiding obstacles and retrieving objects. This program, along with updated versions of our PIC code, will be available to download for all those interested in building or buying one of our IR controllers.

There is only one component, but I designed a schematic diagram for representing the various pin connections (see CS2-4). This diagram works with the supplied program. You could even build this without a PCB and hard wire it if you are handy. There are 8 pins on the PIC12f675 to connect.

If you have any questions about this hack, you can contact Daniel through his Web site: www.alberts-equation.net.

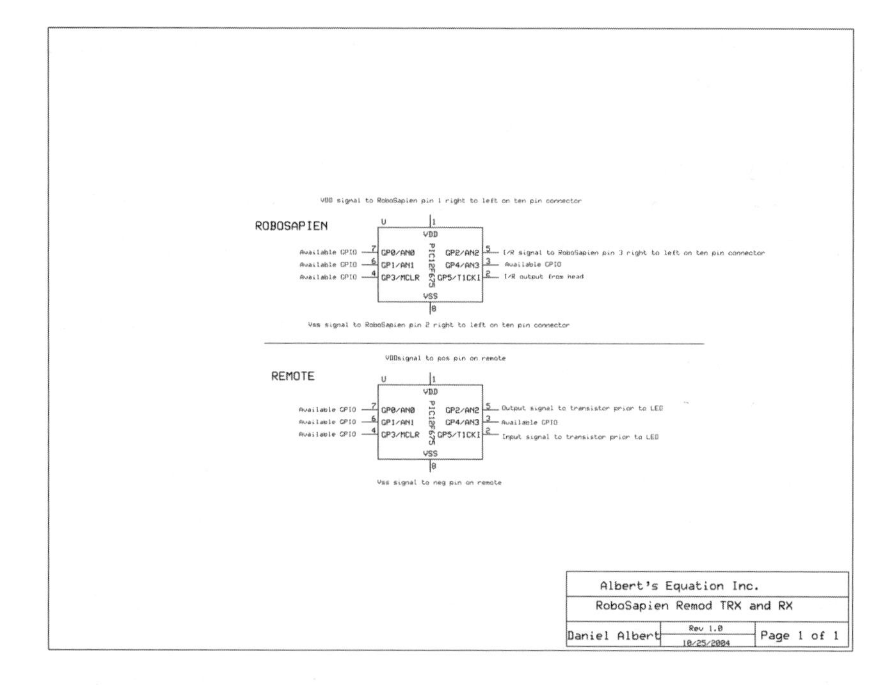

CS2-4 **Although not much of a circuit design, this schematic diagram shows all of the necessary pin assignments for the PIC microcontroller used in this hack.**

Dawn of a New Species—Hack-a-Saurus

How do you "one-up" Robosapien? You take one bold step backwards on the evolutionary timeline, of course. Hot on the heels of Robosapien, WowWee Robotics released a robot that will actually bite the hand that controls it—Roboraptor.™

Casting an impressive 32-inch long and 11-inch high shadow, Roboraptor can snarl, snap, and saunter about on six AA-size batteries. Like its Robosapien ancestor, Roboraptor comes with an IR remote controller; only this controller is vaguely reminiscent of a Sony PlayStation® DualShock2 game controller. By using the IR remote controller, you can guide Roboraptor through movements, modes, and moods. Yes, you read right; Roboraptor can be moody.

Unlike the Robosapien IR remote controller, the one that is shipped with Roboraptor can be used to guide your robotic dinosaur around any prehistoric swamp or through your living room. This feature is called "Laser Targeting."

A special button located on the leading edge of the IR remote controller's right-hand pistol grip is the Laser Targeting button. Just press this button and shine the IR signal at a "target" and Roboraptor will immediately "sense" this target, acquire it, and begin stalking this "irradiated" prey. There is a bright green light adjacent to the Laser Targeting IR emitter called a "tar-

geting assist light" that will help you "see" your target and assist you with aiming Roboraptor.

In addition to targeting objects as Roboraptor prey, you can also select between two distinctly different modes of operation. The first mode is "free roam." During free roam mode, Roboraptor will move around, explore, and interact with its environment, autonomously. You can access free roam mode either via the IR remote controller or by leaving Roboraptor alone for about three minutes. This mode is a very impressive demonstration of robot movement interacting with its onboard sensors.

HANDS OFF

Roboraptor is bristling with sensors. There are touch sensors, sound sensors, and vision sensors. During free roam mode all of these sensors are used in concert for a very believable "live" performance.

In fact, there are dual sound sensors which enable a stereo sensing ability. This ability allows Roboraptor to engage in three vastly different "moods" or reactions to sound. Depending on the location and repetition of a detected sound (e.g., front or either side; one or more sounds), Roboraptor will engage in either a hunting, cautious, or playful mood. Oh, and check this out: When Roboraptor is in one of its "moods" and you accidentally on purpose stick your finger in its mouth, a special mouth touch sensor will allow Roboraptor to "play" with your finger. And I'm not talking about the "pull my finger" gag, either. I'm talking about a very convincing ripping and tearing action that will make you glad that you have more than ten fingers.

Roboraptor has another mode of operation—guard mode. Unlike the autonomous free roam mode, guard mode is a nonmoving activity. Use the IR remote controller to activate guard mode and Roboraptor will use its vision and sound sensors for detecting movement and then reacting to any detected intruder. The guard mode will last about 30 minutes.

HAIL TO THE KING

How did WowWee Robotics manage all of this magic? Well, Roboraptor contains five servo motors, a handful of sensors, a powerful main circuit board, and an exciting array of springs, counterbalances, and articulation points. In fact, there are enough of these "analog" elements to educate all of us. But with Roboraptor you have a totally interactive bot bundled inside one of the most innovative form factors to ever stalk your local toy store shelf.

How about hacking? Sure your first inclination is to "smarten" up Roboraptor with some advanced microcontroller programming, but there's more to this bot than just a brain upgrade. Think collective or pack mentality. Try to organize a "pride" of Roboraptors into a hunting pack.

The Laser Tracking capability would be one avenue to pursue for achieving your own backyard "Jurassic Park." Organize one Roboraptor as the "Alpha Raptor" and equip it with the laser tracking feature. Then when this "king dino" locates prey, the other Roboraptors can circle the victim and let the feeding begin.

While Roboraptor might look intimidating, its bark is worse than its bite. The head movements coupled with the sound effects can make you think that you are being stalked by a living creature. But these jaws aren't strong enough to do any major tissue damage. In fact, the one thing missing from all of this posturing is a large glob of drool dripping from the open jaws and slathering up your workbench.

No Roboraptors were harmed during the course of this book's preparation.

DOGGONE IT—ONE BOT TOO MANY?

Leave it to WowWee Robotics for attempting to tackle the age-old marketing dilemma—how do you sell robots to girls, without painting them pink? While the latest addition to their robotic menagerie, Robopet™, attempts to stake a claim to achieving that goal, it remains to be seen whether this claim will strike gold.

In a continuing departure from the original Robosapien glossy black and white plastic exterior, Robopet sports the new textured flat black and white with silver highlights hide that was first introduced on Roboraptor. Likewise, as we've all come to expect from WowWee Robotics creations, Robopet is equipped with numerous sensors and a sophisticated servo muscle network.

Vision, sound, and a new remarkable edge detection sensor flesh out Robopet's interactive capabilities. Unfortunately, one of the sensors that you would expect to find on your new robotic pet, a touch sensor, is strangely missing.

Unlike Roboraptor and Robosapien, Robopet romps around with all of its servo linkages uncovered and exposed. This undesired exposure is not only unsightly, but also prone to possible damage. And therein lies a new problem for WowWee Robotics—quality control.

I tested five different Robopets and found that two of these pups would occasionally shed plastic parts as readily as Fido casts off fleas. These droppings originated around the two front leg linkages. In each case, the repair

could only be made when I carefully removed the back plate from Robopet. This is a tough repair requirement to levy on unsuspecting parents, especially on Christmas afternoon.

Another bewildering departure from Robosapien is the personality of Robopet, or lack of it. You know, that purported "fusion of technology and personality" that has become the hallmark of all WowWee Robotics toys. In the case of Robopet, there is a mix of truly amazing dog-like attributes (roll over, play dead, shake its paw) along with some bizarre robot-like noises that seem like inappropriate afterthoughts for satisfying the "boy crowd" in its intended audience. While the successful Roboraptor plays the part of a dinosaur to a tee, Robopet is stuck somewhere between the incredible electronic gizmo Hasbro iDOG and the lifelike elitist Sony AIBO.

Robopet is a wonderful crossover robot that has clearly defined appeal to both boys and girls, but limited endurance with parents who might have to master some sophisticated pet maintenance issues. While cute in a technological way, Robopet is missing that certain ambiance that made Robosapien a hit right out of the box. It's tough to argue that this bot isn't a dog in more than just appearance.

HOW TO HACK A ROBOSAPIEN

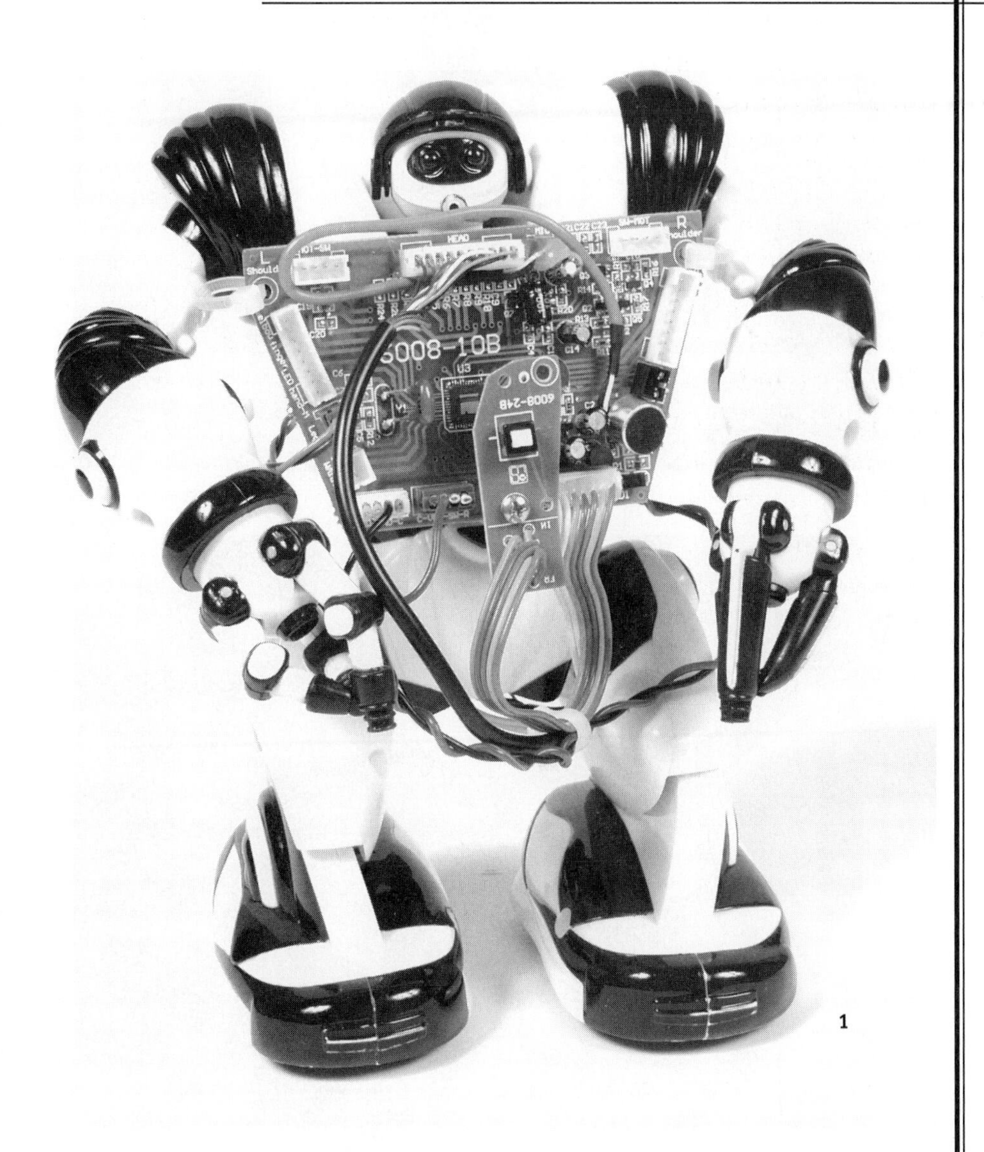

1

1 I added a 3.58-MHz ceramic resonator in this hack which was displayed at RoboNexus 2005 in San Jose, California. You can experiment with ceramic resonators of 2.00–12.00 MHz for a wide variety of slow/fast speed results.

2 A simple No. 0 Phillips screwdriver is all that you need to gain admission to this robot-hacking universe.

3 The Robosapien main circuit board is located on the robot's back. The crystal ceramic resonator is labeled Y1.

4 Snip off the crystal and add a new one for a pepped up Robosapien.

5 Solder in the new ceramic resonator.

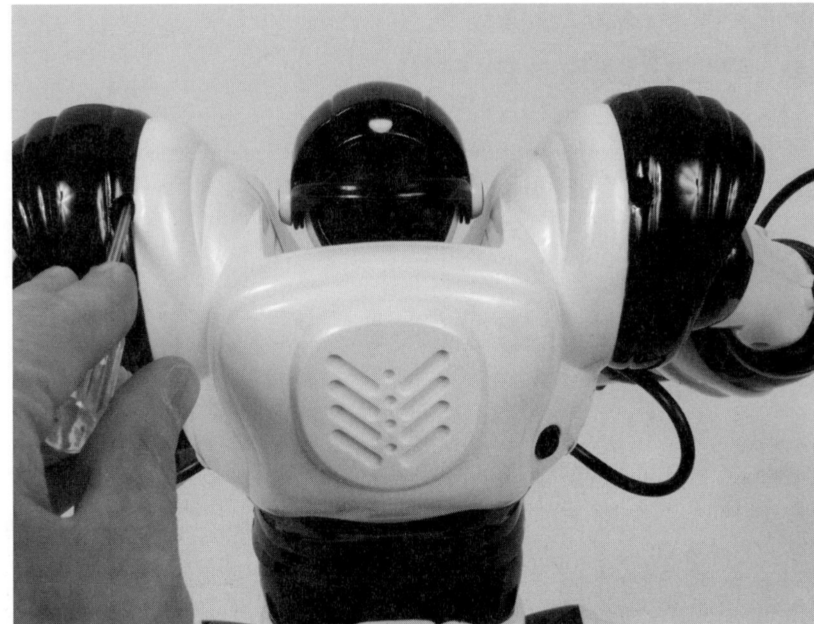

2

3

HEAD
MIC
C21C22 C23
C11
Q3
R14
R20
Q6
Q2
R13
C12
R11
C20
6008-10B
U3
C6
C7
Y1
C2
R23
R1
4

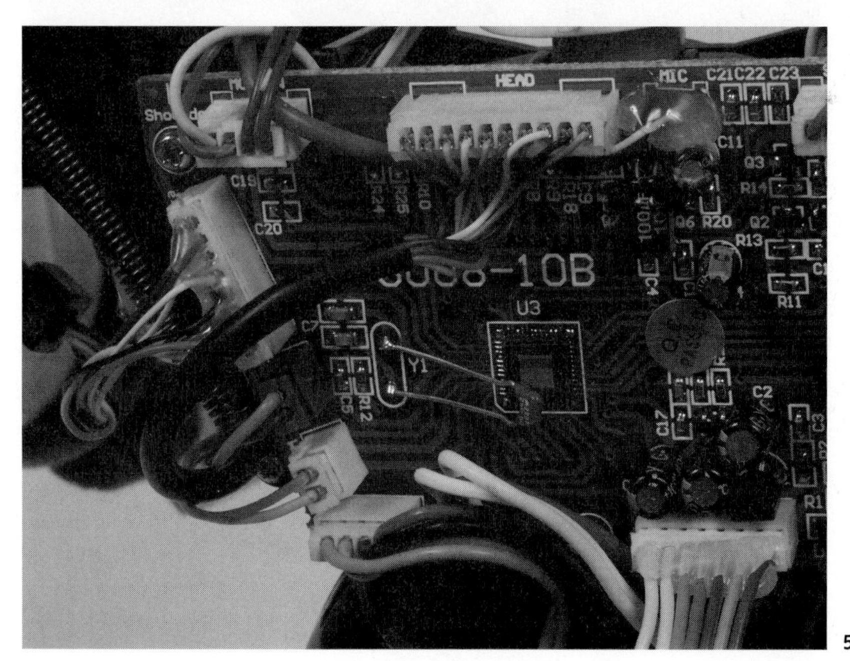
HEAD
MIC
C21C22 C23
C11
Q3
R14
Q6
R20
Q2
R13
R11
C20
-10B
U3
C7
Y1
C2
5

6 Racing stripes are no longer optional.

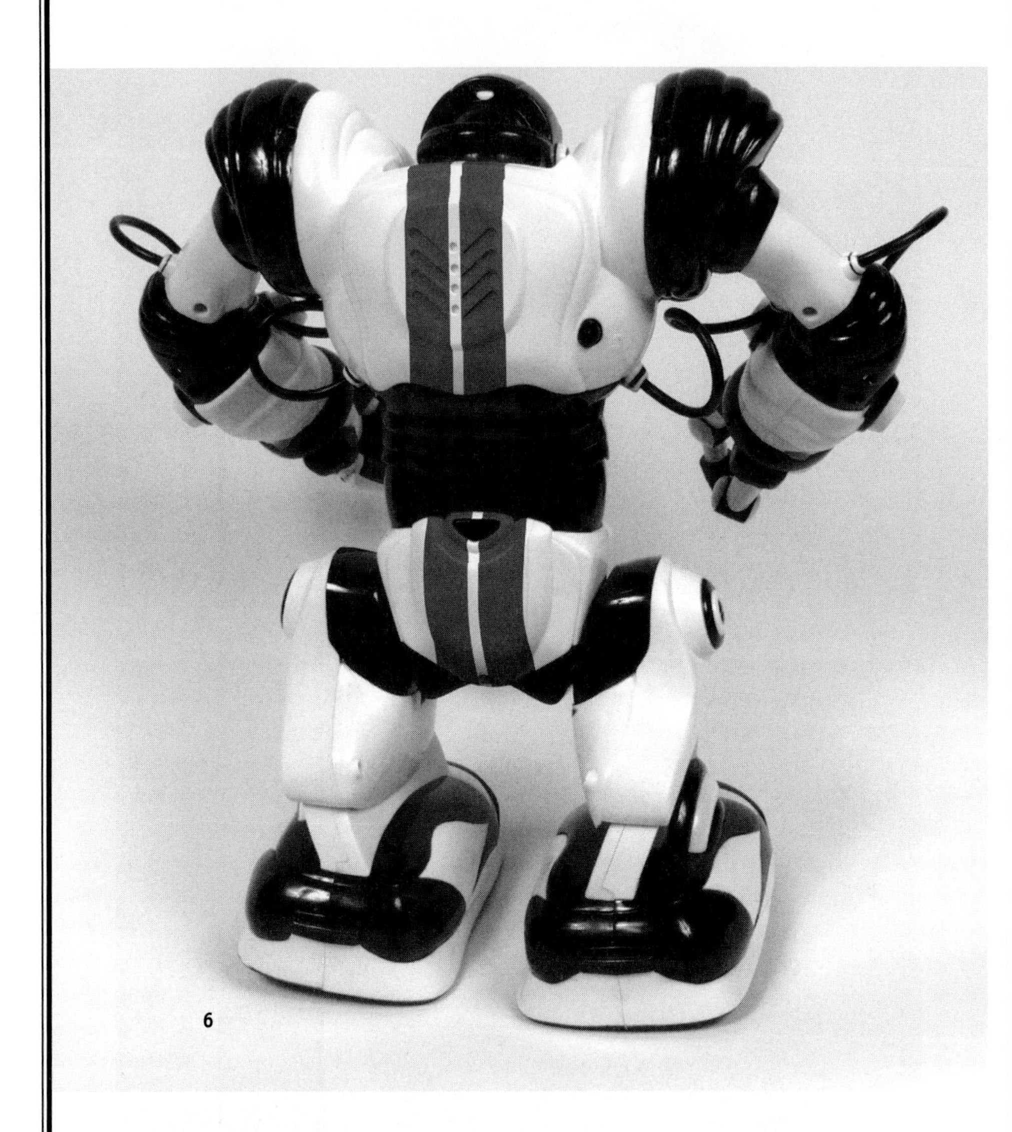

HOW TO HACK A ROBORAPTOR

1

1 The WowWee Robotics Roboraptor stalks the competition feeding on puny wheeled robots and devouring all pod bots.

2 Two heads are better than none. The top Roboraptor head is from a preproduction model and the bottom one was removed from a commercial version.

3 Don't touch that! The tail of Roboraptor is chock-full of touch sensors each cleverly hidden for detecting any type of errant tail grab.

4 Stereo sound sensors are positioned on each side of the head (i.e., one sensor on each side).

5 You can easily connect the Roboraptor head to your favorite robot brain. The benefit is that you get two touch sensors and two sound sensors—housed in an awesome form factor.

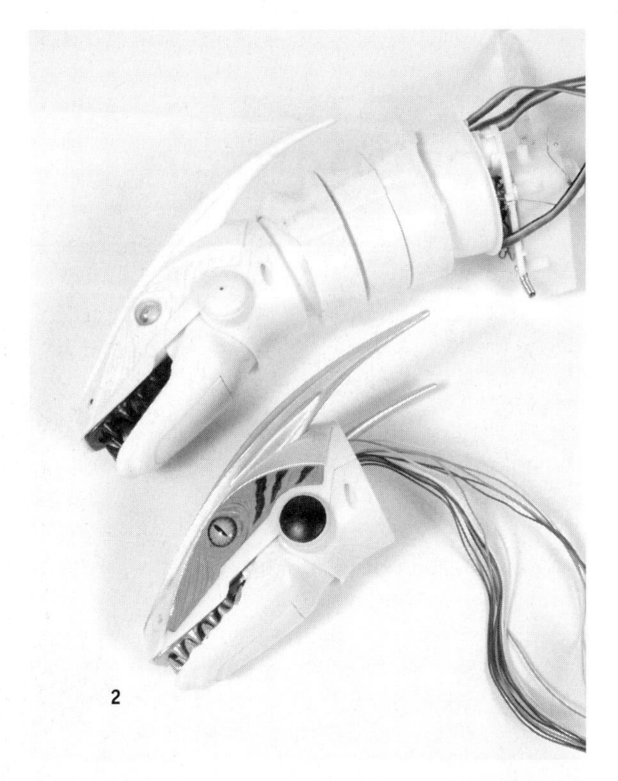

2

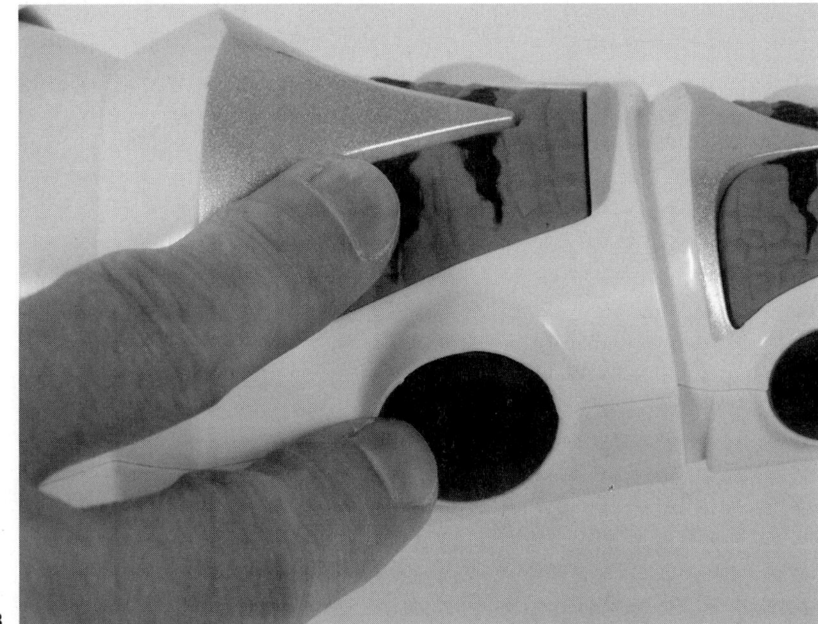

3

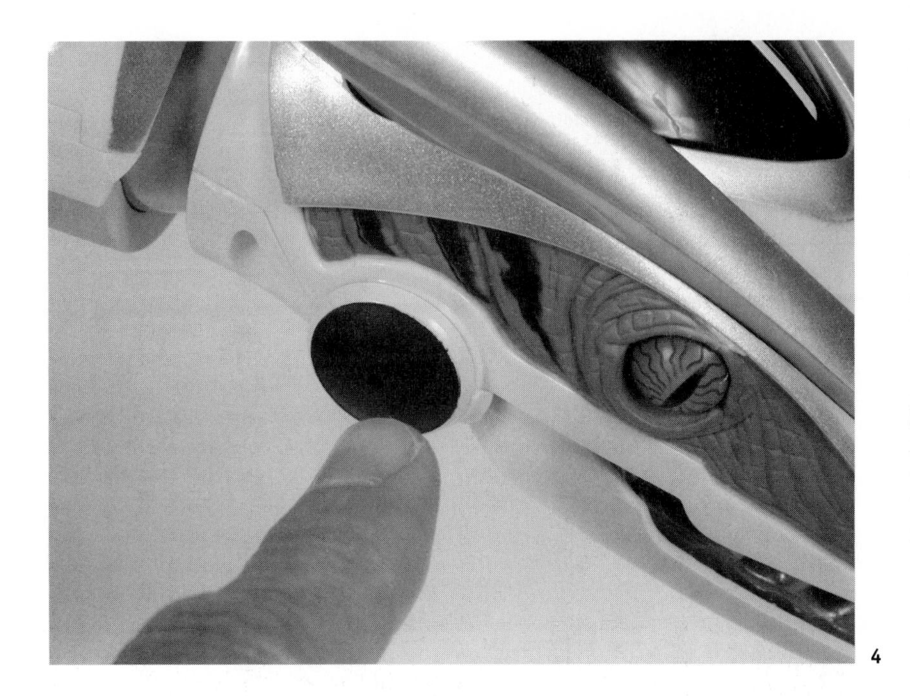

4

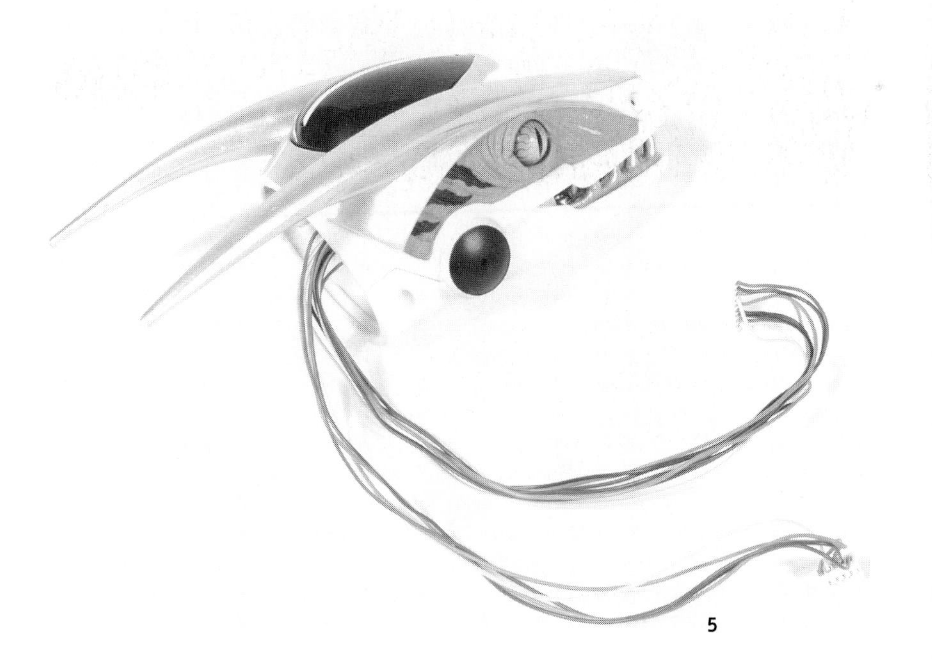

5

6 Watch those fingers. There is a touch sensor under the chin and another one on the dino's hard palate.

7 If you want to see what's inside Roboraptor, you must first remove each leg.

8 If you need a little more support structure with your Roboraptor transplant, just slap a battery box and circuit board onto the beast's head and neck.

9 Just like Robosapien, the Roboraptor main circuit board is a great robot brain that is ready for transplant.

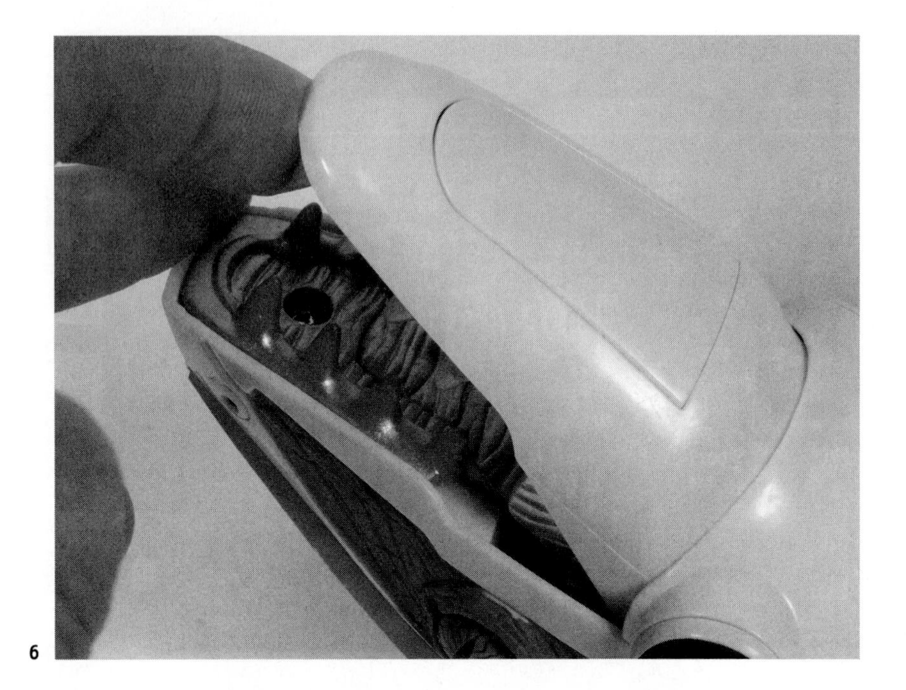

6

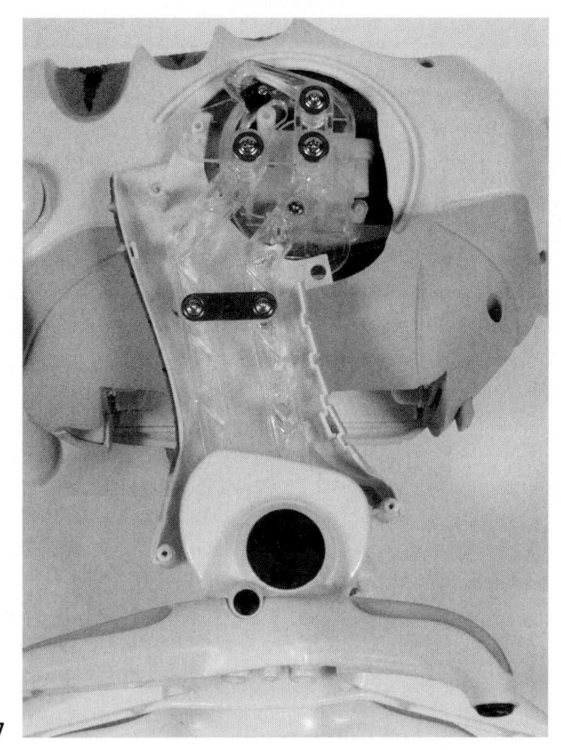

7

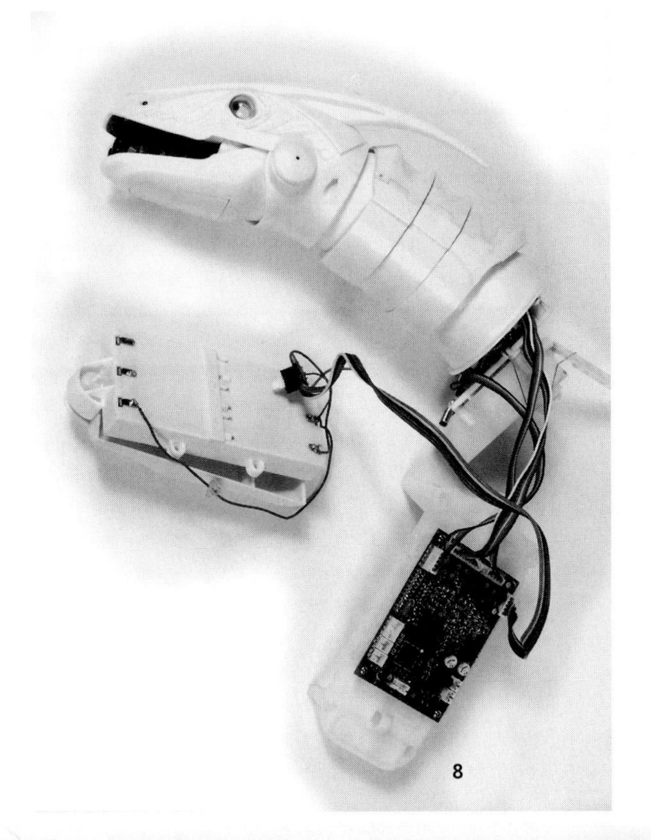

8

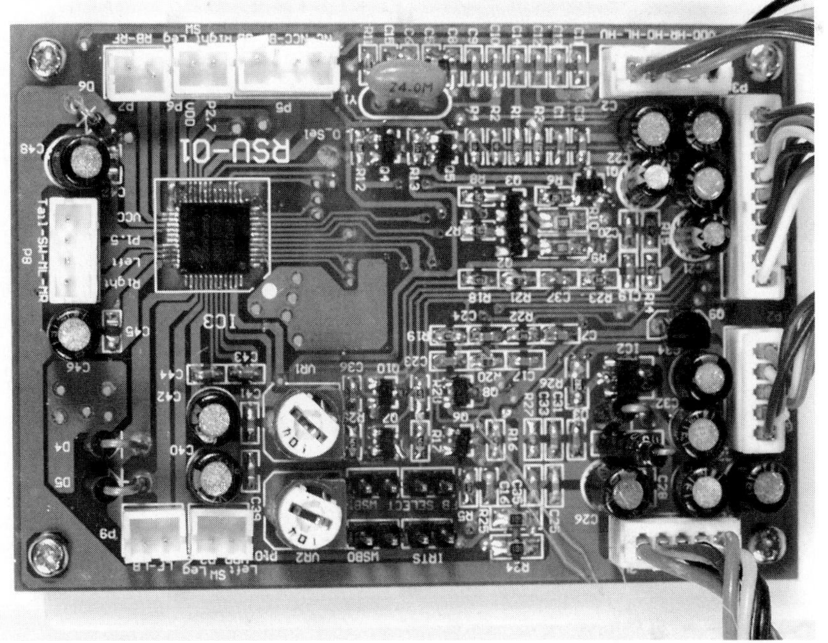

9

HOW TO HACK A ROBOPET

1 WowWee Robotics Robopet™ is a sophisticated robot that is inexplicably missing a batch of obvious touch sensors; along with an absent tail and AWOL ears.

2 Unlike other WowWee Robotics toys, Robopet's leg linkages are exposed. You can also see the new edge detector sensor under this little doggie's chin.

3 This pin could cause parents a ton of grief as I had two Robopets lose theirs within 30 minutes of play. Shortly after the pin fell out, the linkage slipped off the servo and I had two gimpy pups.

4 Robopet is a remarkable breakthrough product for introducing girls to robots.

1

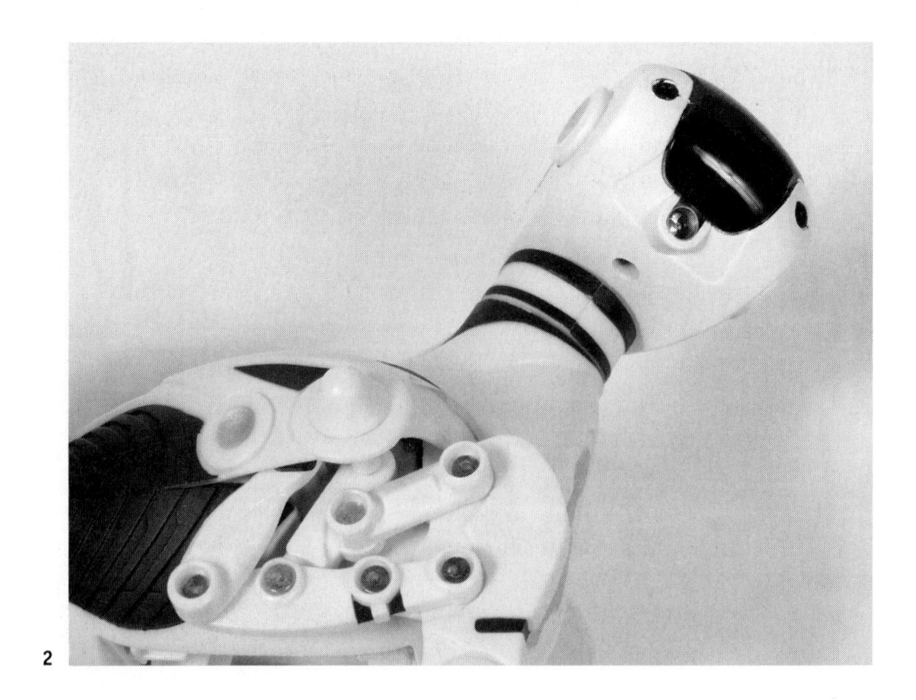

2

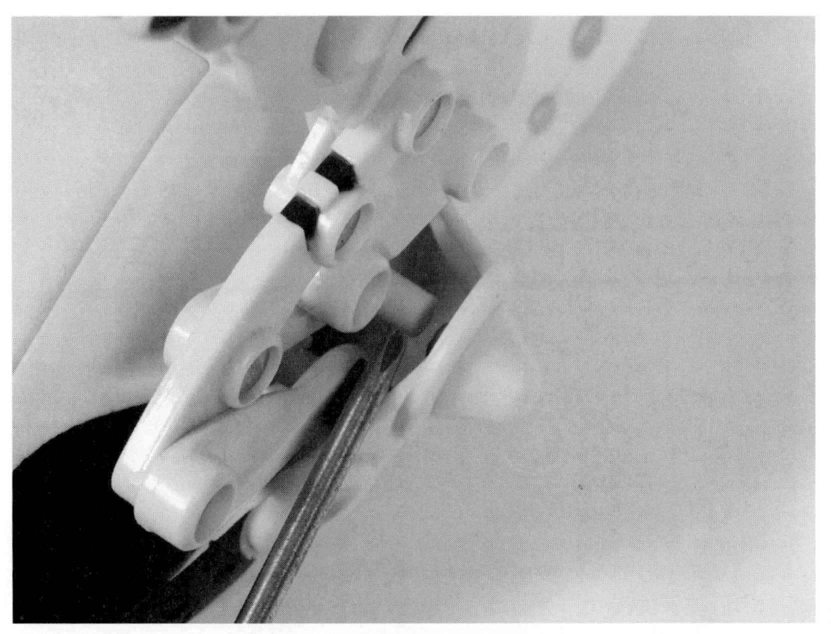

3

4

5 Robopet features a meticulously labeled main circuit board that is ideal for hackers. And for you hackers keeping score at home, you must remove both the front and rear top plastic plates to access the main circuit board.

6 Your first hacking assignment, hacking the JoinMax Robot Dog with Robopet's main circuit board.

5

6

HOW TO HACK A ROBOSAPIEN V2

1 WowWee Robotics Robosapien V2.

2 Two RS V2 Mini gang up on an RS Mini.

3 There are a whole boatload of new capabilities built into RS V2. Other than these sensors, cameras, and natural speech functions, the 24-inch-tall physical presence of this robot is awe inspiring.

4 Each hand has a gauntlet that functions as a touch sensor.

5 A unique color camera coupled to a dual-range IR vision system enable RS V2 to play simple games and sort your laundry.

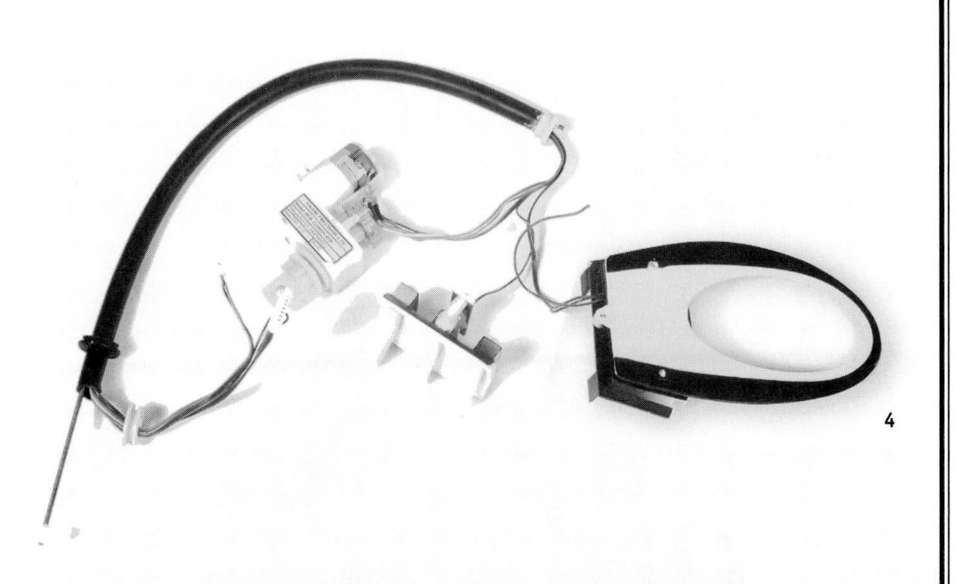

4

5

6 Like its predecessor, Robosapien, each foot of RS V2 is equipped with two touch sensors.

7 To really understand RS V2, you've got to get inside its head, literally. The main circuit board is attached to the back of the neck.

8 This brain is a little trickier than the WowWee Robotics brains for transplanting into other bots.

9 You can easily access the color camera inside the head.

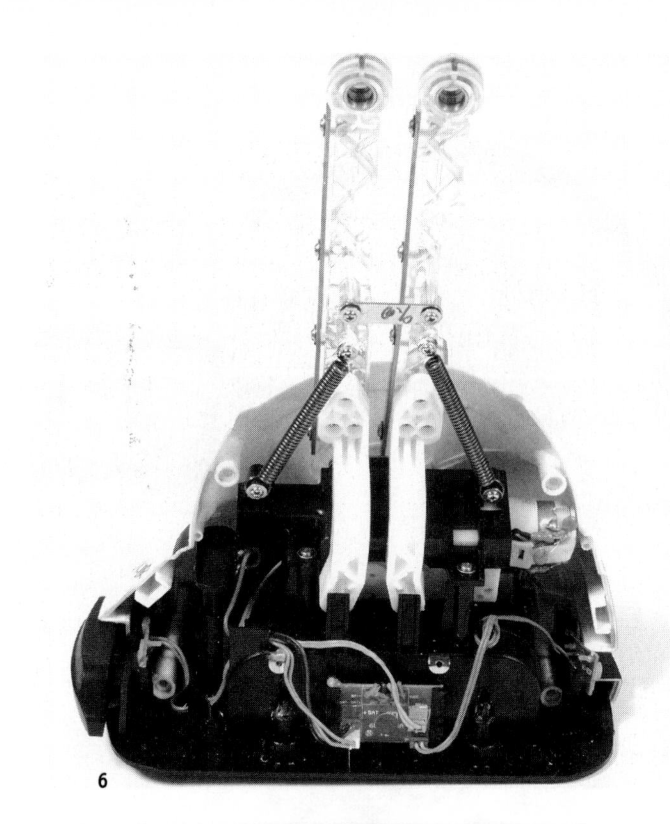

6

7

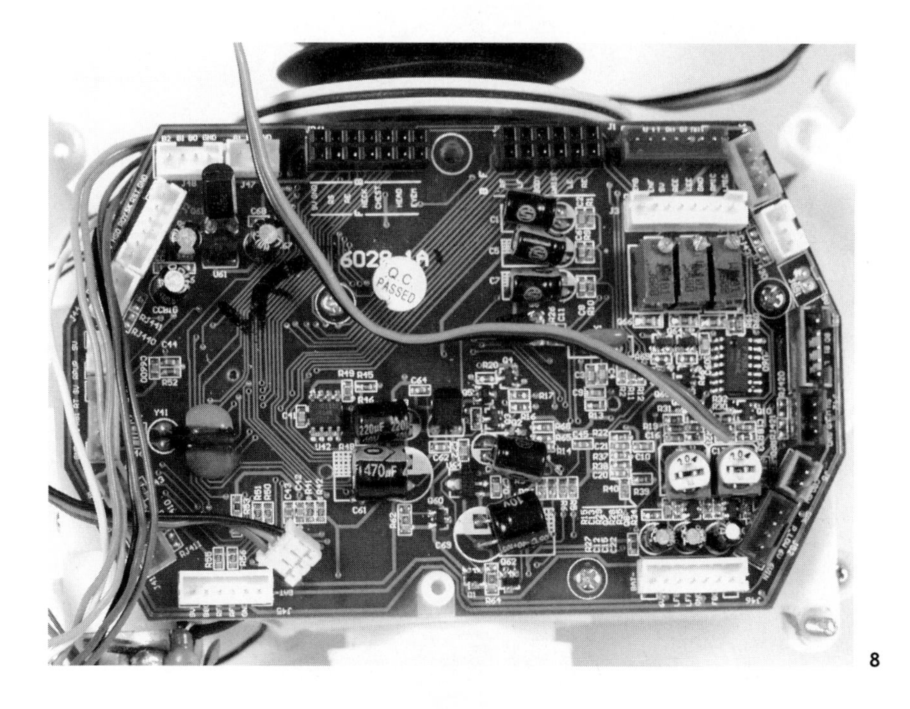

8

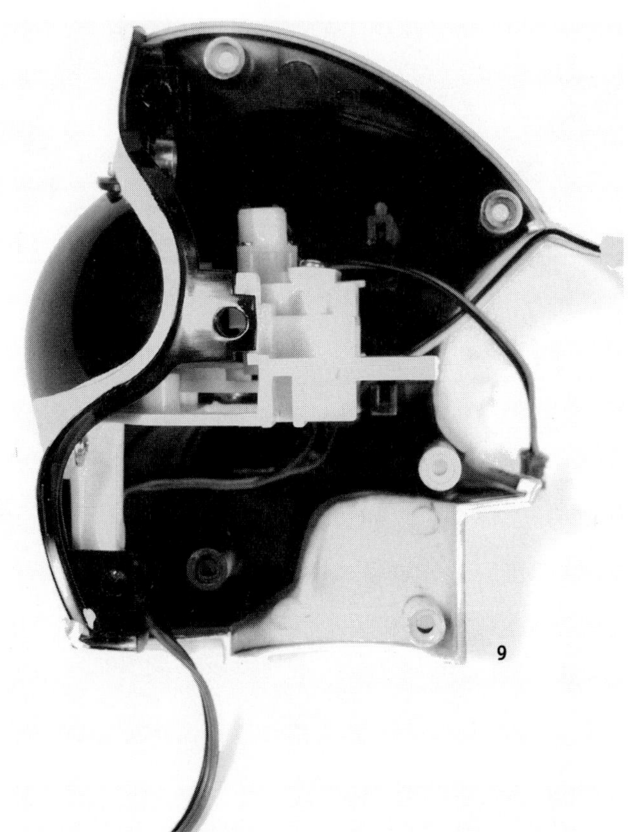

9

10 You can also remove the LED eyes from the head.

11 RS V2 is not only able to interact with you, but it also plays with other WowWee Robotics bots.

12 Remember the Robosapien personality that you like so much? Well, it's all in RS V2, but in clearly spoken English.

13 Equipped with unprecedented flexibility and articulation, RS V2 is able to sit down, lay down, and stand up.

14 RS V3 is coming. Be afraid; be very afraid.

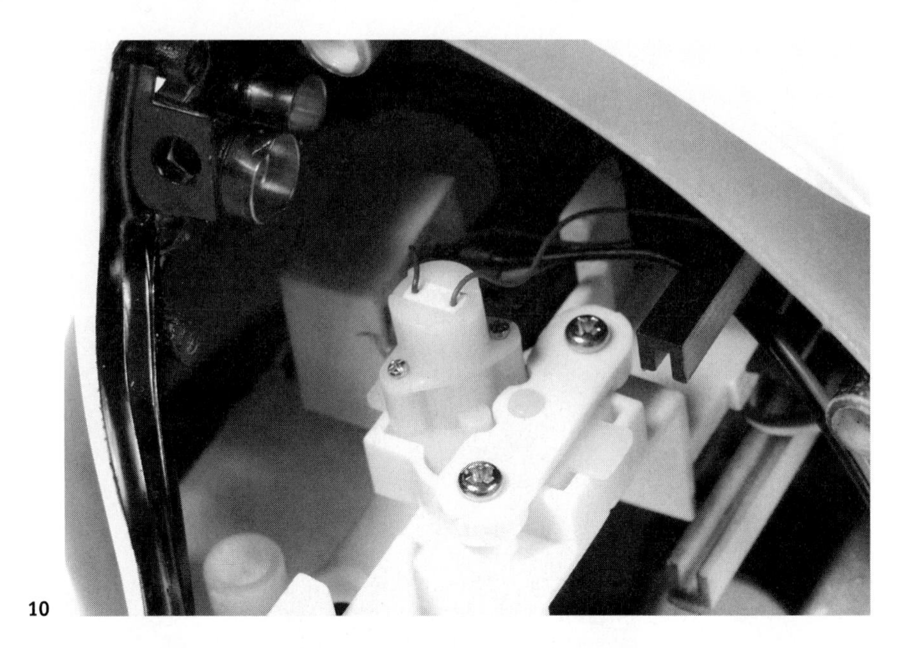

10

11

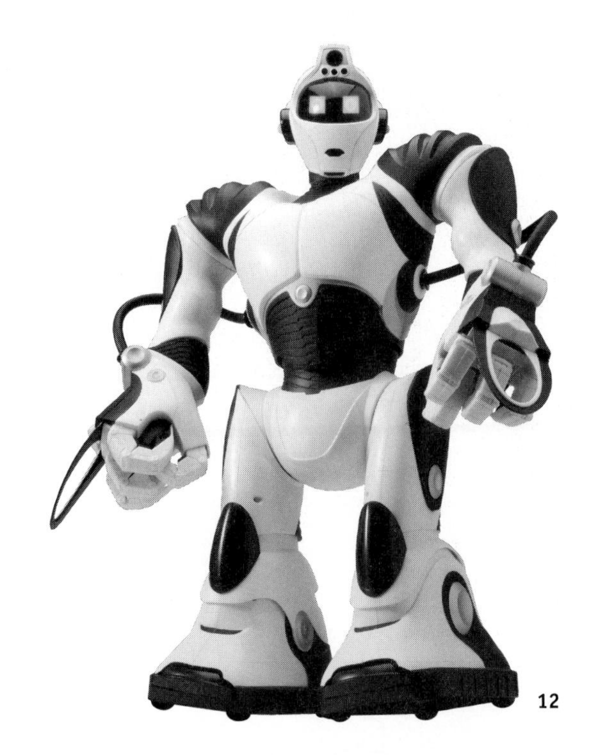

12

13

CASE STUDY 3

Apple Computer iPod

HACK TO A DIFFERENT DRUMMER

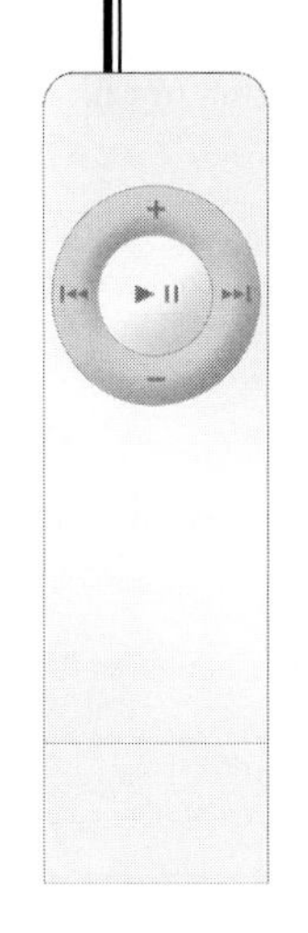

CS3-1
Apple Computer iPod shuffle, 1Gb model.

Who doesn't know about the Apple Computer iPod lineup of music players (see Fig. CS3-1)? Did you know that there are *other* music players on the market?

Even if it does have a goofy name, the iAUDIO X5 will endear you to your heart (and ears and eyes) with an alarm system, time-shift recording, voice recorder, and mobile storage facilities (see Fig. CS3-2). Couple this impressive feature list with Linux, Mac, and Windows support and the iAUDIO X5 is the perfect mate for your PSP.

The iAUDIO X5 sports a 1.8-inch, color LCD that is capable of displaying digital photographs, music videos, and movies. Oh, and did I say that it can play music, too (see Fig. CS3-3)? Available with either 20-Gb or 30-Gb memory capacity (NOTE: This is not a hard disk drive system), the iAUDIO X5 can outperform the entire iPod line leaving Apple's music players behind in a cloud of dust (see Fig. CS3-4).

While holding the aluminum iAUDIO X5 is a sheer joy; its beauty is more than skin deep. After just a short 3-hour charge, the iAUDIO X5 is able to belt out music for over 14 hours of use. Now throw in a multifunction navigation joystick, a USB OTG "host" capability, a high-speed USB 2.0 port, and a direct audio MP3 encoding ability (that's the ability to rip a CD without using a computer) and iAUDIO X5 becomes your perfect companion.

The Name of This Game

You know the event. And you know the outcome. But wait; now *you* are there. At least you are with your new iPod game. "Say what," you ask. A new iPod game? You bet. And I'm going to show you how to hack the next great thing...for iPod's, at least.

CS3-2

CS3-2 COWON America iAudio X5 is a great music player that languishes in the shadow of the over-hyped iPod.

CS3-3

CS3-3 You navigate the iAudio X5 with a push joystick (the round disk located under the right side of the color LCD).

CS3-4

CS3-4 You navigate the iPod with the popular scroll wheel.

First of all, here's a little background. Inside all iPods with iPod 2.x or later software (circa 2003 to more recent), there is a great little application that can display text notes on your Pod's LCD screen. Known as "Note Reader," this app is able to parse rudimentary HTML files that have been stored inside a special "Notes" folder on your iPod, as well as display the Notes folder file hierarchy. All is not bliss with Note Reader, however. There are a couple of limitations, but I'll show you how to overcome these puny obstacles with some clever coding techniques.

Probably the biggest limitation with Note Reader is the restriction on a note's maximum file size. Files can't be larger than 4 Kb in size. Oh, and if you exceed that maximum, the remainder of the file is truncated. Before you throw away your hopes for bringing your adventure game masterpiece to Pod glory, think about this: Note Reader does support the HTML link tag (i.e., <a href>). So with some clever game writing, you could easily link large numbers of notes together into a tapestry of suspense, or whatever else tickles your talents. And Note Reader will allow up to one thousand notes inside the iPod's Notes folder. Before you start coding and linking like a madman, however, let's look at Note Reader's other major limitation.

Navigation on the iPod is an elegant scroll-and-click series of linear steps. While this simplicity is terrific for accessing your favorite playlists, imagine how cumbersome this technique can become with hundreds of linked and cross-linked little text files. Even worse, no matter how sophisticated your linking (e.g., including "Back" and "Start Over" links), you must retrace your file footsteps to exit your game. In other words, if a player clicks on 50 linked text files in your game, the "Menu" button would have to be clicked 50 times to back out and exit the game and return to the iPod's Notes menu screen. Whew! In this context, brevity truly is the soul of your game's source code. Alternatively, you could abort all of this back-clicking by performing a master reset of your iPod (i.e., press and hold the Menu and Select buttons on click-wheel 'Pods and press and hold the Play/Pause and Menu buttons on scroll-pad 'Pods) at the conclusion of the game. But, geesch, that's so, un-Apple.

Other than those two limitations, creating text-based games for your iPod is wide open. In fact, it's as easy as, well, 1-2-3.

First, you must set your iPod up as a hard disk drive. Then you create your masterpiece with any text-oriented software. I used BBEdit, but any program that's able to make a text file is suitable. Finally, you publish all of your hard work on an iPod. Then just sit back, turn on your favorite playlist and play

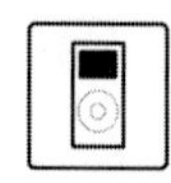

your own custom iPod game. Oh, and while royalties are optional, don't quit your day job quite yet.

THE NAME OF THE GAME

Before you can play your new game, you must set up your iPod to act as a hard disk drive.

1. Jack your Pod into your Mac (or, PC) and iTunes should start automatically.
2. Select your iPod from the iTunes Source list.
3. Click on the iPod Options button in iTunes.
4. Now you can either select "Manually manage songs and playlists" from the Music tab or "Enable disk use" from the General tab. Base your decision on whether you want to continue to automatically update your iPod with your iTunes music library. If so, then make sure that you select the "Enable disk use" option. After making your selection, close the Preferences panel by clicking "Ok."

"Yes, But I Don't Want a Skunk in My Car"

So you think that the idea of making your own iPod game is intriguing, but you don't want to mess around with code writing. Well, then you might want to consider VoodooPad (flyingmeat.com/voodoopad.html). This cross-linking, hypertext library building app is a notepad on steroids that enables you to export your notes to your iPod. There is a "Lite" version that you can experiment with before you purchase the real deal.

Or, maybe you'd rather forego creating your own game and find some text-based games for playing on your iPod. At the iStory Creator Web site (www.ipodsoft.com/istories.aspx) there is a library of user-supplied games that can be loaded onto your iPod. While iStory Creator is a Windows-only program, the games should work equally well on any iPod with the Note Reader.

Cracking the Note Reader Code

NOTE: Reader's 10-tag vocabulary looks deceptively easy to program. Like everything else in life, however, practice makes perfect. So before you can really strut your game's stuff, you must learn to talk the talk.

NOTE: Tags are not case sensitive.

<a href></a> = link to files or songs; links begin at the Notes folder top level or you can link directly to songs (e.g., song=songsnameoniPod) or playlists (e.g., ipod:music?playlist=mypersonalPlaylist)

<body></body> = an optional tag for holding file content; can be toggled on/off with a <meta name> preference tag

 = line break

<errors> = displays all file and link errors detected in Notes folder; use alone in file

<instructions> = displays Note Reader instructions; use alone in file

<meta name> = sets five file preferences with true/false parameters (e.g., content= "true"); LineWrap (name="LineWrap" content="true") controls usage of <br> and <p> tags, ShowBodyOnly (name="ShowBodyOnly" content="true") decides whether to show file content wrapped inside <body> tag, HideAllTags (name="HideAllTags" content="true") removes all Note Reader tags, as well as unrecognized tags, prior to display, NowPlaying (name="NowPlaying" content="true") displays the "Now Playing" screen after clicking on a song link, and NotesOnly (name="NotesOnly" content="true") creates a custom Note Reader user interface that must be globally initialized

<p></p> = paragraph break

<showpreferences> = displays global preferences for notes; global preferences which are overridden with <meta name> tags are displayed instead

<title></title> = specifies different title for note; notes, by default, display the file name for the title

<?xml encoding> = foreign language and non-ASCII text Unicode; always place the encoding tag at the top of the file

You can download the complete iPod Note Reader User Guide from the Apple Computer Web site (developer.apple.com/hardware/ipod/).

5. Your iPod icon is now ready on your computer's desktop.
6. Double-click on the icon.
7. Your future game files will be copied into the now-visible iPod Notes folder.

LIGHTS, CAMERAS, AND TEXT-BASED ACTION

This is the step where you separate the Macs from the hacks. Every game is only as good as the story that surrounds the play. This mandate is even more clear in text-based games that can't rely on the drool factor of sound effects, graphics, and animation. Don't expect to become an overnight sensation—remember to work your Mac and learn your craft. And, of course, share your games with the rest of us.

1. Develop a game concept and organize your background research for creating character biographies, location settings, time constraints, objectives, and goals. In other words, all of that journalistic "who," "what," "where," "when," and "why" stuff.
2. Make a game timeline. This document is your "bible" for keeping the game on track and your fellow game players satisfied.
3. OK, with all of that preliminary work taken care of, it's now time to start turning your timeline into code.
4. Working on your Mac, set up a folder hierarchy for holding all of your game's text files.
5. Use your favorite text editor, like BBEdit, for making short (remember, each file must be smaller than 4 Kb in size) text files. Your first file should be named "Main.linx." This file will be the opening screen for the game. Unlike other Note Reader text files, Main.linx can only contain two elements: a title and links to other files. If you opt to omit the Main.linx file (and you can, if you like), your game's opening screen will display a file list of the Notes folder's contents. Ugh, that's not a very professional greeting to game players, now is it?
6. Start writing, naming, and saving each of the text files that are needed for your game. Keep this physical dimension in mind when writing these files: The Note Reader's screen measures approximately 8 rows by 21 columns. Now this is a rough dimension that varies slightly between iPod mini, iPod 3rd generation, iPod 4th generation, etc. Regardless, the point is to keep your verbiage tight and line feeds to a minimum.

PUBLISH IT!

After you've gathered all of your text files and folders together into a manageable structure, it's time to publish your game to your iPod. The initial phase of this publishing step should include game testing and debugging. Thankfully, Note Reader includes a tag that makes this laborious process almost painless. The <errors> tag fills a file's contents with a comprehensive listing of all errors that were detected in all of your game's text files—automatically. Neat, huh? Just add this tag to any file and you have instant error checking. I make a special file for this purpose and include a link to it from inside Main.linx. Once you have completed your testing and debugging, you can remove this link and delete the <errors> file.

1. With your iPod on your desktop, double-click on its icon.
2. Navigate to the Notes folder inside your iPod.
3. Drag your Main.linx file and your game's entire folder/file hierarchy into the Notes folder.
4. **NOTE:** The file inside your Notes folder named "Instructions" is a short text file that describes how to use the Note Reader. You don't need to delete this file. It won't affect your game's play.
5. Bask in the glory of being a published game designer.

About the Game

I chose the dramatic sinking of the RMS Titanic as the backdrop for this text-based iPod game hack. In this context, I used real passenger names and actual events to lend an air of accuracy to this otherwise totally fictional portrayal of this tragedy. If you would like to locate additional information about the sinking of the Titanic, you can learn a lot at the Encyclopedia Titanica Web site (www.encyclopedia-titanica.org/index.php).

NOTE: The character Henriette Yvois was an actual passenger aboard the RMS Titanic, but very little is known about her. The Encyclopedia Titanica Web site would welcome any information that you could provide about this real life soul.

HOW TO CREATE AN IPOD GAME

1 Halo 2? Nah, better; it's your own custom-made iPod game.

2 Activate your iPod hard disk drive capability by selecting "Manually manage songs and playlists."

3 Or, you can activate your iPod hard disk drive by selecting "Enable disk use" on the General tab.

4 You will copy your future game to your iPod's Notes folder.

Yvois
April 14, 1912, 11:40 PM
You are playing Patience
with William Harbeck in
the main lounge.

Crow's nest signals

1

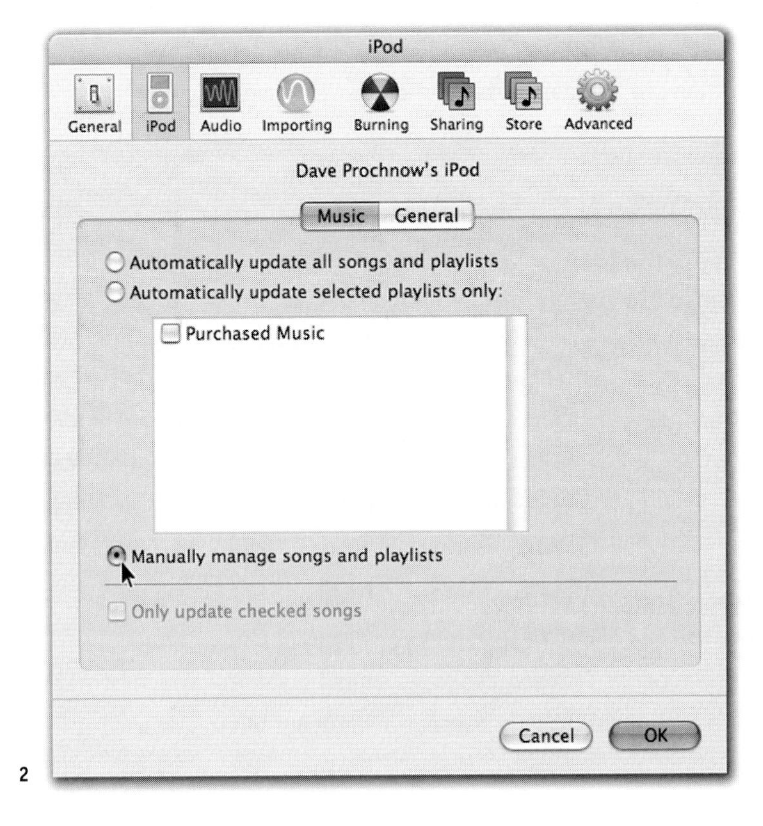

2

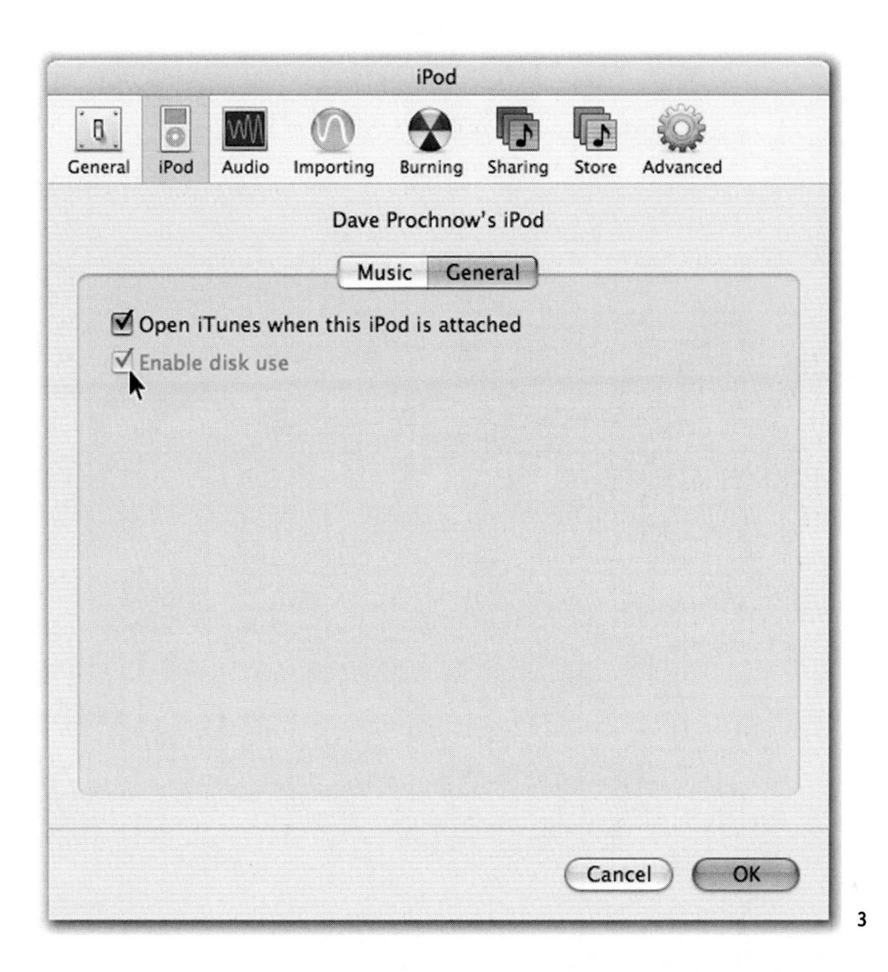
iPod
General
iPod
Audio
Importing
Burning
Sharing
Store
Advanced
Dave Prochnow's iPod
Music
General
Open iTunes when this iPod is attached
Enable disk use
Cancel
OK

3

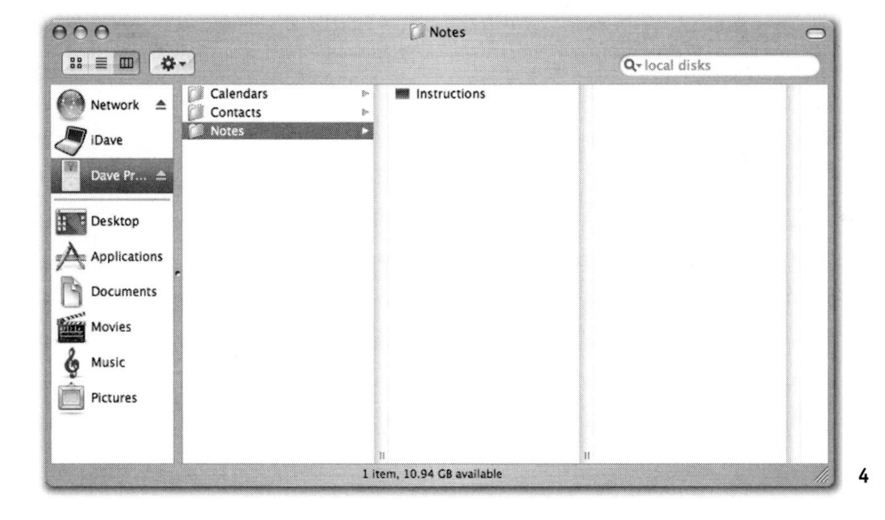
Notes
local disks
Network
iDave
Dave Pr...
Desktop
Applications
Documents
Movies
Music
Pictures
Calendars
Contacts
Notes
Instructions
1 item, 10.94 GB available

4

5 Make a “working” file/folder hierarchy on your Mac for holding all of your game’s text files.

6 A Main.linx file contains a title (I used the title for our game) and links to the beginning files for your game (I used the names of our game’s passengers).

7 Writing an iPod game is deceptively easy and unusually terse. I opted for setting the “LineWrap” <meta> tag preference to “false” so that I could eliminate <br> and <p> tags and keep my file sizes to a bare minimum.

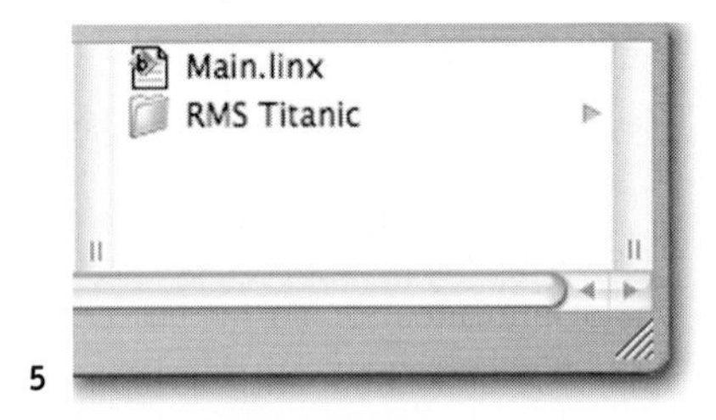

5

6

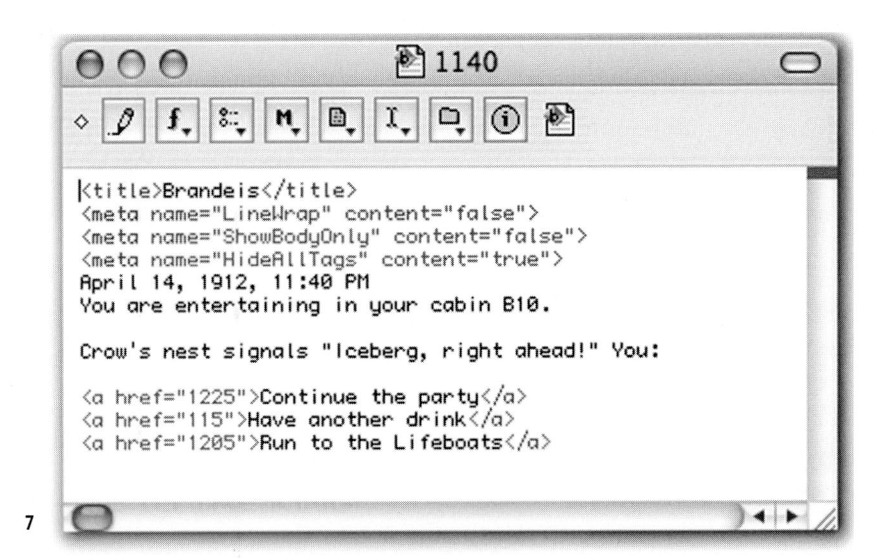

7

8

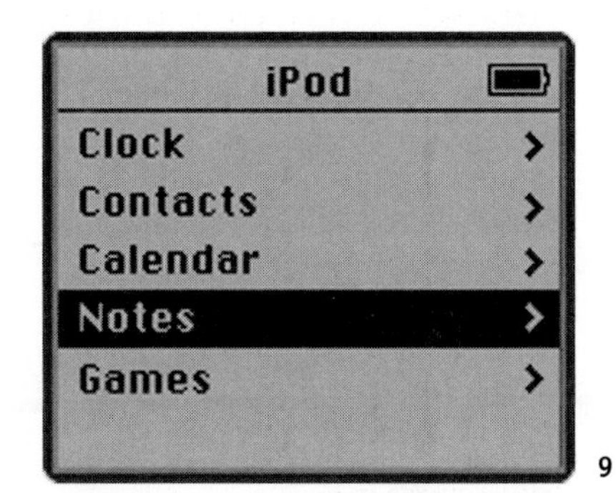

9

RMS Titanic
Emil Brandeis
Henriette Yvois

10

Yvois
April 15, 1912, 2:05 AM
The last lifeboat leaves.
William films the departure. You:
Help with the shot
Head toward the bow

11

1:13 PM
Link Error in file Note\RMS Titanic\Yvois\1225.
<A HREF="1230"> at char 129.
Bad link, file not found. (5)

12

8 You will find the Note Reader app inside the Extras menu on your iPod.

9 The app itself is called "Notes" inside the Extras menu.

10 By using the nifty Main.linx opening file, your game takes command of the entire iPod Notes folder.

11 Navigating through Note Reader links can be tricky. A dark underline highlights the selected link, while a gray underline represents "other" link options. You can toggle through various links by rotating the iPod's click wheel or scroll wheel.

12 By inserting the <errors> tag in a file, you will receive a comprehensive listing of all errors within your game. Use this error listing to quickly debug your game and then delete the file when your game is ready to be published.

HOW TO ADD LINUX OS TO YOUR IPOD

1 Jack your 'pod into your computer, enable hard disk drive capability, and run the iPod-Linux Installer.

2 Decide whether or not you want Linux to be the default OS when you start up your iPod. You can always toggle between the 'Pod's native OS with a simple keypress during a master reset of your music player.

3 The root file hierarchy for Linux on your iPod. Please note that the PSP and DCIM folders are not part of the Linux installation.

4 Reset your iPod by pressing and holding the Menu and Play buttons for several seconds.

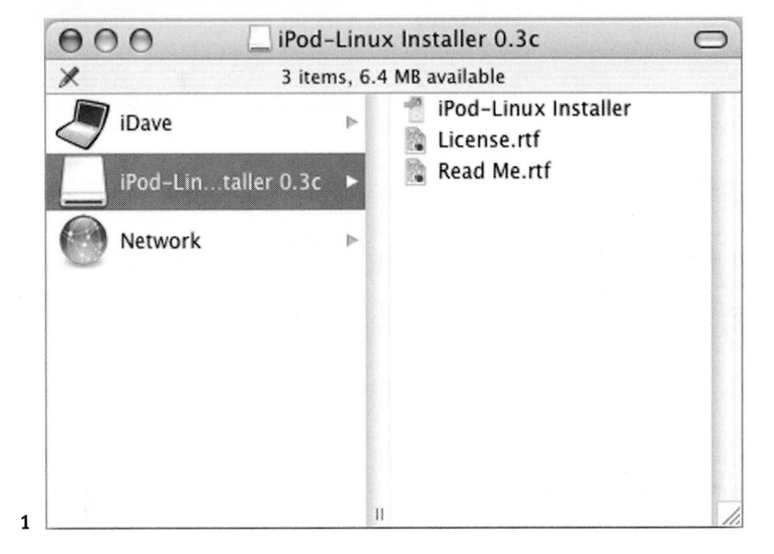

1

iPod-Linux Installer

iPod: Dave Prochnow's iPod

Install
Make Linux Default OS
Uninstall

Click the Install button to install Linux on this iPod.

Eject iPod
Check For Updates

2

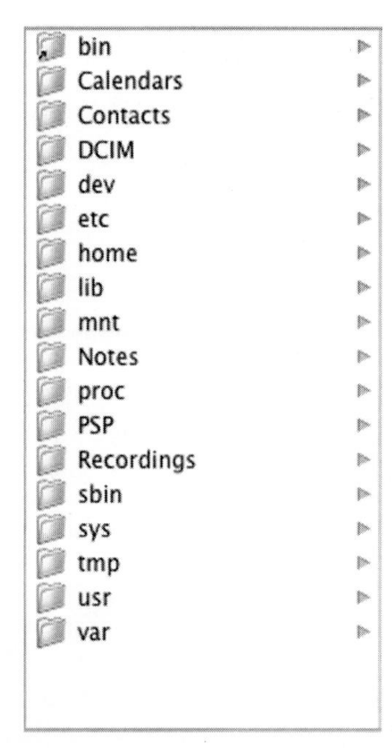
bin
Calendars
Contacts
DCIM
dev
etc
home
lib
mnt
Notes
proc
PSP
Recordings
sbin
sys
tmp
usr
var

3

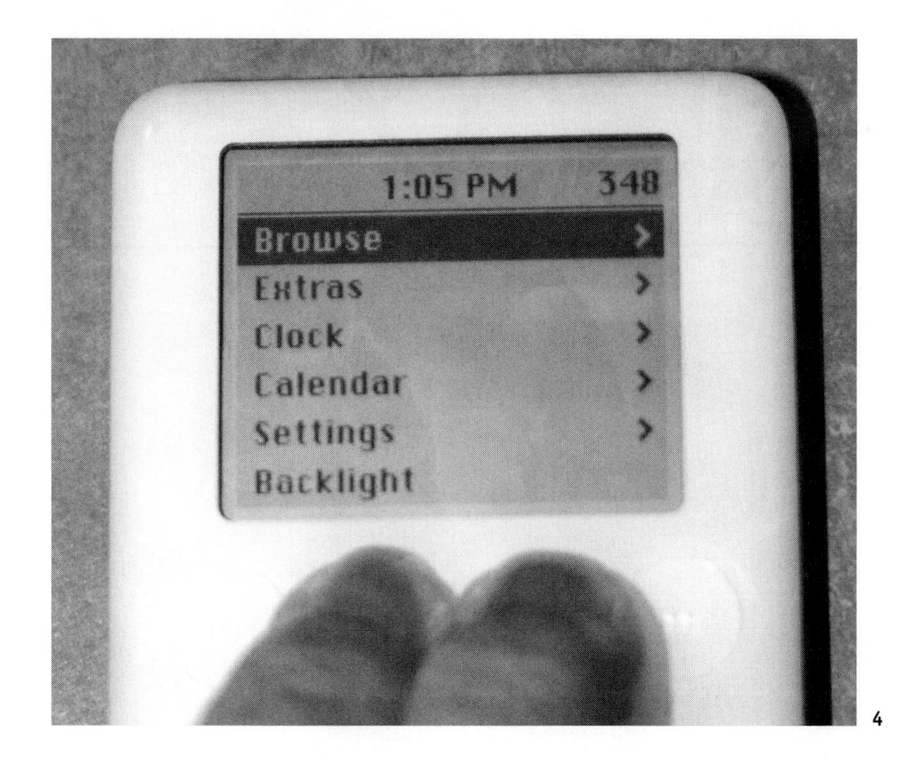
1:05 PM 348
Browse
Extras
Clock
Calendar
Settings
Backlight

4

5 During startup, you can toggle between the non-default OS (i.e., in this case Linux is the non-default OS) by pressing and holding the iPod's Rewind button.

6 Welcome to *podzilla*, the Linux OS for iPod.

7 When you perform an iPod reset from podzilla, a command line interface is displayed on the iPod LCD screen. This stuff is music to geeks' ears.

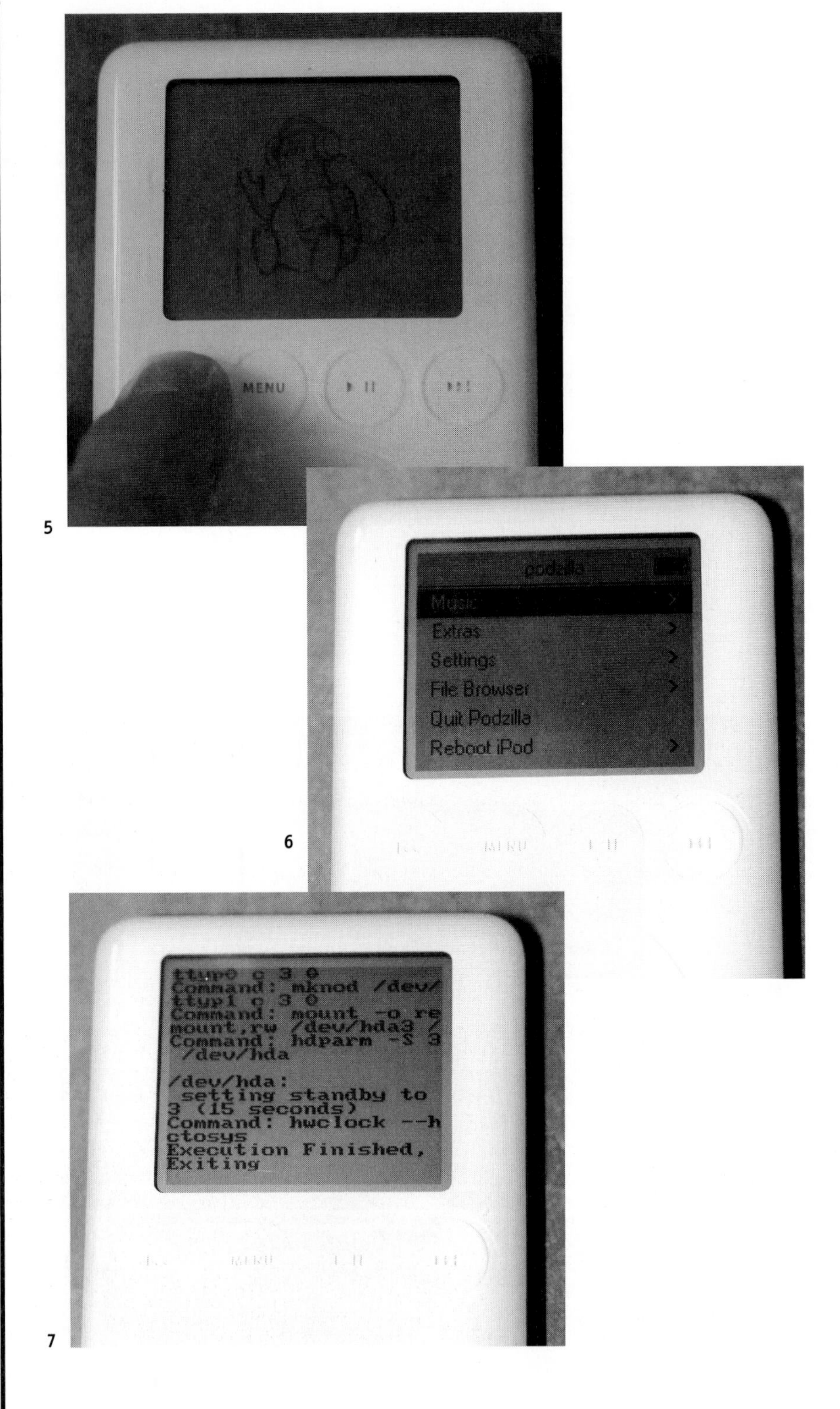

CASE STUDY 4

Sony PSP™

MAKE YOUR PSP HAPPY, HACK IT INTO A SONY MAC OS PSP AND LOCATIONFREE, LOCATIONFREE, LOCATIONFREE TV

On the heels of the publication of my bestselling book about hacking the Sony PlayStation Portable handheld entertainment system, *PSP Hacks, Mods, and Expansions* (McGraw-Hill, 2006), I was interviewed by a reporter with *BusinessWeek*.

This Tokyo-based reporter had a lot of great questions about hacking electronic gadgets, integrating the PSP into a digital lifestyle, and Sony Corporation's treatment of hackers (see Fig. CS4-1). Unfortunately, very little of this interview was published in the final article.

CS4-1 Sony PSP™ handheld entertainment system.

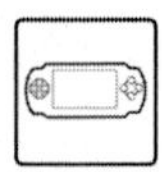

Two important topics that were discussed in the interview have been addressed by the two hacks that are presented in this Case Study for the Sony PSP. In order to provide you with some background perspective on these hacks, here is the entire text from that interview.

KENJI HALL, Correspondent with *BusinessWeek*: "I'm a Tokyo-based reporter with *BusinessWeek* magazine. When you're free, I'd be interested in chatting with you about your hacker's guide to the PSP."

DAVE PROCHNOW: "It is a pleasure to meet you, Kenji."

KH: "A company that makes chat software says it's working with game machine makers, including Sony and Nintendo, to offer users of handheld devices a whole new way of creating a community."

DP: "Like yourself, I've heard rumors of a 'Wi-Fi' IM chat capability, but actual confirmation of this feature has been impossible to obtain. Likewise, I've heard that there *might* be a limited test or evaluation program being conducted in a Korean market. At present, however, this rumor is unsubstantiated."

KH: "Have you already found a way to create a chat function on your PSP?"

DP: "Alas, no I haven't, nor have I heard of any reliable "hacker" who has been able to succeed in adding this capability to a stock PSP. You are correct that a pseudo-chat function could be jury-rigged with the help of a Wi-Fi hotspot, the 2.x Web browser, and a couple of dedicated Web pages. Making an IM chat capability, however, like the one found in the awesome Aeronix ZipIt Wireless Messenger is still just a rumor.

"Granted, the Internet Web browser feature introduced in the PSP Firmware 2.0 update was a significant capability that Sony added to the PSP; but, at what price? What sounded like a good enough reason for every Firmware 1.5x user to upgrade, became a digital line drawn in the sand for owners of Firmware 1.5 PSPs. Basically, if a 1.5 user upgrades to the 2.x Firmware he/she loses access to the homebrew hackers underground. And that loss translates into the inability to play emulated ROM games, shareware PC games, and Lua-scripted games. That's a loss that few, if any, 1.5 owners are willing to accept. Therein lies one of Sony's most glaring failures with the PSP: Understanding that these really outstanding capabilities, like the browser and this new IM chat feature are shunned by the most coveted PSP owner demographic—the die-hard gamer."

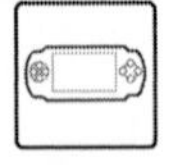

KH: "...they'll make the iPod video look downright prehistoric."

DP: "Actually, Kenji, I'm one of those few writers who actually thinks that the iPod video is already prehistoric—dead right out of the box. Here you had a revolutionary breakthrough music-playing product that is now being kludged into a video player, a photo scrapbook, a voice recorder, and, of course, still a music player. Even in its initial incarnation, iPod wasn't that great of a music player. Heck, COWON America makes music players that can run circles around the iPod. What made the iPod into a killer product was the brilliant Apple Computer marketing ploy of bundling a remarkably powerful and simple-to-use software interface with each iPod—iTunes. This FREE interface to the iPod made music collection management a breeze. Unfortunately, it took all of us a while before we realized that we had sold our souls to the devil—the DRM devil, that is.

"At any rate, I am a complete PSP convert. Yes, I have a couple of iPods, but the PSP is the complete package for me. The video playback from a UMD movie (or even a homebrew movie off of a Memory Stick Pro DUO media card) is breathtaking. Likewise, onboard stereo speakers, a great arrangement of movie control buttons, and battery life that would make every iPod owner envious, makes the PSP into a viable video playback platform...in spite of Sony. Why this qualifier? Where's the user interface software that could make PSP into a definitive killer product? Where's the PSP equivalent of iTunes? No where; and that might be the reason that PSP lurks in the shadow of even the anemic iPod video."

KH: "How hard is it to hack into PSP?"

DP: "Ha, ha, Kenji—Hey, buy the book. Actually, once again, in spite of Sony, the PSP can be hacked by virtually any type of user/owner. 'Oh sure,' you say, 'the solder-heads and the code hackers can find something inside the PSP to entertain themselves. But what about regular users like myself?' There are many hacking opportunities for exploiting the PSP that don't require solder or code. Exploring USB OTG, manipulating Memory Stick Pro DUO folder contents, and modding a PSP case are each avenues worthy of 'hacking' by every user. Depending upon your technical prowess and your creative instincts, the PSP is a great canvas for anyone to express her/his newfound digital lifestyle."

KH: "I think I want to change the angle of this story, if you don't mind. I think it's interesting how Sony is so dead-set against allowing this [hacking] to happen, and has changed the firmware to make sure future PSP buyers can't do to

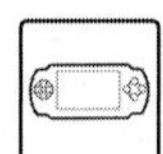

their machines what you've done to yours. What drove you to hack into your own PSP?"

DP: "That's a good question. Just what would make someone void the warranty on everything that he owns? (Tee-he.) From my perspective, a product, like the Sony PlayStation Portable, for example, is as stimulating as a blank canvas is to an artist. I see an opportunity to 'make' this product into 'my own' personal digital lifestyle product. Rather than relying on the operation and functions defined by the manufacturer, I see the possibilities of integrating this device into my life...but I don't want to alter my entire life just to be able to view movies, play games, and listen to music. Therefore, I hack it."

KH: "When did you buy your PSP? Did you immediately set about modifying it? Or did you test it first?"

DP: "I was lucky enough to get my hands on one of the first PSP models released in Japan. As I held that exquisitely crafted device in my hands, it was like an epiphany or a digital head slap. Whoa. Here in my hands was the digital product that I had always longed for. Games, movies, music; everything together in one discrete, highly portable form factor. It was digital lifestyle lust at first sight. Then the reality, the Sony reality, hit me. My PSP was crippled...deliberately crippled—forced to use a proprietary media format (UMD), limited to handheld viewing (no video output port), and lacking a practical method for accessing my current game and DVD libary. My Sony PSP was an 'almost' perfect product."

KH: "What do you think is wrong with the PSP?"

DP: "Don't get me wrong, the stock Sony PSP is a wonderful device. If you don't already own a PlayStation 2 game library or a bookshelf full of DVD movies, then buying UMD media for your PSP is an expected and agreeable facet of staking your claim for a piece of the new digital lifestyle. There are probably very few people who live inside this kind of digitally-deprived world, however. Owners of PSP should expect, if not demand, that previous media investments be leveraged against this newfound portable product. Heck, it is called the 'PlayStation' Portable, right? Wrong! Why didn't Sony include a DVD input port and a video output port on the PSP? I'm not talking about crippling the PSP with ancient technology I/O like VHS video tape or floppy disks. Just let my new toy work with my current media."

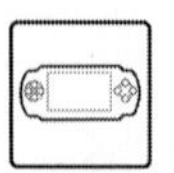

KH: "Internet browser, MP4 compatibility, LocationFree Player. These are a few of the new features that users can get with firmware v2.0 and v2.5. You don't care that your PSP doesn't get these?" (See Fig. CS4-2.)

CS4-2 Sony Firmware update 2.6 added the LocationFree™ Player to the PSP.

DP: "Actually, I'm a different type of PSP owner. I have several PSPs. Therefore, I can keep a couple of PSP examples that have the old firmware, as well as install the newer, more potent versions on other handhelds for sampling these new features. For my money, it is very tough to convert all of my PSPs over to the latest and greatest firmware. I enjoy the homebrew games that, as of today, can only be played on firmware 1.5–equipped PSPs. That isn't to say that I don't find the new features, especially the Web browser tool, to be valuable...I do. In fact, when I travel I regularly take one of my newest PSPs with me. Sitting in an airport terminal with a couple of UMD games and movies, as well as being able to surf the Web, all on the same device. Now that's a good fit into *my* digital lifestyle."

KH: "What can your PSP now do that it couldn't do before? In other words, what games do you play or movies do you watch? Any other cool functions that normal PSP users won't have?"

DP: "Homebrew games, like the shareware version of DOOM, and emulator mods for my already-owned NES game cartridges are the most high-profile attributes that one of my PSPs with the older firmware share which are absent on my newer PSPs. Is that a big deal? I'm not so sure. There are a lot of terrific things that you can do with a PSP, any PSP, regardless of the firmware version. Printing hard copies of photographs on a PSP, sharing music within your entire collection of MP3 devices (including your Apple Computer iPod), and installing and watching your own videos on a PSP are all activities that any PSP owner can do. The only real trick is knowing how to perform these tasks. And that's where any, and every, PSP owner can become a hacker. You don't have to be a coding wizard to be a hacker, just be willing to try some unconventional experimentation with your PSP."

KH: "How much programming went into this?"

DP: "I hack with a soldering iron and screwdriver. Other hackers rely on modifying the PSP through software engineering. For example, if you want to play homebrew games on your 1.5 version PSP, then you need to utilize the numerous programming hacks for exploiting the PSP firmware. On the other hand, if you want to mod your PSP with some electroluminescent wiring, regardless of its firmware version, then all you need is some carefully crafted soldering iron and screwdriver hacking."

KH: "The word 'hacker' has good or bad connotations, depending on your view. For some people, hackers are the guys who break into people's e-accounts and rob them of their life savings. For others, they're the guys who outsmart the system. How do you see yourself?"

DP: "Oh, I'm glad that you asked that, Kenji.

"Not too long ago, around 1620 to be precise, the term 'hacker' was used to denote a person who was unskilled or inexperienced at a particular activity. For example, you were a golf hacker; or, I was a painting hacker. Now fast forward approximately 350 years, and the term 'hacker' has undergone a radical facelift.

"Around 1970, rather than being unskilled or inexperienced at an activity, a hacker was considered to be a highly skilled, clever programmer or a technology expert. What? You thought a hacker was a criminal, a ne'er-do-well, a malcontent who thrives on crippling computer systems? Wrong-o, bucko. Let's set the record straight, once and for all.

"As coined in the early 1970s, a hacker was a genius, a gifted individual who was the master of a software or hardware domain over which digital dominion was exercised. Ah, the good ol' days. Unfortunately, in the 1980s, a different definition for hacker came into mainstream usage. This metamorphosis didn't have to happen, however.

"As far as hackers are concerned, the 1980s started in a heroic manner with the publication of Steven Levy's book *Hackers: Heroes of the Computer Revolution*. In 1989, however, the hyperanimated Clifford Stoll (author of *The Cuckoo's*

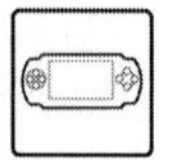

Egg) penned a different take on hackers—as computer criminals. Soon, public awareness and media propagation embedded the hacker-nee-criminal concept into colloquial English. And, the negative connotation for hacker stuck.

"Now, if you ask a hacker what to call a computer criminal, the most common reply would be a "cracker." According to hackers, a cracker is a thief, subversive, or criminal who is determined to break, cripple, or vandalize a computer network or system. Additionally, the term "black hat" might also be applied to this form of computer criminal. In this case, the appellation is derived from a negative nod to a popular flavor of Linux (i.e., Red Hat). And Linux is the most popular weapon in the arsenal of the hacker."

KH: "When did you start writing your book? How long did it take? Why did you decide to write it?"

DP: "I started working on my book as soon as I held that brand new shiny PSP in my hands. In fact, I had the proposal for the book project completed in less than one hour—not written into a word processor exactly, rather it was all 'written' inside my head. Now I just had to put my ideas down on paper. The presentation or 'pitching' of the book project took longer than writing the proposal, however. The publisher, McGraw-Hill, was very skeptical about producing a book dedicated to 'just another game machine.' It took me roughly two months of relentless pushing, prodding, and selling to convince the publisher that a book about hacking the PSP would be as valuable to this handheld as a UMD disc. I even went so far as to supply dummy copies of PSPs to all of the key players at McGraw-Hill. Once the contract was inked, the actual writing and illustrating of the book took less than two months. A lot of people worked many long and tireless hours contributing to this book's development. Even the destructive aftermath of Hurricane Katrina couldn't stymie this book. I lived right in the path of this epic storm and had to drive over 300 miles to reach an airport for mailing the final manuscript to the publisher. Once McGraw-Hill saw this manuscript, however, they were convinced about the merits for this book. So much so, that the printing and marketing of this book became a top priority for this publisher. Why all of this unprecedented interest? I feel that it's the power of the PSP—once you see it or have someone show you what you can do with it, you're hooked."

KH: "If you and Sony had to switch places, would you try to stymie the hackers? Or, would you let them be?"

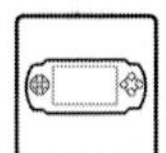

DP: "There are three significant marketing moves that Sony could make which might enhance the sale of the PSP:

"1. Open up the hardware and software for empowering the hacker community. In turn, this 'open' system would produce an incredible explosion in the development of follow-on after-market products. By embracing third-party accessories (beyond the case, grip, and skin markets), Sony would benefit from the advertising overhead that these outside manufacturers would generate.

"2. Minimize the digital rights management (DRM) impact on the user. Learn from the sales secret of the iPod—package DRM in a tasty FREE software product that insulates the user from arcane computer actions and you will have a killer product. You don't need to resort to rootkit viral infections, either. Just enable PSP owners to bring their current DVD and PS2 game libraries to the PSP and you will have a product that your mother will want as much as you do.

"3. Take the strengths of the PSP to the core demographic—gamers. Why would any gamer want a PSP? Good question, ask Sony. Develop a FREE software package that can convert PS2 games to run on a PSP Memory Stick Duo and you will have a product that could steal important thunder from the impending Microsoft Xbox 360 release."

KH: "I've heard there's a thriving PSP hacking community."

DP: "Yes, there is a PSP hacker community...it is a small, vocal, and fiercely defiant group. I haven't 'tapped' this community, but I do regularly recognize and acknowledge its contributions through the articles that I write, as well as in my PSP book. While perceived as a thorn in Sony's side, the PSP hacker community's greatest strength is in its loyal camaraderie which, if properly channeled, could serve to exploit more than just 'holes' in firmware. This hacker community could be a healthy 'think tank' for fostering the commercial exploitation of this potent handheld entertainment system into becoming a 'must-have' digital lifestyle product."

KH: "How many PSPs do you own?"

DP: "During the preparation of the manuscript the exact number of PSPs fluctuated wildly—between two to five at one time. Typically, I would give 'test' models away when I was finished with 'em...mainly to area kids. At present, my PSP stable contains three thoroughbreds."

KH: "Do you literally have to take apart your PSP to get at the motherboard inside?"

DP: "Yes, a PSP was completely disassembled. In fact, there is an excellent photo spread in the book that illustrates *all* of the PSP internal details in gorgeous photographs. I'm a meticulous hardware hacker...I won't even tolerate scratched screw heads. And that kind of diligence comes from many years of hardware hacking. Likewise, my hacking talents then pay dividends to my readers who are able to revel in all of my hacks, mods, tips, and tricks without ever having to get their hands dirty, or, gasp, risk damaging their own PSP."

KH: "You identify the incompatibility of PSPs with PS consoles as one of the shortcomings of PSPs. Don't the PSP updates now let users do many of these things?"

DP: "Actually, Kenji, I might take issue with the premise of this question. Yes, the firmware updates provide *some* of this capability, but learning how to convert a movie into the proper PSP format, for example, can be a daunting task. The book provides all of the necessary tutelage that is required for mastering this technique, as well as accomplishing a whole host of other PSP functions and capabilities. I was able to obtain the newest firmware release prior to the submission of the final manuscript. And I can assure you that the material in the book is neither compromised nor outdated by any of the Sony PSP firmware updates. As an example, learning how to implement USB OTG in your PSP will enable you to leave your notebook computer at home and still shuttle music tracks, photographic images, movie footage, and homebrew games back and forth between your PSP and a USB host. In my opinion, this is a 'killer' capability and it is painstakingly described in the book. Even better, this type of hack works with *any* and *every* firmware update. But, best of all, *any* user with *any* experience level can perform this hack with perfect results. Everyone that I've demonstrated this hack to has exclaimed, 'Wow, I didn't know that you could do that!' Hmm, maybe that should be a subtitle for the book?"

KH: "If you had your way, what else would you want your PSP to do?"

DP: "Ha, ha, ha...other than seamlessly mesh with my DVD collection and access my PS2 game library...I can't think of anything else." (See Fig. CS4-3.)

CS4-3 The LocationFree Player enables you to watch your TV, DVD, or PS2 (watch not play) games on your PSP.

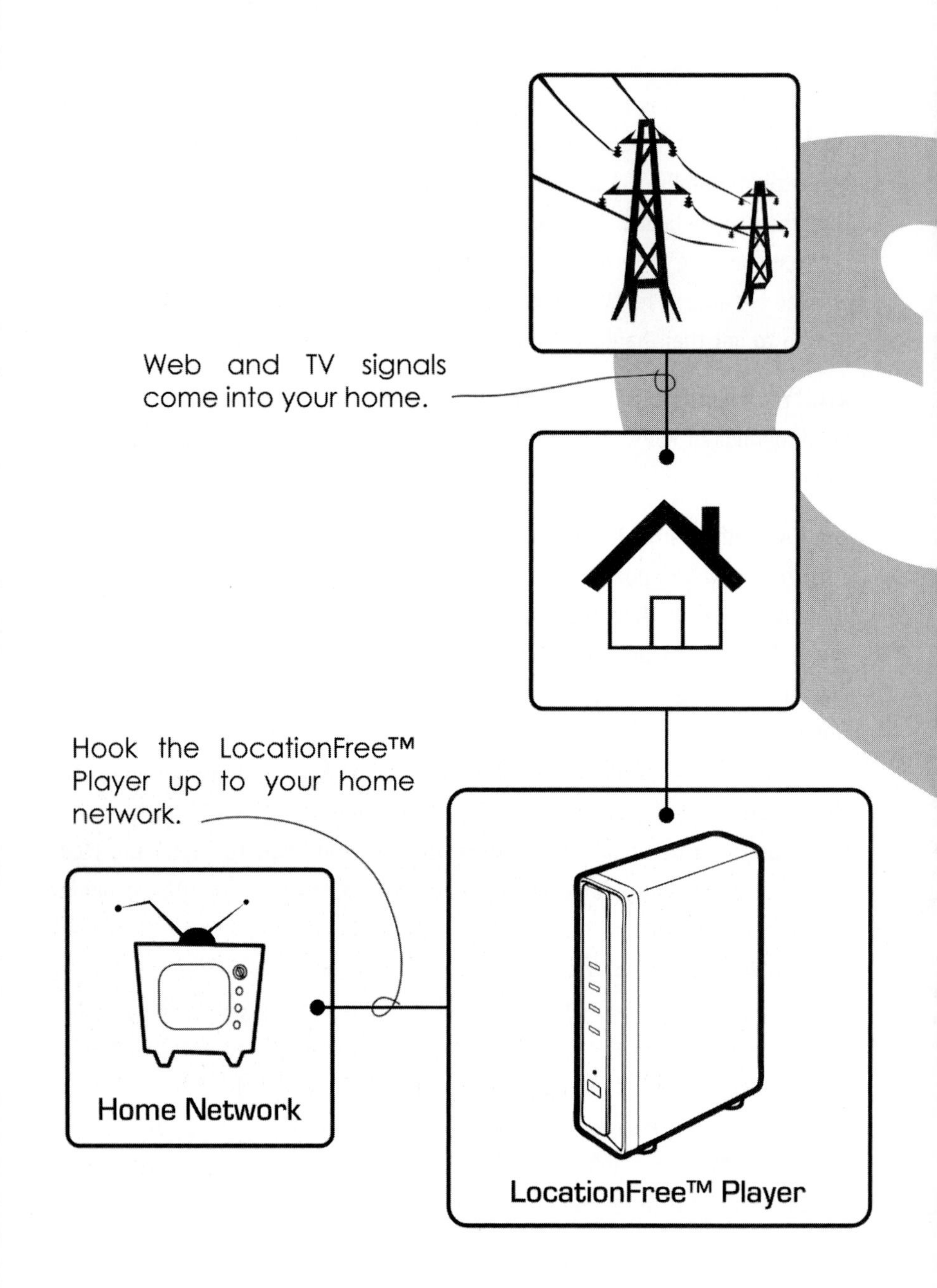

Web and TV signals come into your home.
Hook the LocationFree™ Player up to your home network.
Home Network
LocationFree™ Player

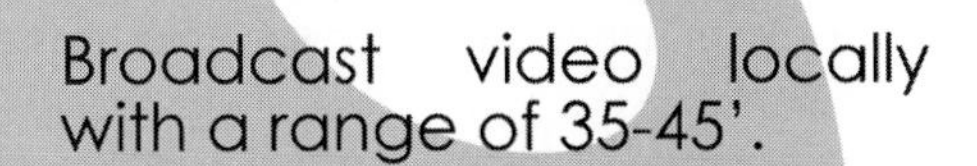

Broadcast video locally with a range of 35-45'.

Sony PSP™

Watch your video anywhere any time. Use a static IP for watching your video at any WiFi hot spot.

PDF Your PSP

One of the neatest hacks that emerged on the heels of the publication of my PSP book was the release of *Basilisk II PSP Port*. Started by Kiera Mohr and enhanced by ChaosKnight, this Mac emulator program is so powerful that you can readily load virtually *any* Mac OS app on your PSP and it will work...just like on a real Mac.

Able to support Mac OS System 7.5.5 (yes, some PSP hackers, like ChaosKnight, have been able to use Mac OS System 8.x on the PSP, but the file-size overhead might not be worth the space that Basilisk II PSP Port will use on your Memory Stick), this hack will only run as a homebrew application. Therefore, in order to use this hack, you *must* have a 1.5 Firmware PSP. (I have not tried this hack on the new *Grand Theft Auto* homebrew hack, yet.)

Once you get Basilisk II PSP Port running on your 1.5 Firmware PSP, you can load your Mac OS System 7.5.5 apps provided they are in an HQX compression format. In addition to numerous games, I was able to load Adobe Acrobat 2.1.

Once Acrobat is up and running inside Basilisk II PSP Port, you will be able to convert the entire contents of my PSP book into a 2.1-compatible PDF, load the PDF onto one of my 1.5 Firmware PSPs, and read the book, anytime, anywhere.

HOW TO HACK THE MAC OS ONTO YOUR PSP

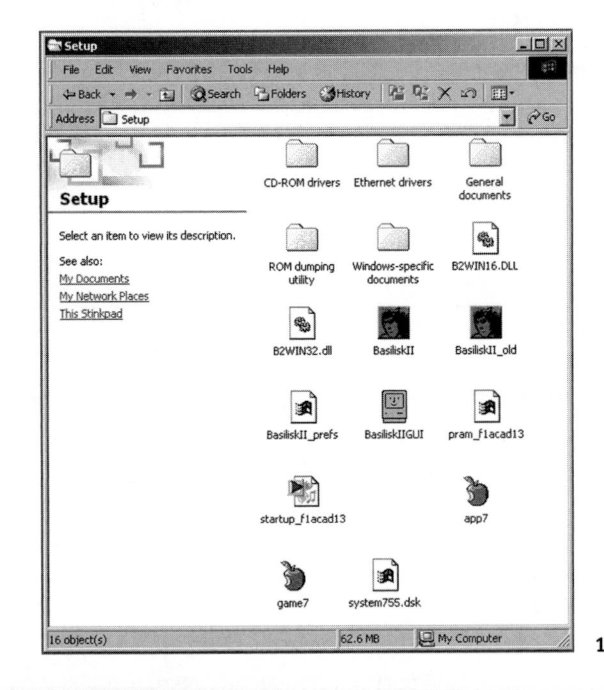

1

2

1 *Basilisk II for Windows*. Ironically, using a PC is the *best* way for installing the Mac OS emulation program, Basilisk II for PSP.

2 This is my "working" folder for holding Mac OS files and Basilisk II for PSP.

3 An example HFS folder that I used for installing and testing Basilisk II for PSP.

4 *HFVExplorer* for Windows is an essential utility for setting up Basilisk II for PSP.

5 The root directory structure for a Mac OS System 7.5.5 disk in HFVExplorer.

6 A "blessed" Mac OS System 7.5.5 folder in HFVExplorer. In order to get a "blessed" system folder (i.e., the Finder icon looks like a Mac SE *and* the System folder icon looks like a Mac SE inside a suitcase), you *must* have a valid Mac ROM image. You can either download this image from one of your old Macs *or* you can find a pirated copy on the Internet.

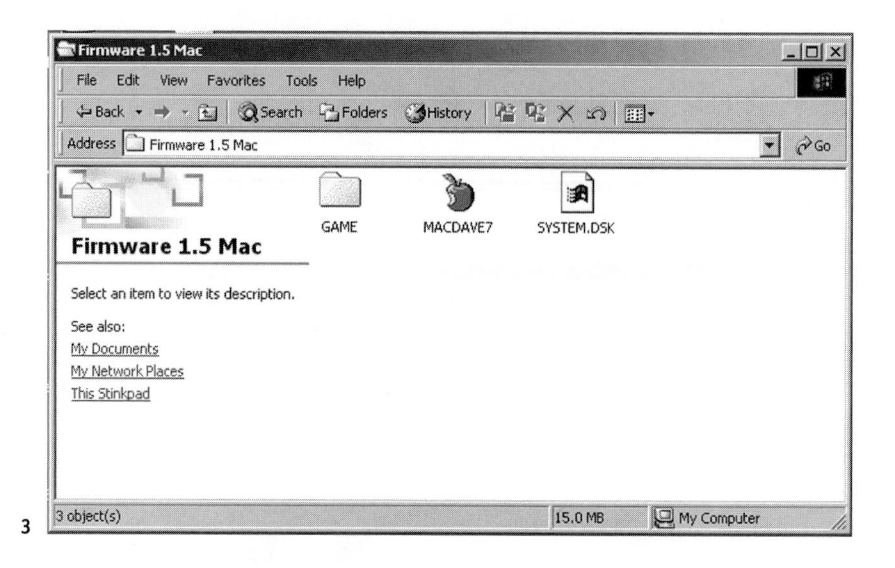

3

4

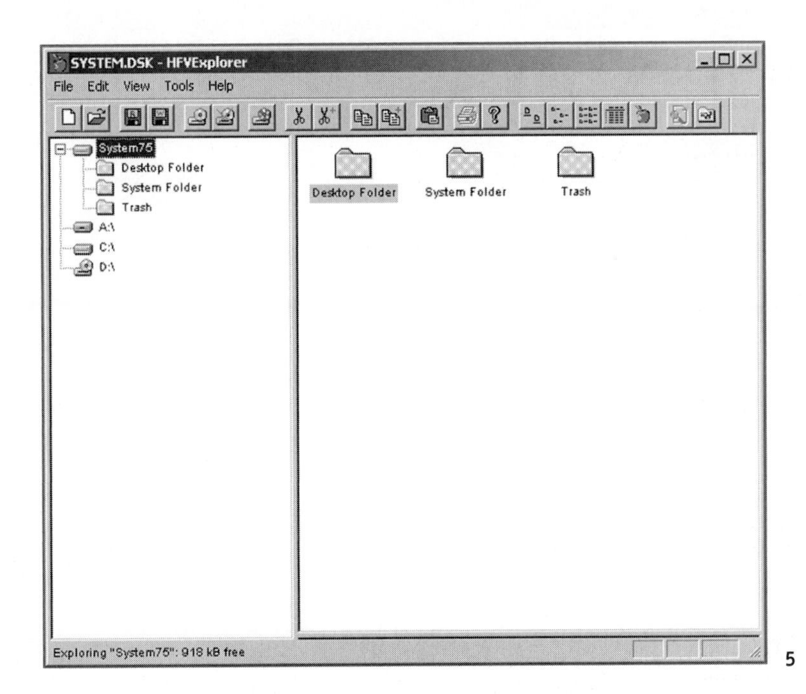
SYSTEM.DSK - HFVExplorer
File Edit View Tools Help
System75
Desktop Folder
System Folder
Trash
A:\
C:\
D:\
Desktop Folder
System Folder
Trash
Exploring "System75": 918 kB free

5

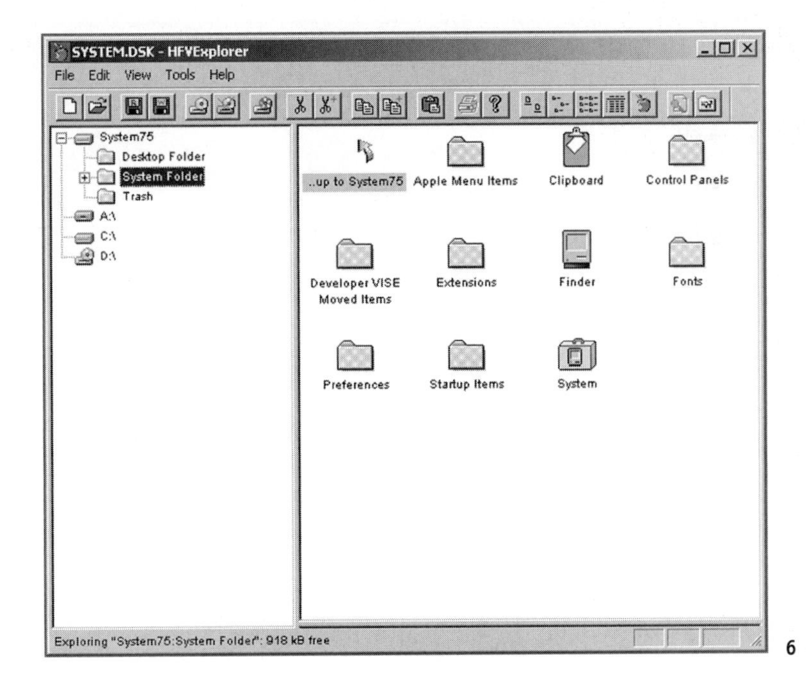
SYSTEM.DSK - HFVExplorer
File Edit View Tools Help
System75
Desktop Folder
System Folder
Trash
A:\
C:\
D:\
..up to System75
Apple Menu Items
Clipboard
Control Panels
Developer VISE Moved Items
Extensions
Finder
Fonts
Preferences
Startup Items
System
Exploring "System75:System Folder": 918 kB free

6

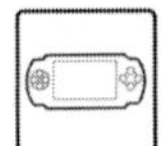

7 A valid game folder in HFVExplorer. You can tell that these games are valid (and, therefore, playable) by the presence of an icon.

8 Make sure that you have Stuffit Expander™ 5.5 installed inside Basilisk II for PSP *before* you try to load any other software, games, or applications.

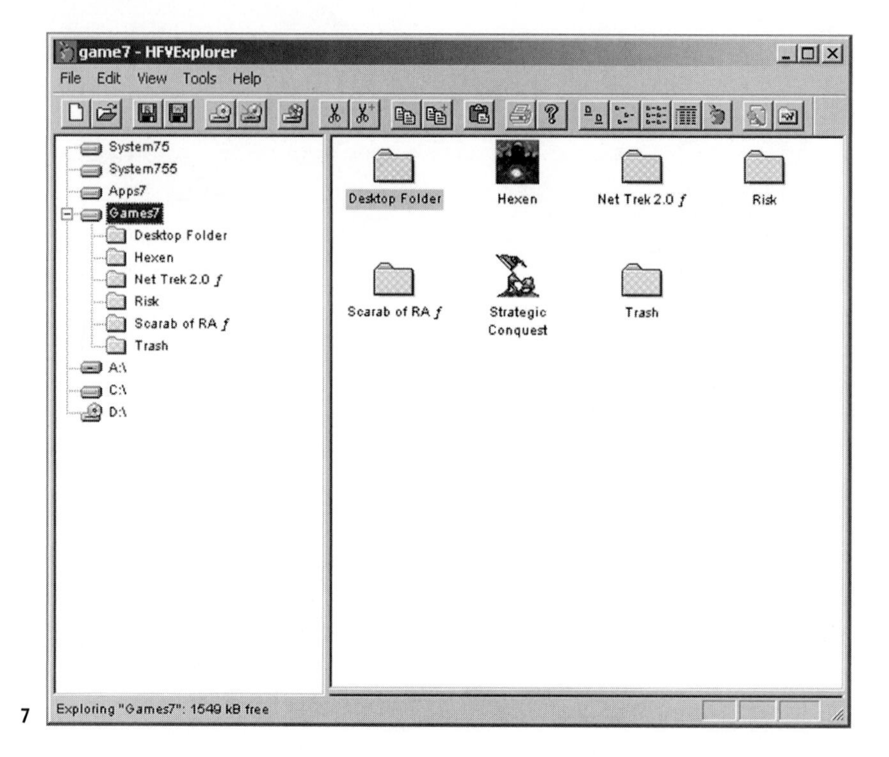

7

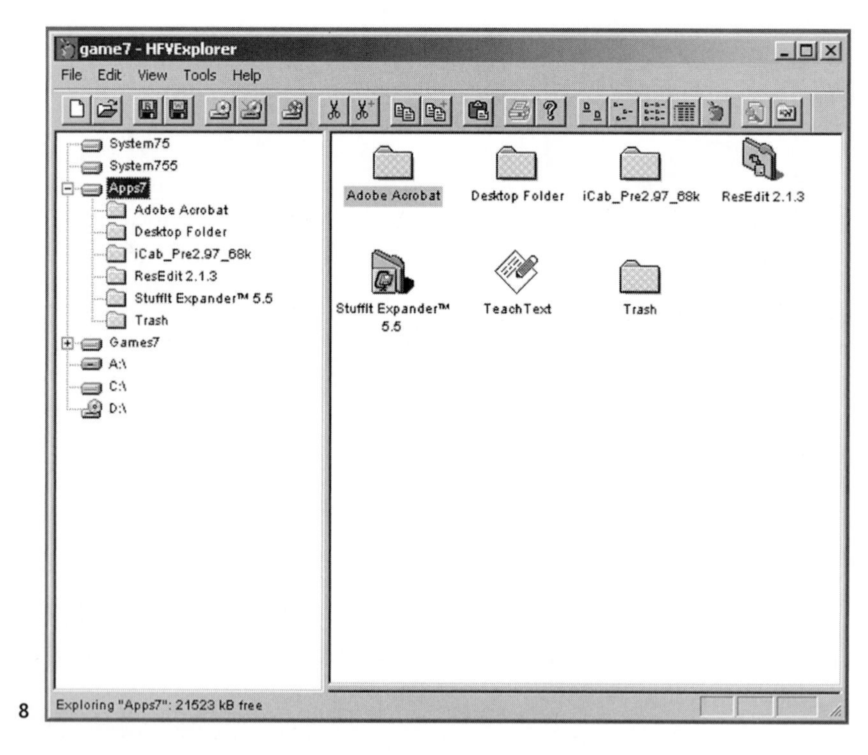

8

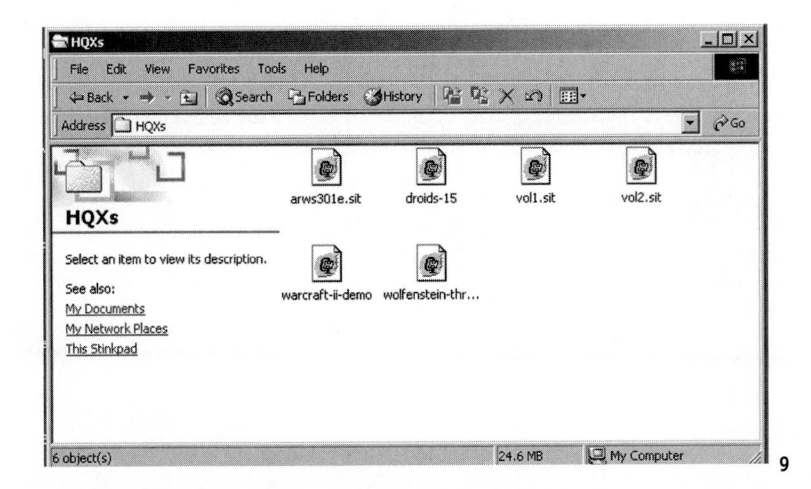

9

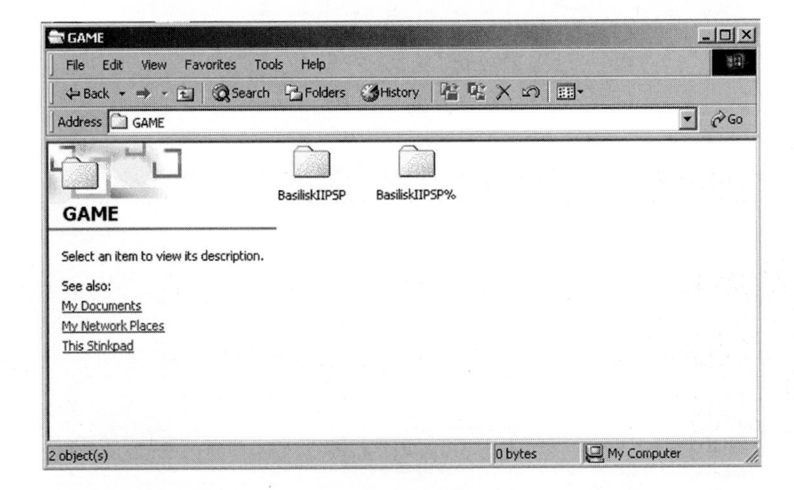

10

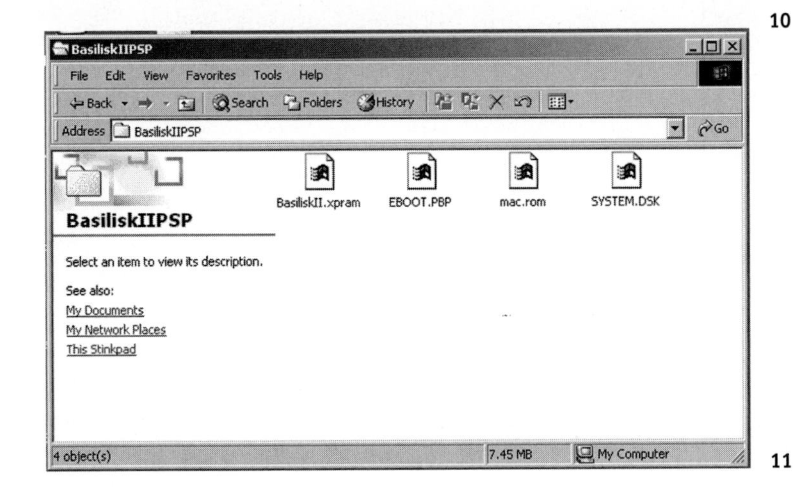

11

9 You *must* load Mac software as HQX files when copying from your PC to Basilisk II for PSP. Sources for finding suitable games, applications, and utilities include: old issues of *MacAddict* CD-ROMs (e.g., issues 1-4; yeah, that's real old), copies of old Mac OS books (e.g., *The Mac Bible Goodies Pack*, featuring *The Mac Bible 6 CD-ROM* and *The Fall 1996 BMUG PD-ROM*), and the Internet.

10 Copy these two folders to the Game folder on your PSP.

11 The contents of the BasiliskIIPSP folder inside your PSP Game folder. It's a good idea to use HFVExplorer *and* Basilisk II for Windows on your PC for setting up and configuring this BasiliskIIPSP folder *prior* to installing it on your PSP. Please note the correct naming for "mac.rom" (this is the Mac ROM image file) and "SYSTEM.DSK" (this is the complete Mac OS System 7.5.5; including all system, application, and game folders).

12 Success! A Sony PSP is running Mac OS System 7.5.5 as a homebrew application under Firmware 1.5. The load time for Basilisk II for PSP is approximately 20 seconds.

13 In my opinion, this is the last, great operating system—Mac OS System 7.5.5. No command line interface, simple, small footprint, easy to use, and rock solid with Ethernet functionality, and color display, this OS was the "high water mark" for all computer operating systems. And now you can have it on your PSP. Please note that I have three separate HFS disk images for system, applications, and games.

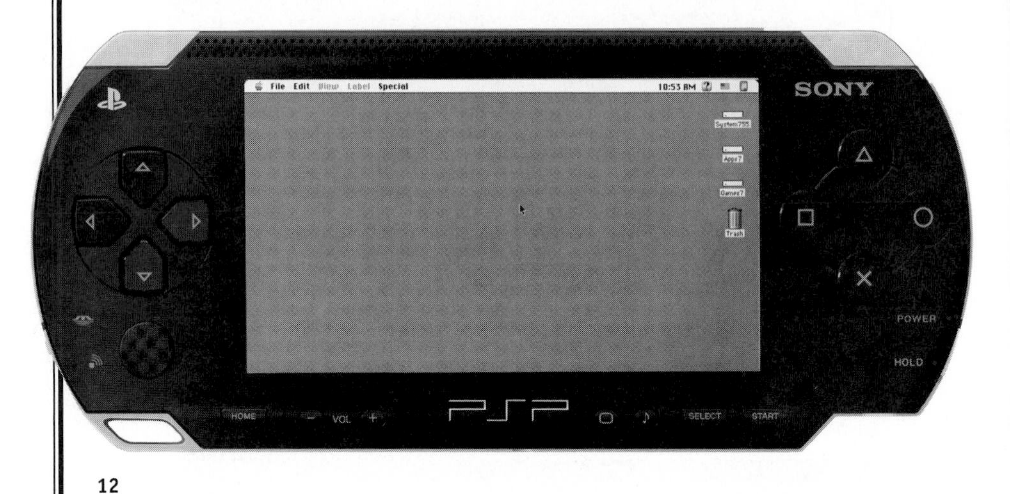

12

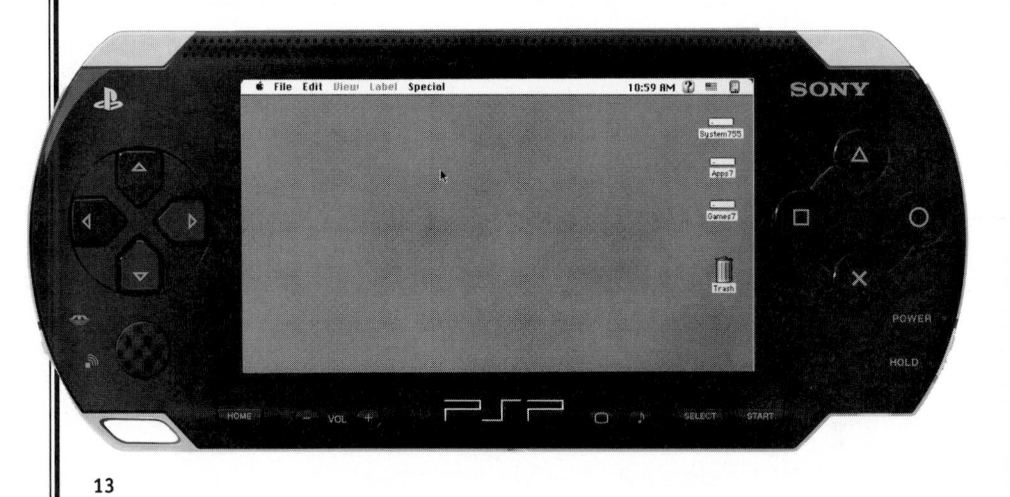

13

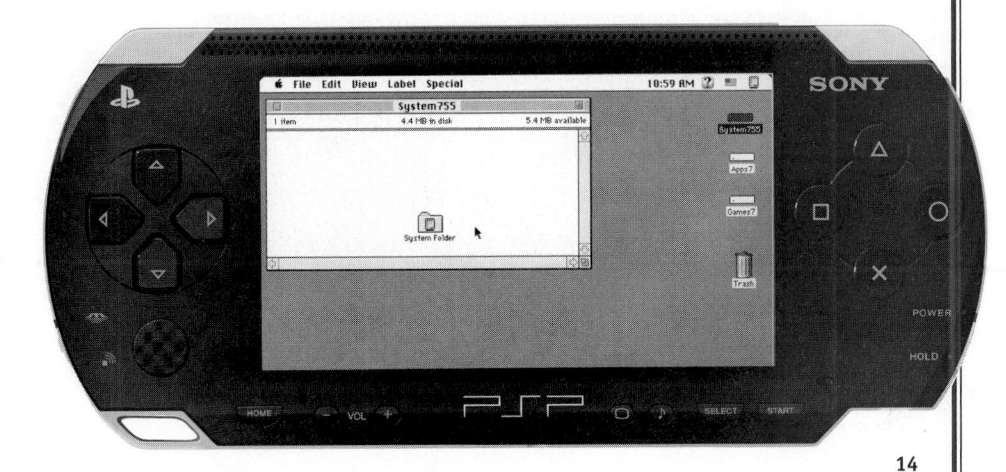

14

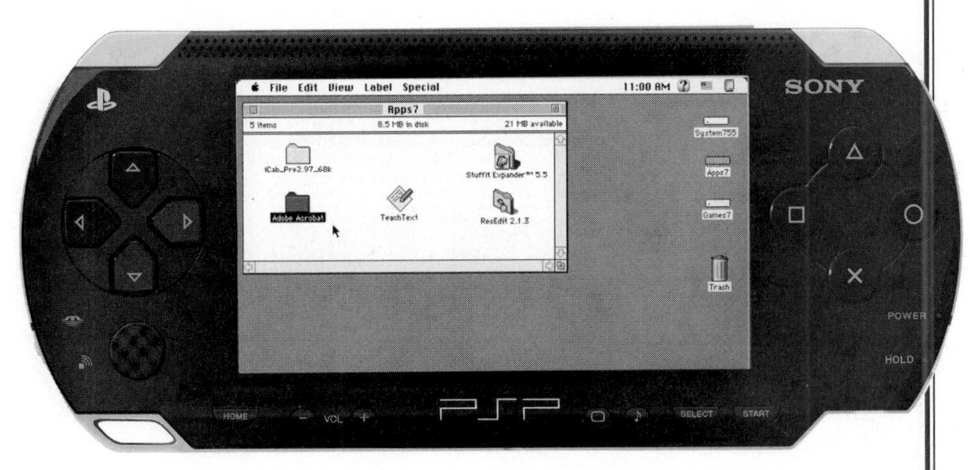

15

14 The system folder sits inside the system disk image.

15 I installed a Web browser (iCab), text program (TeachText), a resource editor (ResEdit), and, most importantly, Adobe Acrobat Reader inside my applications disk image.

16 These games all work on Basilisk II for PSP. Games like *Wolfenstein 3D could* work (it does inside the Windows version), but the Basilisk II for PSP implementation can't accept the mandatory display change that is executed by this game. Also, *Strategic Conquest* by Delta Tao Software (in my opinion, this is the best strategy game ever written) does work, even though its icon isn't properly shown. Be forewarned, these games do run slowly and there are some color errors, but they are playable.

17 Yes, you can read PDF documents (version 2.1 or older) with Acrobat Reader on your PSP.

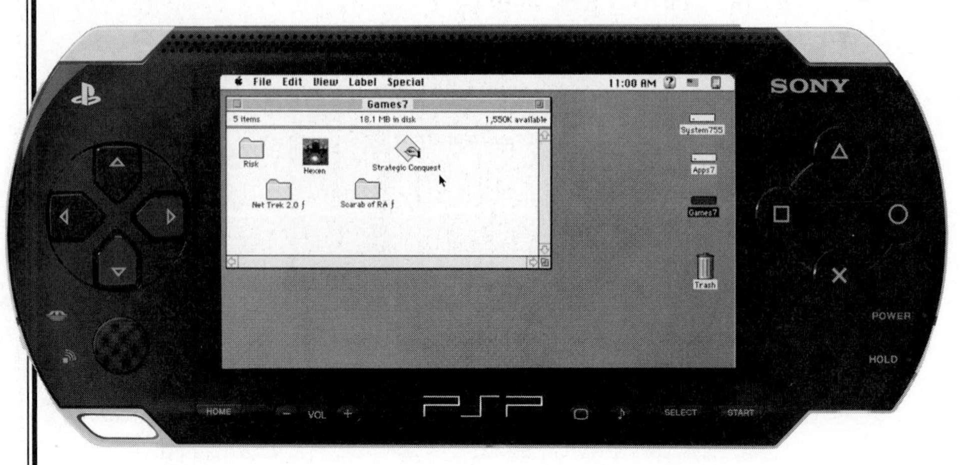

16

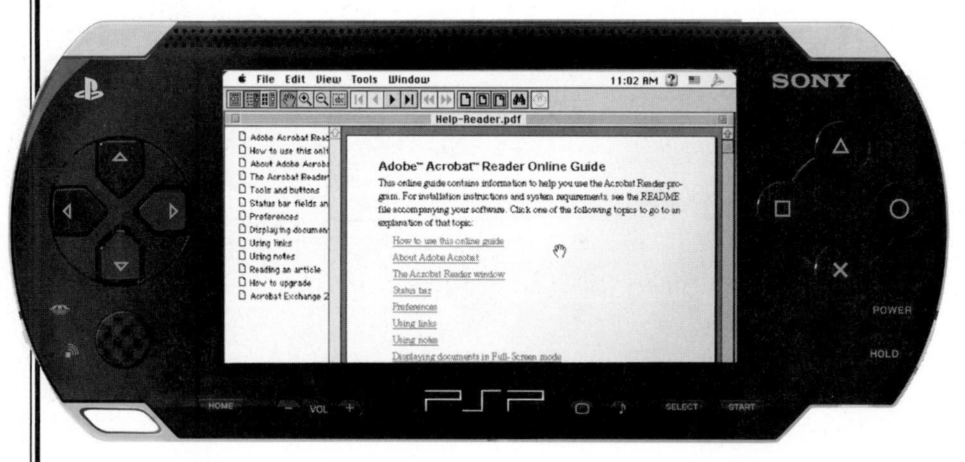

17

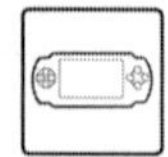

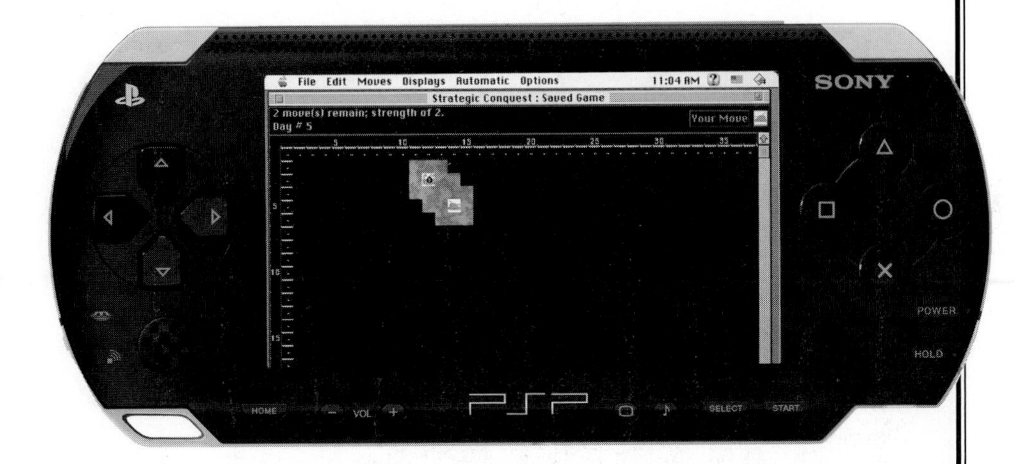

18

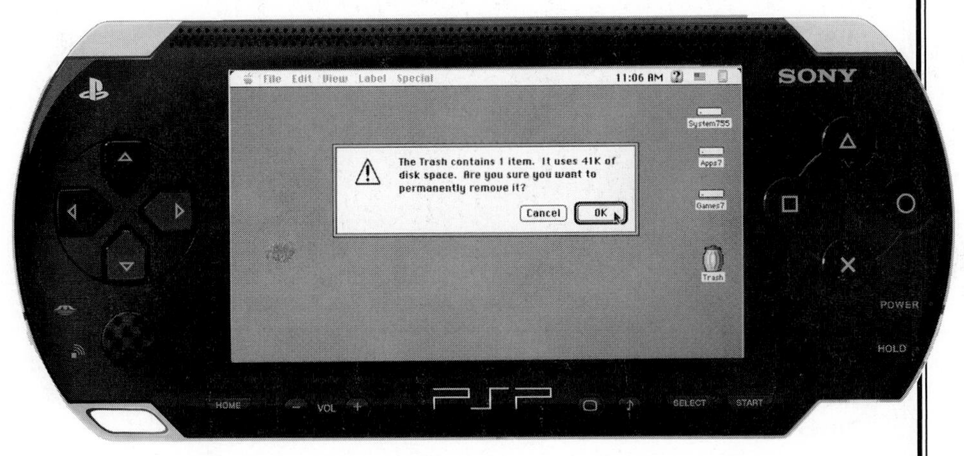

19

18 A game of Strategic Conquest is underway.

19 You can only delete items from inside your Mac OS disk images with the Trash and not files from the PSP.20. Use the Control Panels inside the System Folder to configure the Mac OS.

20 Select the Shut Down menu command before you exit your homebrew Basilisk II for PSP with the PSP Home button.

21 Select the Shut Down menu command before you exit your homebrew Basilisk II for PSP with the PSP Home button.

20

21

EPILOGUE

Hacking Now for Your Future

So why on Earth would you ever want to be a hacker?

Well, there are several plausible reasons: *curiosity* in the way stuff is built, accepting the *challenge* to combine dissimilar objects into the creation of one single über object, or enjoying the *fun* of recovering a dead or obsolete piece of techno-garbage and restoring a new and different function to it.

On an even more primal level, however, you might become a hacker out of necessity.

Remember when your older third-generation iPod's battery died and you couldn't bear the expense of shipping it to an Apple authorized repair center for installing a new battery? So, instead, guided by some instructions that you downloaded from the Web, you pried the back off of your trendy MP3 player, removed the dead battery, and installed a new one. Then you put it back together and...it worked. Oh, the joy. Oh, the wonder. Oh, you hacker, you.

Now that didn't hurt, did it?

The trouble is most people don't recognize this effort as hacking. Rather, these folks will use a more generic or politically popular label like: DIYer, tinker, (Mr.)/(Ms.) Fix-it, repairman/repairperson, dabbler, or experimenter—actively avoiding a term like "hacker."

Semantics aside, it doesn't really matter what name you call this rose, it still smells so sweet when your project succeeds. And succeed, it must. Why? Our world's landfills are not a good resting place for our dead or obsolete electronics. A much better "final" resting place for our gadgets, gizmos, and goodies is inside a "parts box" underneath our workbench.

Likewise, this hacking shouldn't be the restrictive domain of the homeowner. Business owners, IT professionals, and Web workers would all benefit from an organized e-recycling plan.

In fact, one of the first programs that I regularly established at the various jobs that I've held during my career was the creation of an e-junk box, cable drawer, or hangar queen closet. Hangar queen? An old US Air Force term for an aircraft that is typically not flight worthy, but can still be used for salvaging parts to keep other aircraft flyable.

How does a hangar queen relate to techno-stuff recycling?

For example, the owner at a graphic design company once told me that he needed a RAID (Redundant Arrays of Independent Disks). Prior to placing the order for this costly peripheral I made a short trip through my hangar queens. In less than one day, I had located a set of SCSI drives that I was able to install inside a dead Power Computing case. This reutilization of resources that he already owned made a valuable impression on the owner and helped to ensure my continued employment.

So before you pitch it, pinch it from the garbage can. Then secret it away inside your own e-junk box, cable drawer, or build a squadron of hangar queens. The reward will be more than a financial return, it'll contribute toward environmental renewal and, who knows, maybe even ensure your own job security.

APPENDIX A

Things to Know, Places to Go

Here is a collection of every reference, document, product, and Web site that is mentioned in this book. Live and learn.

1999 *Surgeon General's Report on Physical Activity and Health*—www.cdc.gov/nccdphp/sgr/contents.htm
ACG—www.acg.de
Aeronix GNU/Linux—www.zipitwireless.com/linux.html
AiboHack Zipit Software Reflashing Web site—www.aibohack.com/zipit/reflash.htm
Aiptek—www.aiptek.com
American Buffalo Products—www.americanbuffaloproducts.com
American College of Sports Medicine—www.acsm.org
American International Toy Fair—www.toy-tia.org/Content/NavigationMenu/TIA_Trade_Shows_and_Events/American_International_TOY_FAIR/American_International_TOY_FAIR1.htm
American Pet Products Manufacturers Association, Inc.—www.appma.org
Arcata Community Bike Program—www.arcata.com/greenbikes/
AvrFlasher—www.ipnas.org/garnir/avr
Basel Action Network (BAN)—www.ban.org
Basilisk II Emulator—psp-news.dcemu.co.uk/basilisk.shtml
Basilisk II PSP Port (Mac Emulator)—www.digital-yume.net/neisha/
Basilisk II for Windows—emulation.net/basiliskII/system753_tutorial/index.html

BEAM Bicores Online—www.beam-online.com/Robots/Circuits/bicore.html

BG Micro—www.bgmicro.com

Bike Nashbar—nashbar.com

Bikes Not Bombs—www.bikesnotbombs.org

Blue Sky Alpacas—www.blueskyalpacas.com

"Bored or Lonely Cat? There's Help for That" (*New York Tails: A Magazine for the People and Pets of New York*, January 2005)—www.newyorktails.com/boredcat.htm

BoxWave Corporation—www.boxwave.com

Bawdsey Radar Group—www.bawdseyradargroup.co.uk

"CA Assemblywoman Clarifies $7 Bike Surcharge, Welcomes Industry Comment" (*Bicycle Retailer and Industry News*, February 25, 2005)—www.bicycleretailer.com/bicycleretailer/search/article_display.jsp?vnu_content_id=1000817554

"California announces new solar power plan" (PhysOrg.com; Science, Technology, Physics, Space News, January 25, 2006)—www.physorg.com/news10247.html

California Energy Commission—www.energy.ca.gov

California Manufacturers and Technology Association—www.cmta.net/index.php

California Public Utilities Commission California Solar Initiative Program—www.cpuc.ca.gov/static/energy/051214_solarincentive.htm

CBS Corporation—www.cbscorporation.com

Centers for Disease Control and Prevention: Computer Workstation Ergonomics—www.cdc.gov/od/ohs/Ergonomics/compergo.htm

C. Crane—www.ccrane.com

Coalition for Appropriate Transportation—www.car-free.org

Convergence Technologies—www.econvergence.net/electro.htm

"Could Your Cat Be Craving Companionship?" (Arm & Hammer®)—www.armhammerpets.com/catpan_press.aspx

"Cutting Back On Lawn-Mower Pollution" by Davide Castelvecchi *(Santa Cruz Sentinel*, November 29, 2003)—www.santacruzsentinel.com/archive/2003/November/29/local/stories/10local.htm

CVS PV2 Disposable Digital Camera—www.maushammer.com/systems/dakotadigital/lcd.html

Dale Wheat—www.dalewheat.com

"Dead man's phone rings inside coffin" (Expatica News; reprinted from *Gazet van Antwerpen*, November 18, 2003)—ww.expatica.com/source/site_article.asp?subchannel_id=24&story_id=2374

"'Digital Dumps' Heap Hazards at Foreign Sites" by Elizabeth Grossman (Special to *The Washington Post*, Monday, December 12, 2005)—www.washingtonpost.com/wp-dyn/content/article/2005/12/11/AR2005121100664.html

"Dogs Home Alone" (PetPlace®.com)—www.petplace.com/dogs/dogs-home-alone/page1.aspx

Easy Linux CDs—www.easylinuxcds.com/catalog/index.php

Echowell—www.echowell.biz/

Edmund Scientific—www.scientificsonline.com

EIA Environment: Consumer Education Initiative (CEI)—www.eiae.org/

Element Products—www.elementinc.com/

"Eleven Days Gone" by Dave Prochnow (*MAKE* Web Extra September 12, 2005)—www.makezine.com/extras/25.html

elinux.org—elinux.org/wiki/JuiceBox

EM Microelectronic-Marin SA—www.emmicroelectronics.com/

Encyclopedia Titanica—www.encyclopedia-titanica.org/index.php

An Empty Spacesuit Becomes an Orbital Experiment (NASA)—www.nasa.gov/mission_pages/station/expeditions/expedition12/26jan_suitsat.html

ENERGY STAR—www.energystar.gov

Environmental Caucus of the California Democratic Party—www.environmentalcaucus.org

"EPA: Old Computers No Longer Junk" by Kendra Mayfield (*WIRED News*, 2002-06-03)—www.wired.com/news/technology/0,1282,52876,00.html

EPA Plug-In To eCycling—www.epa.gov/epaoswer/osw/conserve/plugin/

EPA WasteWise—www.epa.gov/epaoswer/non-hw/reduce/wstewise/index.htm

Familiar Quotations Compiled by John Bartlett (10th ed.), Boston, MA, Little, Brown, 1919

Federal Communications Commission—www.fcc.gov

Federal Trade Commission—www.ftc.gov

Feng Shui News—www.fengshuinews.com

Fuji—www.fujibikes.com

"Google's Great Works in Progress" by Burt Helm (BusinessWeek.com—News Analysis, December 22, 2005)—www.businessweek.com/

technology/content/dec2005/tc20051222_636880.htm?campaign_id=technn_Dec22&link_position=link13

"GOP ex-EPA chiefs bash Bush policies" (CNN.com, January 19, 2006)—www.cnn.com/2006/POLITICS/01/18/global.warming.ap/index.html

Grand Idea Studio—www.grandideastudio.com

Grass Cutting Beats Driving in Making Air Pollution, ENS 31 May 2001—www.mindfully.org/Air/Lawn-Mower-Pollution.htm

Hand Crank Generator—www.hometrainingtools.com/catalog/physical-science-physics/electricity-electronics/motors-generators/p_el-genhand.html

Hantronix—www.hantronix.com/2_2.html

Harbor Freight Tools—www.harborfreight.com

"Healthy Lawn, Healthy Environment: Caring for Your Lawn in an Environmentally Friendly Way" (EPA, 700-K-92-005; June 1992)

HFS Utilities—www.mars.org/home/rob/proj/hfs/

Home Training Tools, Ltd.—www.hometrainingtools.com

"ID Chips Not Just for Pets Any More"—(CNN.com, February 13, 2006)—www.cnn.com/2006/TECH/02/13/security.chips.ap/index.html?section=cnn_topstories

Identifying American Furniture: A Pictorial Guide to Styles and Terms, Colonial to Contemporary by Milo M. Naeve (W. W. Norton & Company, 3rd ed., 1998)

International Bicycle Fund—www.ibike.org/index.htm

International Council of Shopping Centers 2002 Holiday Watch—www.icsc.org

iPod-Linux Installer—www.ipodlinuxinstl.sourceforge.net

iPod Note Reader User Guide—developer.apple.com/hardware/ipod/

iRobot Corporation Hacker—www.irobot.com/sp.cfm?pageid=198

ISO-TIP—www.iso-tip.com/html/soldering_tools.htm

iStory Creator—www.ipodsoft.com/istories.aspx

Is Your Cat Infected with a Computer Virus? by Melanie R. Rieback, Patrick N. D. Simpson, Bruno Crispo, Andrew S. Tanenbaum (Department of Computer Science, Vrije Universiteit Amsterdam, March 15, 2006)—www.rfidvirus.org/index.html

jpbConverter.jar—epoxy.mrs.umn.edu/~john4390/jbpConverter.jar

J Schatz Egg Bird Feeders—www.jschatz.com/eggbirdfeeders/index.html

Junkbots, Bugbots & Bots on Wheels by Dave Hrynkiw and Mark W. Tilden (McGraw-Hill, 2002)

Juice Box—www.juicebox.com/home.aspx

Juice Box SD/MMC Hack—www.elinux.org/wiki/JuiceBoxMMCHack

K-Byte—www.kbytememory.com/pages/zipit.shtml

Kenpo Jacket for iPod—kenpofashion.com/kenpo.html

The Knitter's Handy Book of Patterns: Basic Designs in Multiple Sizes & Gauges by Ann Budd (Interweave Press, 2002)

"Legislator peddles bicycle tax but idea is as welcome as a flat tire," by K. Lloyd Billingsley (American City Business Journals Inc.; *Silicon Valley/San Jose Business Journal,* April 22, 2005)—sanjose.bizjournals.com/sanjose/stories/2005/04/25/editorial3.html

Levi's® RedWire™ DLX Jeans—www.levistrauss.com/news/pressrelease.asp?r=0&c=1&cat=2&pr=764&area=

LinuxDevices.com—www.linuxdevices.com

Linux Journal—www.linuxjournal.com

Linux Kernel Archives—www.kernel.org

Linux Online—www.linux.org

Mac OS X software for Slow-Scan TV (SSTV), Narrow-Band TV (NBTV) and weather satellites—homepage.mac.com/kd6cji

Mac ROM Image—www.the-underdogs.org/nonpc.php

MAREX-MG ISS SpaceCam1 Project—www.marexmg.org/spacecam/spacecam.html

Mattel Juice Box—www.juicebox.com/home.aspx

"Men & Boys Knitting Up A Storm" (CBS News, February 4, 2005)—www.cbsnews.com/stories/2005/02/04/national/main671644.shtml

Mercury-Containing Equipment Classified as Universal Waste Fact Sheet—www.epa.gov/epaoswer/hazwaste/recycle/electron/mce-fs.htm

Mind Control—www.elementdirect.com/product_info.php?products_id=28

Mister Feng Shui—www.misterfengshui.com

Monster iCruze for iPod—www.monstercable.com/icruze

Moog Music—www.moogmusic.com

National Do-Not-Call Registry—www.fcc.gov/cgb/donotcall/

National Federation of Coffee Growers of Colombia—www.juanvaldez.com

National Gypsum Properties—www.nationalgypsum.com

National Hurricane Center's Tropical Cyclone Report TCR-AL122005 Katrina—www.nhc.noaa.gov/2005atlan.shtml?

National Institute of Standards and Technology: Manufacturing Extension Partnership—www.mep.nist.gov/rfid/rfid.htm

National Museum of American History's *America on the Move*; Volkswagen Beetle—americanhistory.si.edu/onthemove/collection/object_1316.html

National Priorities Project: Cost of War—nationalpriorities.org/index.php?option=com_wrapper&Itemid=182

New Bright—www.newbright.com

Occupational Safety & Health Administration: Computer Workstations—www.osha.gov/SLTC/etools/computerworkstations/

The Official Robosapien Hacker's Guide by Dave Prochnow (McGraw-Hill, 2006)

"One small step for trash is giant leap for ham-kind" (CNN.com, February 3, 2006)—www.cnn.com/2006/TECH/space/02/02/recycled.spacesuit.reut/index.html?section=cnn_topstories

Ontrak Control Systems, Inc.—www.ontrak.net

Organizations Recycling Bicycles—www.ibike.org/encouragement/recycling/recycling-orgs.htm

OS Heaven—www.osheaven.net

PAiA Electronics, Inc.—www.paia.com

Parallax, Inc.—www.parallax.com

Parallax RFID Reader Module—www.parallax.com/detail.asp?product_id=28140 and www.grandideastudio.com/portfolio/index.php?id=1&prod=36

Patte International Furniture—www.stefan-patte-design.de/html/patte_en.html

Peck-Polymers—www.peck-polymers.com

Pepper Pad—www.pepper.com

Hugo Perquin's Roomba PCB connectors—prj.perquin.com/roomba/pcb.php

Pfaff Sewing Machines—www.pfaffusa.com

Polar Heart Rate Monitors—www.polarusa.com

PowerMax: new, used, and certified pre-owned Macs—www.powermax.com

Preemptive Media—www.preemptivemedia.net

The President's Challenge—www.presidentschallenge.org

Presidential Champions Rules—www.presidentschallenge.org/the_challenge/presidential_champion_rule.aspx

President's Council on Physical Fitness and Sports—www.fitness.gov

PSP Hacks, Mods, and Expansions by Dave Prochnow (McGraw-Hill, 2006)

Qsent: Wireless 411 Service for Consumers—www.qsent.com/wireless411/index.shtml

"Radio ID Tags: Beyond Bar Codes" by Kendra Mayfield (*WIRED News*, May, 20, 2002)—www.wired.com/news/technology/0,1282,52343,00.html

RFID Journal—www.rfidjournal.com

RFID in Japan—ubiks.net/local/blog/jmt/stuff3/

Resource Revival—www.resourcerevival.com

Roomba Dev Tools—www.roombadevtools.com

Roomba SCI Specification—www.irobot.com/sp.cfm?pageid=198

SANE—Scanner Access Now Easy—www.sane-project.org

The Scanner Photography Project—www.scannerphotography.com

Sears Brands LLC (Craftsman brand)—www.sears.com

SFBags Waterfield Designs—www.sfbags.com

Shreve Systems: The oldest and largest recycler of Macs in the world—www.shrevesystems.com

Simplehuman—www.simplehuman.com

Skype Limited—www.skype.com

SmokeHouse Plans: Build Your Own Meat Smoker—www.smokehouseplan.com

Sokymat—www.sokymat.com/index.php?id=2

Solarbotics—www.solarbotics.com

The Solar Cooking Archive—www.solarcooking.org

Solar Electric Power Association—www.solarelectricpower.org

Sony LocationFree® Owner's Lounge—products.sel.sony.com/locationfreetv/owners/psp.html

Spielwarenmesse International Toy Fair Nürnberg—www.spielwarenmesse.de

Stuffit Standard Edition for System 7.x —www.stuffit.com/mac/standard/updates.html

Suitsat-1 RSORS—www.amsat.org/amsat-new/articles/BauerSuitsat/index.php

System 7.5.5—www.the-underdogs.org/nonpc.php

Tamiya—www.tamiyausa.com

Telephreak—www.telephreak.org

Theremin World—www.thereminworld.com

"They've Got Your Number ..." by Annalee Newitz (*WIRED*, Issue 12.12, December 2004)—www.wired.com/wired/archive/12.12/phreakers.html

Tim Riker's Zipit Experiments and Hacks—elinux.org/wiki/ZipIt

Transformed: How Everyday Things are Made by Bill Slavin (Kids Can Press Ltd., 2005)

Traxxas—www.traxxas.com

Trek—www2.trekbikes.com

"TV SPORTS; CBS Loses Battle With Clock" by Michael Goodwin (*The New York Times*, September 16, 1987)—query.nytimes.com/gst/fullpage.html?res=9B0DE3D91630F935A2575AC0A961948260

Universal Waste—www.epa.gov/epaoswer/hazwaste/id/univwast.htm

U.S. Census Bureau Facts for Features & Special Editions—www.census.gov/Press-Release/www/releases/archives/facts_for_features_special_editions/index.html

U.S. Department of Homeland Security, Hurricane Katrina: What Government Is Doing?—www.dhs.gov/interweb/assetlibrary/katrina.htm

USG Corporation—www.usg.com:80/index.jsp

VoodooPad—flyingmeat.com/voodoopad.html

VW Trends Magazine—www.vwtrendsweb.com

Wacom Technology—www.wacom.com

Wahl Clipper Corporation—www.wahl.com

"Warming Tied to Extinction of Frog Species" by Juliet Eilperin (*The Washington Post*, January 12, 2006)—www.washingtonpost.com/wp-dyn/content/article/2006/01/11/AR2006011102121.html

"What's Lurking in That RFID Tag?" by Olga Kharif (*BusinessWeek Online*"Your Yard and Clean Air" (EPA, Fact Sheet OMS-19; May 1996)

APPENDIX B

DIY Death

Glued cases, tamperproof security fasteners, obliterated, omitted, and obfuscated circuit markings, and limited replacement parts—these are all the unfortunate birthmarks of a new age in technology. The dawn of products that are designed *not* to be hacked.

It wasn't too long ago when you could readily obtain wiring diagrams, exploded parts views, and even thorough battery replacement instructions for just about anything that you purchased. Not so, anymore.

Want to replace the battery in your dead iPod? Forget it. Return it to an authorized repair center, fork over around $90, and wait for *your* iPod's return. All of this hassle for just a simple rechargeable battery replacement. That's crazy. But your iPod isn't the only product that has been designed not to be hacked.

Just look out in your garage. Chances are that your car, truck, or SUV has a whole slew of parts inside of it that are sold only as "assemblies" and not as individual components. Even worse, if you can "debug" your auto's ailment, chances are the parts assembly that you must purchase for repair will cost a king's ransom. That's right, even though a simple two-bit component could fix your problem, you'll end up buying a much bigger, and much more expensive, parts assembly instead.

Luckily, some manufacturers are very upfront about their ambivalence toward end-user repair or hacking of their products. These types of products will usually proclaim something like "No user serviceable parts inside." There ya go; no guess work, if it breaks, you can't fix it. Generally speaking, this type of manufacturer declaration should be read as, "Don't buy this gizmo."

This type of user-unfriendliness is the hardware equivalent of digital rights management (DRM) with software and media files. Even though you buy DRM stuff, you really don't own it... ever. So, to me, DRM products really mean "Don't Reach for My wallet." Oh sure, I will buy a book or a music CD any day, but I will *never* purchase an iTunes song. It's just bad business sense—no cents from my business.

Now please don't consider me to be just some cranky old Luddite. 'Cause I'm not. For example, I regularly cut my CD music into MP3 files where I then physically load them onto my various iPods—Linux-based iPods. I didn't used to be this cagey, however.

When the iTunes phenomenon began, I took the bait; hook, line, and sinker. That is until one day when I was trying to finish an iMovie project. I wanted to drag a Green Day song onto an audio track of my movie and I couldn't grab the song. In fact, I couldn't even see the song from inside the iMovie app.

Thinking that this goofy action was a response to some odd file format conflict, I attempted to convert the song into an AIFF track. No go. And so my lengthy research into the growing incestuous relationship between DRM and the entertainment industry began.

Remarkably, that event was several years ago and billions of iTunes song downloads later. What's wrong with this picture? It sounds like a broken record—Doesn't the consumer get it?

DRM fast-food for thought like iTunes keep serving billions of songs like a digital McDonald's. Now DRM-controlled video is gaining steam and e-books are being "printed" with DRM ink. The day is quickly coming when you will own nothing, yet you will regularly fork over large amounts of money for just the "right" to listen, watch, or play something.

Everything that you own will really just be rented. In a country that was founded on ownership, we are slowly being lead into the fiscal irresponsibility of "rentalship." The only problem is that none of us are "kicking or screaming" about this transition.

What, you don't like all of these limitations on your gadgets and entertainment media? Well, do something about it. Refuse to buy, err, rent *any* DRM media. Never buy a product that has been designed *not* to be hacked. Don't worry, you won't be cut off from civilization. There are plenty of user-friendly options available out there. You just have to be a diligent shopper and not cave into popular trends or crazes.

Don't super-size your digital debt. For example, the COWEN America iAudio X5 is just as capable of an MP3 player as the iPod. It just doesn't have the same cachet. So buy an iAudio X5 and shun the iPod.

So could a ban on DRM media really work? Well, just look at that other company who boasted about serving over one billion products—McDonald's. Feeling increased pressure from other "quick serve" sandwich restaurants, McDonald's had to change their strategy, marketing, and, most importantly, their menu. The result was a "fast-food" restaurant being successfully transformed into a revamped "quick serve" sandwich (and salads) restaurant.

Now if we can only clean up this DRM mess, we'll all be saying that "we're lovin' it."

Index

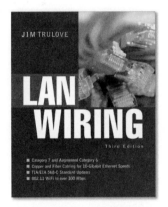

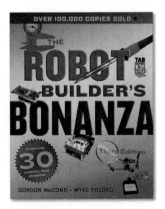

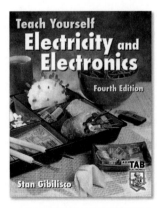